Sustainable Solid Waste Collection and Management

Ana Pires • Graça Martinho • Susana Rodrigues
Maria Isabel Gomes

Sustainable Solid Waste Collection and Management

Ana Pires
Faculty of Sciences and Technology
Universidade NOVA de Lisboa
(FCT NOVA)
Caparica, Portugal

Graça Martinho
Faculty of Sciences and Technology
Universidade NOVA de Lisboa (FCT NOVA)
Caparica, Portugal

Susana Rodrigues
Faculty of Sciences and Technology
Universidade NOVA de Lisboa
(FCT NOVA)
Caparica, Portugal

Maria Isabel Gomes
Faculty of Sciences and Technology
Universidade NOVA de Lisboa (FCT NOVA)
Caparica, Portugal

ISBN 978-3-319-93199-9 ISBN 978-3-319-93200-2 (eBook)
https://doi.org/10.1007/978-3-319-93200-2

Library of Congress Control Number: 2018949367

© Springer International Publishing AG, part of Springer Nature 2019
This work is subject to copyright. All rights are reserved by the Publisher, whether the whole or part of the material is concerned, specifically the rights of translation, reprinting, reuse of illustrations, recitation, broadcasting, reproduction on microfilms or in any other physical way, and transmission or information storage and retrieval, electronic adaptation, computer software, or by similar or dissimilar methodology now known or hereafter developed.
The use of general descriptive names, registered names, trademarks, service marks, etc. in this publication does not imply, even in the absence of a specific statement, that such names are exempt from the relevant protective laws and regulations and therefore free for general use.
The publisher, the authors, and the editors are safe to assume that the advice and information in this book are believed to be true and accurate at the date of publication. Neither the publisher nor the authors or the editors give a warranty, express or implied, with respect to the material contained herein or for any errors or omissions that may have been made. The publisher remains neutral with regard to jurisdictional claims in published maps and institutional affiliations.

This Springer imprint is published by the registered company Springer Nature Switzerland AG
The registered company address is: Gewerbestrasse 11, 6330 Cham, Switzerland

Preface

Waste collection is one of the most relevant parts of the integrated solid waste management system in technical, economic, environmental, and social terms. However, the vital role of waste collection has not been recognized, and the support given by the quadruple helix – academics, industry, state, and citizens – is reduced when looking for the other operational units of an integrated solid waste management system. To exemplify the missing interest of academia for waste collection, in a Google Scholar search, the results for "waste collection" are about 157,000, for incineration are 375,000, and for landfill are 639,000. This example shows that collection is seen only as the way to make waste to get into high-tech infrastructures, like incineration plants, where real science is applied. The practitioners of the waste collection are the range of workers with lower income backgrounds, making waste collection not attractive enough to be devoted to high-tech solutions for waste collection problems. Many times, waste managers working in waste collection systems are more focused on the trucks and containers and the costs involved in the collection process, which is a considerable amount of local authorities' budget.

In a meeting in 2014 between academia and waste collection sectors at the Nova University of Lisbon, waste collection professionals expressed the need for technical skills and knowledge based on waste collection and management practice, in a bottom-up approach. Technical skill areas include cost-efficiency, recycling behavior, environmental impacts, and technical operation. Waste collection professionals all over the world have the same necessities when implementing and managing a waste collection system to know more about the subject, but in such a way that knowledge could be affordable in technical and economic terms.

The waste collection professionals' call to fulfill the technical background needs required experts in several fields of waste collection and management. This book results from the collaboration from different science areas and experiences in waste collection and management. Dr. Ana Pires, from whom the original idea was formed, has a scientific role in analyzing solid waste management from a system analysis perspective, where the application of life cycle assessment and multi-criteria decision-making are the techniques applied by her to make waste collection and management more sustainable. Prof. Graça Martinho has a long career in solid waste management and recycling behavior studies. Prof. Isabel Gomes has scientific expertise in operational research applied to waste collection and reverse logistics. To keep us focused on the goal of this book, we required someone from the waste collection professional field. Dr. Susana Rodrigues is a waste collection manager, which brought the vision and expertise from the technical operation of waste collection in practice.

This book intends to provide those who work in the scientific field of waste management and who are practitioners the backgrounds of waste collection and its incredible role in the success of an entire waste management system. Bringing the most recent developments on the subject to people who are not keen in searching for scientific articles to obtain knowledge and apply it to its professional life is the challenge of this book. We do not intend to define the best technology to implement waste collection. We want to give readers the tools to improve waste collection by integrating their work within the entire waste management system.

A particular interest for graduate students: this book shows the recent technology tendencies in the field, which will help students finding new directions their study and graduation at waste collection systems. This book will allow students to understand the applicability of system analysis through case studies.

The timeliness of this book is justified by the current context of essential changes in the waste management sector and the critical role that legal aspects and organizations have on the promotion of the sustainable development on a sector essential for well-being and population health. We are facing challenges in developing countries where integrated solid waste management systems are being built, and the scientific background of this book can help them to direct waste collection to become more efficient and sustainable. In developed countries, we are shifting away from a waste collection system, with no value product collection and management, to a resource collection and management system. This shift is a real challenge, where the consumer (the product owner) has the ability to make the collection and management system work as a provider of raw materials to the economy.

Last but not least, we would like to acknowledge the precious help given by Springer in supporting all the questions, issues, and delays that occurred during the writing of this book. Writing a book in English by nonnative speakers required extra effort from us. Springer was very helpful in the entire process of writing and editing this book.

Caparica, Portugal
May 2018

Ana Pires
Graça Martinho
Susana Rodrigues
Maria Isabel Gomes

Contents

Part I Fundamental Background

1 Introduction 3
   1.1 The Concept of Sustainable Development 3
   1.2 Sustainability in the Context of Solid Waste Collection and
       Management 5
   1.3 The Framework for Sustainability Assessment 6
   1.4 The Structure of This Book 8
   References 10

**2 Prevention and Reuse: Waste Hierarchy Steps Before Waste
Collection** 13
   2.1 Waste Hierarchy Principle: Saving Materials Before Becoming
       Waste 13
   2.2 Waste Prevention 16
   2.3 Products and Goods Reuse 19
   2.4 Final Remarks 21
   References 21

3 Technology Status of Waste Collection Systems 25
   3.1 Waste Collection 25
       3.1.1   Waste Collection Role 25
       3.1.2   Waste Collection Systems 27
   3.2 Waste Collection System Classification 28
       3.2.1   State of the Art 29
       3.2.2   Waste Collection System Types 32
   References 42

**4 Preparation for Reusing, Recycling, Recovering, and Landfilling:
Waste Hierarchy Steps After Waste Collection** 45
   4.1 Waste Hierarchy Principle: All After Becoming Waste 45
   4.2 Preparing for Reuse 47

4.3	Recycling		48
	4.3.1	Upcycling	49
	4.3.2	Recycling	50
	4.3.3	Downcycling	50
	4.3.4	Recycling Challenges	51
	4.3.5	Remarks	52
4.4	Other Recovery		53
4.5	Disposal		55
4.6	Final Remarks		56
References			57

5 Economic Perspective 61

5.1	International Legislation on Waste		61
	Basel Convention		61
	5.1.1	European Union Waste Policies	63
5.2	National Waste Regulation in European Union Countries		63
	Batteries and Accumulators		65
	5.2.1	End-of-Life Vehicles (ELV)	66
	5.2.2	Packaging Waste	67
	5.2.3	Waste Electrical and Electronic Equipment	68
	5.2.4	Waste Oils	69
5.3	Final Remarks		69
References			70

6 Psychosocial Perspective 73

6.1	Contributions of Social Psychology to Source Separate Waste Collection		73
6.2	Determining Factors of Recycling Behaviors		74
6.3	Understanding and Predicting Models of Recycling Behaviors		76
	6.3.1	Schwartz Model of Altruistic Behavior	77
	6.3.2	Theory of Reasoned Action and Theory of Planned Behavior	78
6.4	Strategies to Change Behaviors and Their Evaluation		81
6.5	Current Limitations and Future Perspectives for Social Psychology		86
References			89

7 Economic Perspective 95

7.1	Waste Collection Costs		95
	7.1.1	Investment Costs (CAPEX)	103
	7.1.2	Operating and Maintenance Costs (O&M)	106
7.2	Financial Concerns of Waste Management Systems and Instruments of Waste Policy		108
7.3	Public and Private Sector Financing		116
References			119

Contents ix

8 Environmental Context 123
    8.1 Environmental Context of Twenty-First Century 123
        8.1.1   Globalization and Economic Growth 124
        8.1.2   Megacities, Eco-cities, and Industrial
                Symbiosis at Cities 126
        8.1.3   Climate Change 128
    8.2 Sustainability and Circular Economy Considerations 129
    8.3 Adaptive Management Strategies for Waste Collection
        Systems 131
    8.4 Final Remarks 134
    References 135

Part II Models and Tools for Waste Collection

9 Design and Planning of Waste Collection System 141
    9.1 Waste Generation Estimation 141
        9.1.1   Time Series 143
        9.1.2   Forecast Accuracy 144
        9.1.3   Linear and Multiple Linear Regression Models 146
        9.1.4   Advanced Forecast Models 146
        9.1.5   Case Study: Using Time-Series Models
                to Estimate MSW in Kaunas, Lithuania 147
    9.2 Waste Collection System Planning and Selection 148
        9.2.1   Factors to Consider When Planning a Collection
                System 149
        9.2.2   Factors to Consider When Selecting WCS Devices 150
    9.3 The Role of GIS in Waste Collection Planning 152
        9.3.1   Routes Definition 153
        9.3.2   Case Study: Minimizing Operational Costs
                and Pollutant Emission in Collection Routes Using GIS 156
    9.4 Conclusion 157
    Appendix A: Forecasting Methods 158
        A.1: Naïve Forecast Model 158
        A.2: Moving-Average Models 158
        A.3: Exponential Smoothing Model 159
        A.4: Holt's Model 159
        A.5: Holt-Winters Method 160
        A.6: ARIMA Models 160
    Appendix B: Measures of Accuracy 161
        B.1: Mean Absolute Error (MAE) 161
        B.2: Geometric Mean Absolute Error (GMAE) 161
        B.3: Mean Square Error (MSE) 162
        B.4: Mean Absolute Percentage Error (MAPE) 162
        B.5: Geometric Mean Relative Absolute Error (GMRAE) 162

Appendix C: Linear Regression Models		162
C.1: Simple Linear Regression Model		162
C.2: Multiple Linear Regression Model		163
References		164

10 Operation and Monitoring ... 167
10.1 Descriptive Indicators ... 167
10.2 Performance Indicators ... 168
    10.2.1 Technical-Operative and Logistics Indicators ... 168
    10.2.2 Case Study: Calculating Waste Volume
             Weight Inside the Container and Emptying
             Time in Greater Lisbon Area, Portugal ... 171
10.3 Economic Indicators ... 175
10.4 Environmental Indicators ... 178
10.5 Social Indicators ... 179
10.6 Final Remarks ... 179
References ... 180

11 Assessment and Improvement ... 183
11.1 Life Cycle Assessment and Carbon Footprint ... 183
    11.1.1 Goal and Scope Definition ... 184
    11.1.2 Life Cycle Inventory ... 188
    11.1.3 Life Cycle Impact Assessment ... 189
    11.1.4 Interpretation ... 191
    11.1.5 LCA Software ... 192
    11.1.6 Carbon Footprint ... 192
11.2 Life Cycle Costing ... 193
11.3 Social Life Cycle Assessment ... 194
11.4 Behavior Studies and Awareness Campaigns ... 197
11.5 Final Remarks ... 198
References ... 199

Part III Sustainable Solid Waste Collection: Integrated Perspective

**12 Optimization in Waste Collection to Reach Sustainable Waste
Management** ... 207
12.1 Introduction ... 207
12.2 Single Objective Models ... 208
    12.2.1 Linear Programming Model ... 208
    12.2.2 Mixed-Integer Linear Programming ... 210
    12.2.3 Stochastic Programming ... 212
    12.2.4 Nonlinear Programming ... 214
    12.2.5 Solving a Linear Programming Model ... 215
12.3 Some Special Problems ... 215
    12.3.1 Traveling Salesman Problem ... 216
    12.3.2 Vehicle Routing Problem ... 218

Contents   xi

	12.3.3	Chinese Postman Problem	220
	12.3.4	Transportation Problem	222
	12.3.5	Location Problem	223
12.4	Multiple Objectives		224
	12.4.1	Lexicographic Method	226
	12.4.2	Weighted Sum Method	227
	12.4.3	Distance Minimization to the Ideal Point	227
	12.4.4	ε-Constraint Method	229
	12.4.5	Iterative Methods	230
12.5	Case Study 1: Integrated Assessment of a New Waste-to-Energy Facility in Central Greece in the Context of Regional Perspectives		230
12.6	Case Study 2: A Recovery Network for WEEE – A Sustainable Design		231
	12.6.1	The Current Network	232
	12.6.2	The Optimal Network	233
	12.6.3	Scenario Comparison	234
	12.6.4	Multi-Objective Analysis	234
12.7	Final Remarks		235
References			236

13 Multi-criteria Decision-Making in Waste Collection to Reach Sustainable Waste Management 239

13.1	Introduction		239
13.2	Generic Multi-Criteria Analysis Methodology		241
13.3	Multi-Criteria Decision Aid Methods		242
	13.3.1	Simple Additive Weighting (SAW)	243
	13.3.2	Analytic Hierarchy Process (AHP)	243
	13.3.3	TOPSIS	244
	13.3.4	PROMETHEE	245
	13.3.5	ELECTRE	245
	13.3.6	Comparison of MCDA Methods	246
13.4	Sensitivity Analysis		246
13.5	MCDA Software		250
13.6	Dealing with Multiple Stakeholders in the Decision-Making Process		251
13.7	MCDA Case Studies		252
	13.7.1	Multi-criteria Decision Analysis for Waste Management in Saharawi Refugee Camps	253
	13.7.2	Multi-Criteria Analysis as a Tool for Sustainability Assessment of a Waste Management Model	254
	13.7.3	Ranking Municipal Solid Waste Treatment Alternatives Based on Ecological Footprint and Multi-criteria Analysis	255

| | | 13.7.4 | An AHP-Based Fuzzy Interval TOPSIS Assessment for Sustainable Expansion of the Solid Waste Management System in Setúbal Peninsula, Portugal | 256 |

13.7.4   An AHP-Based Fuzzy Interval TOPSIS Assessment
         for Sustainable Expansion of the Solid Waste
         Management System in Setúbal Peninsula, Portugal . . 256
13.7.5   Assessment Strategies for Municipal Selective Waste
         Collection Schemes 257
13.8   Final Remarks 258
References 259

14   A Sustainable Reverse Logistics System: A Retrofit Case 261
14.1   Introduction 261
14.2   Sustainability Objectives 263
14.3   Modeling and Solution Approach 264
14.4   Results and Analysis 266
       14.4.1   Routes Generation 266
       14.4.2   Sustainable Collection System 267
14.5   Conclusion 272
Annex A: Multi-objective Formulation for the MDPVRPI 273
Annex B: Solution Procedure 277
       B.1 Step 1: Routes Generation Procedure 277
       B.2 Step 2: Solution Method for the Multi-objective Problem . . 282
References 284

**15   Collection of Used or Unrecoverable Products: The Case
     of Used Cooking Oil** 287
15.1   Introduction 287
15.2   Company Current Operation Mode 289
15.3   The New Collection Network 291
       15.3.1   If the Current Network Is Optimized 291
       15.3.2   Network Expansion 293
15.4   Conclusions 299
Annex A 299
References 303

**Part IV   Challenges and Perspectives for Sustainable Waste
           Management Through Waste Collection**

16   The Evolution of the Waste Collection 307
16.1   Definition of Integrated Waste Collection Concept 307
16.2   The Functioning of the Integrated Waste Collection (IWC) 310
       16.2.1   How Collection Interacts with Waste Prevention 310
       16.2.2   How Collection Interacts with Preparation for Reuse . . 311
       16.2.3   How Collection Interacts with Sorting for Recycling . . 312
       16.2.4   How Collection Interacts with Biological Treatment . . 313
       16.2.5   How Collection Interacts with Energy Recovery 316
       16.2.6   How Collection Interacts with Disposal 318

Contents xiii

16.3 Sustainability in Integrated Waste Collection 318
16.4 Final Remarks 319
References 320

17 Trend Analysis on Sustainable Waste Collection 323
17.1 Reverse Logistics 323
17.2 Crowd Logistics 328
17.3 Physical Internet 329
17.4 Freight on Transit 330
17.5 Final Remarks 331
References 331

18 Technical Barriers and Socioeconomic Challenges 335
18.1 Developed Countries 335
18.1.1 Advancements in Environmental Informatics 335
18.1.2 Advancements in Information and Communication
Technology 336
18.1.3 Waste Infrastructure Synergies 337
18.1.4 Reaching All-in-One: Citizens Satisfied
and Participative, Cost Affordable, and Low
Environmental Impact of the Waste Management
System 338
18.2 Developing Countries 339
Basics on Waste Collection System Are Still
in Development 339
18.2.1 The Conversion of Informal Sector into Formal
Waste Management Sector 340
18.2.2 The Importation of Hazardous Waste and Trade
of Hazardous Waste 342
18.2.3 Public Health Related to Mismanagement of Waste
and Its Dependents 343
18.2.4 Social Apathy for Participation 344
18.3 Final Remarks 345
References 345

19 Future Perspectives 349
19.1 The Goals of the 2030 Agenda for Sustainable Development . . . 349
19.1.1 SGD 1 "No Poverty" and SWM 350
19.1.2 SDG 2: Zero Hunger 350
19.1.3 SDG 3: Good Health and Well-being 351
19.1.4 SDG 6: Clean Water and Sanitation 351
19.1.5 SDG 7: Affordable and Clean Energy 352
19.1.6 SDG 8: Decent Work and Economic Growth 352
19.1.7 SGD 9: Industry, Innovation, and Infrastructure 354
19.1.8 SDG 11: Sustainable Cities 354
19.1.9 SDG 12: Responsible Consumption and Production . . 355

	19.1.10	SDG 13: Climate Action	356
	19.1.11	SDG 14: Life Below Water	357
	19.1.12	SDG 15: Life on Land	358
	19.1.13	SDG 17: Partnerships	358
19.2	Final Remarks		359
References			359

Index 361

Part I
Fundamental Background

Chapter 1
Introduction

Abstract The statistics are precise: the population is increasing, and, consequently, the amount of waste generated in the entire world is increasing, ending at open dumpsites, with reduced recycling and recovery. The missing of integrated solid waste management systems that ensure controlled management, where environmental and health risks are reduced and where the waste system drives economic growth and social progress, are major challenges for science and engineering. This chapter intends to emphasize the essence of sustainable development in the collection and management of waste, how to make part of the waste management, and how it can constitute the framework of a usual integrated solid waste management. Case studies that show how to promote the sustainable waste collection and, consequently, solid waste management are introduced in the subsequent chapters.

Keywords Sustainable development · Waste collection · ISWM · Waste flows · SDGs

1.1 The Concept of Sustainable Development

The Sustainable Development Goals are the most recent initiative from the United Nations to leverage sustainable development in the world until 2030. The SDGs are present at the report "Transforming Our World" (United Nations 2015), predicted at "The Future We Want," the document resulting from the Rio+20 Conference, 20–22 June 2012, organized by the United Nations, where all countries were called to renew their commitment to reach a sustainable future.

The United Nations started to call countries to the sustainable development cause on 16 June 1972, when the United Nations Conference on the Human Environment occurred, held in Stockholm, where the "rights" of the human families to a healthy and productive environment were delineated (UN-DESA 2015a). The step was crucial to bring environment into the agenda of industrialized countries, and the theme was again brought into the spotlight with the publication of the "World Conservation Strategy" by the United Nations Environment Programme (UNEP), World Wide Fund for Nature (WWF), and International Union for Conservation of

© Springer International Publishing AG, part of Springer Nature 2019
A. Pires et al., *Sustainable Solid Waste Collection and Management*,
https://doi.org/10.1007/978-3-319-93200-2_1

Nature and Natural Resources (IUCNNR). This strategy was the precursor to the concept of sustainable development, which aimed (UNEP/WWF/IUCCNR 1980):

- To maintain fundamental ecological processes and life-support systems, vital for human survival and development
- To preserve genetic diversity, on which depend the breeding programs needed for the protection of plants and domesticated animals, as well as much scientific innovation, and the security of the many industries that use living resources
- To ensure the sustainable utilization of species and ecosystems (notably fish and other wildlife, forests, and grazing lands), which supports millions of rural communities as well as significant industries.

The discussion on the accelerated degradation of the environment and its effects on the economic development led the United Nations to discuss "The World Commission on Environment and Development" in 1983. In 1987, the Brundtland Report "Our Common Future" defined "sustainable development" as (WCED 1987):

The development that meets the needs of the present without compromising the ability of future generations to meet their own needs.

The first UN Conference on Environment and Development in Rio de Janeiro occurred in 1992, where the document of "Agenda 21: A Programme of Action for Sustainable Development," also known as *The Rio Declaration on Environment and Development*, was adopted (UN-DESA 2015b). The Agenda 21 establishes 27 principles around the 3 pillars of sustainability: economy, society, and environment. After Agenda 21, several other documents and programs have been elaborated by the United Nations, which are all being reaffirmed by "The Future We Want" (UN-DESA 2018):

The Programme for the Further Implementation of Agenda 21, the Plan of Implementation of the World Summit on Sustainable Development (Johannesburg Plan of Implementation) and the Johannesburg Declaration on Sustainable Development of the World Summit on Sustainable Development, the Programme of Action for the Sustainable Development of Small Island Developing States (Barbados Programme of Action) and the Mauritius Strategy for the Further Implementation of the Programme of Action for the Sustainable Development of Small Island Developing States.

The attempt to bring sustainability into practice is a real challenge, and continual removal of the goals is needed to keep the subject on the agenda. However, the definition of the goals has been considered to be vague, weak, or meaningless (Hopwood et al. 2005; Stafford-Smith 2014; Stokstad 2015). Holden et al. (2017) criticize the three pillars: the economic, social, and environmental, sustaining that the critical dimensions of sustainable development should be:

The moral imperatives of satisfying needs, ensuring equity and respecting environmental limits. The model reflects both moral imperatives laid out in philosophical texts on needs and equity, and recent scientific insights on environmental limits.

The concept of sustainable development by Holden et al. (2017) goes further, by developing a model that quantifies how sustainable the development of the country or groups of countries is. The model includes critical themes, indicators, and thresholds, which would be far way more practicable than the SDGs defined by the United Nations (2015).

The intention of this book goes much further in the establishment of sustainable development definitions. The way how solid waste collection and management should include sustainable development and contribute to the SDGs is, in fact, one of the goals of this book.

1.2 Sustainability in the Context of Solid Waste Collection and Management

Paragraph 218 of "The Future We Want" devotes to the development and enforcement of comprehensive national and local waste management policies, strategies, and regulations, regarding a life cycle approach and promotion of policies of resource efficiency and environmentally sound waste management (UNEP and UNITAR 2013).

Before defining any sustainability waste management policies, the definition of waste should be discussed. The need to define what waste is from what is not influences the need to control or not the output material resulting from a process or from the urban metabolism. According to the European Parliament and Council (2008):

Waste means any substance or object which the holder discards or intends or is required to discard.

All materials that can be considered as waste according to the definition can also be classified by source, nature, physical and mechanical properties, chemical and elemental properties, biological/biodegradable properties, and combustion properties (Chang and Pires 2015). Concerning the source, waste can be classified as municipal solid waste (which includes commercial and services waste), construction and demolition waste, medical waste, industrial waste, and other wastes that can require a specific identification. Concerning nature, waste can be classified in hazardous waste (presenting one hazard (at least) to humans or environment); inert waste, which has no transformation at physical, chemical, or biological levels; and nonhazardous waste, which has no hazardous features. Physical, mechanical, and chemical properties include physical composition, density, moisture content, particle size and size distribution, pH, chemical composition, C/N ratio, calorific value, and biological features (Chang and Pires 2015).

All the waste can also be divided by waste stream, i.e., by the product that gave origin to the waste. The need to define waste streams started at municipal solid waste, which presents high heterogeneity on materials due to the diversity of products consumed in the urban system. Waste streams present in municipal solid waste can be packaging waste (which is even divided by materials like paper, cardboard, glass, plastic, liquid carton beverages packaging, ferrous metals, nonferrous metals), batteries, food waste, biodegradable waste, green waste, waste of electrical and electronic equipment, construction and demolition waste, and domestic hazardous waste, and many others may appear. The definition of waste by streams is devoted to the need of being separately managed, in the beginning, because the recycling industry wanted the source-separated materials to be recycled in their process, i.e., an internal value market exists for those wastes. If waste streams with potential market value are removed, the residual fraction (or stream) can be managed in such way that only waste can be managed to ensure its processing more efficiently and straightforwardly.

The segregation of waste through its properties, mostly, can impulse its environmental sound management of waste, with financial revenues and positive social impacts. The segregation of waste leads to its management without being contaminated with hazardous (or nonhazardous) materials or allowing its maximum use. Segregation helps to increase the value of recyclables and recyclates, resulting value-added by-products. The use of recyclates contributes to a green gross domestic product (GDP), an index of economic growth that corrects the environmental consequences from GDP (Chang and Pires 2015). The social well-being reached with an integrated solid waste management and with the source separation of waste is notable, although source separation of waste requires citizens' participation which can be demanding and challenging for waste system managers. The public participation in the decision-making process on waste management is also a reality nowadays, where waste players, from products life cycle, can be brought to deliver strategic plans and actions plans to prevent and manage waste.

1.3 The Framework for Sustainability Assessment

Several authors (CSLFG/STSP/PGA/NRC 2013; Sala et al. 2015; Sonnemann et al. 2015) have elaborated their framework for sustainability assessment, with the aim of being used to support decision-making and policy development at any field. Sustainability assessment can have different definitions:

> The goal of sustainability assessment is to pursue that plans and activities make an optimal contribution to sustainable development. (Verheem 2002)

Alternatively,

> Sustainable assessment refers to the interaction of different methodologies in such a way that is geared toward obtaining an analysis, an evaluation, or a plan that approaches several management aspects in which the sustainability implications may be emphasized and illuminated. (Chang et al. 2011)

The proposed sustainability framework by Sala et al. (2015) (Fig. 1.1) presents the comprehensive approach required to impulse the sustainability in the waste

1.3 The Framework for Sustainability Assessment

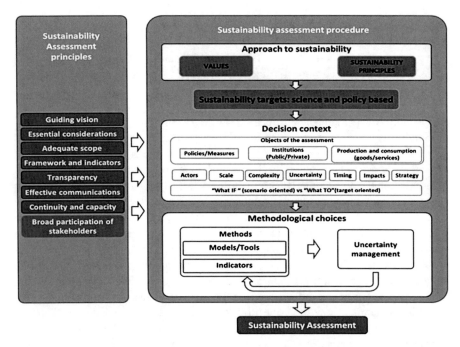

Fig. 1.1 Scheme of the conceptual framework for sustainability assessment. (Source: Sala et al. (2015))

sector. The principles of sustainability defined by Sala et al. (2015), i.e., guiding vision, essential considerations, adequate scope, framework and indicators, transparency, effective communications, continuity and capacity, and broad participation of stakeholders, can constitute the guide for practitioners to perform the assessment. The sustainability assessment procedure comprises several steps (Sala et al. 2015):

- Approach to sustainability: the values and sustainability principles are to be defined by the organization requiring the assessment.
- Sustainability targets: to define the level of sustainability intended to be accomplished.
- Decision context: where information from sustainability assessment will be translated in practical terms.
- Methodological choices for the assessment: the core of the sustainability analysis, which involves the identification of the most suitable assessment methodologies (methods, models, tools, indicators), sensitivity and uncertainty analysis, and definition of the monitoring strategies to track progress toward sustainability targets.

1.4 The Structure of This Book

The interactions between all pillars of sustainability and the urban, agriculture, and industry activities are complicated and often difficult to quantify, manage, and give a rapid response. Technologies, related to machinery or related to computational solutions, are called to contribute significantly to the resolution of sustainability issues in the waste management sector.

Solid waste management complexity occurs like in any other human-based processes. Making decisions in complex systems requires following principles, processes, and practices to proceed from information and desires to choices that inform actions and outcomes (Lockie and Rockloff 2005). The decision process is different in each phase of the waste collection and management. In the begging, the design, planning of the waste collection, and management must consider not only the international, national, and local strategies devoted to waste management (like collection rates, recycling rates) but also the local context where the waste management system is to be implemented, which will define and constraint the type of infrastructure to be employed. During the operation of a waste collection system, decision process requires information on the amounts of waste being collected for several destinations, how the collection routes and vehicles allocated to them are defined, and the waste quality reached, which will influence the following waste management processes. The collection system requires constant redefinition, not only due to the changes in the background where the system is occurring but due to the waste collection system itself, where decisions to improve its performance have to be taken, to not disregard the economic, environmental, and social impacts (positive and negative) of the waste collection system. The book intends to give highlights to waste collection practitioners on the view of their system and is intended to be used in conjunction with existing literature and other relevant guidance, firstly by academic researchers, policymakers, and waste researchers in public and private sectors. It also aimed to challenge the research for an interdisciplinary view where climate change, economic growth, environmental pollution, and social impact are variables which need to be brought to the waste management decision-making.

This book proposes a systemic decision framework where the waste collection has the leading role to leverage solid waste management at a more sustainable level. The processes inherent to waste collection operation are discussed and treated through science-based analysis with various perspectives of sustainable solid waste management. A comprehensive bibliography is provided at the end of each chapter, and some case studies are presented to describe how the system thinking can promote the needed management to reach sustainability in the waste management sector. The integrated approach is reflected in the structure of the four parts as follows.

Part I: Fundamental Background. The context on waste hierarchy upward to waste collection like waste prevention, reduction and reuse, the waste collection itself, and afterward waste hierarchy steps like preparation for recycling, recycling, treatment, and landfilling is provided. The related sustainability science background regarding the environment, social, and economic perspectives of sustainable solid waste management will be addressed. The following chapters lead to the integrated discussion on the role of waste collection in the solid waste management and on the waste hierarchy principle:

- Introduction (Chap. 1)
- Prevention and Reuse: Waste Hierarchy Steps Before Waste Collection (Chap. 2)
- Technology Status of Waste Collection Systems (Chap. 3)
- Preparation for Reusing, Recycling, Recovering, and Landfilling: Waste Hierarchy Steps After Waste Collection (Chap. 4)
- Economic Perspective (Chap. 5)
- Psychosocial Perspective (Chap. 6)
- Economic Perspectives (Chap. 7)
- Environmental Context (Chap. 8)

Part II: Models and Tools for Waste Collection. The waste collection in solid waste management is an operation unit that requires, at first, design and planning and then the operation of the collection itself. During operation, it is required to monitor the operation, to understand if it occurs according to the plan. At last, the assessment and improvement of the waste collection system are needed, to find the constraints to be solved to help the collection to be more sustainable and integrated into the solid waste management system.

- Design and Planning of Waste Collection System (Chap. 9)
- Operation and Monitoring (Chap. 10)
- Assessment and Improvement (Chap. 11)

Part III: Sustainable Solid Waste Collection: Integrated Perspective. The role of sustainability in the way how waste is collected and consequences to the solid waste management system is discussed in this part. The following chapters are organized to provide information on the use of systems analysis methods – optimization and multi-criteria decision-making – as well as case studies where those methodologies are used to improve the waste collection systems regarding economic, environmental, and social perspectives.

- Optimization in Waste Collection to Reach Sustainable Waste Management (Chap. 12)
- Multi-Criteria Decision-Making in Waste Collection to Reach Sustainable Waste Management (Chap. 13)
- A Sustainable Reverse Logistics System: A Retrofit Case (Chap. 14)
- Collection of Used or Unrecoverable Products: The Case of Used Cooking Oil (Chap. 15)

Part IV: Challenges and Perspectives for Sustainable Waste Management Through Waste Collection. The waste collection requires new approaches to face the challenges of the future to make economic growth which is capable of satisfying the needs of the citizens, ensuring equity on accessing the waste collection and the solid waste management service, and respecting the environment and public health.

- The Evolution of the Waste Collection (Chap. 16)
- Trend Analysis on Sustainable Waste Collection (Chap. 17)
- Technical Barriers and Socioeconomic Challenges (Chap. 18)
- Future Perspectives (Chap. 19)

References

Chang NB, Pires A (2015) Sustainable solid waste management: a systems engineering approach. IEEE Wiley, Hoboken

Chang NB, Pires A, Martinho G (2011) Empowering systems analysis for solid waste management: challenges, trends, and perspectives. Crit Rev Environ Sci Technol 41:1449–1530

Committee on sustainable Linkages in the Federal Government, Science and Technology for Sustainability Program, Policy and Global Affairs, National Research Council (CSLFG/STSP/ PGA/NRC) (2013) Sustainability for the nation: resources connection and governance linkages. National Academy Press, Washington, DC

European Parliament, Council (2008) Directive 2008/98/EC of the European Parliament and of the Council of 19 November 2008 on waste and repealing certain directive. Off J L312:3–30

Holden E, Linnerud K, Banister D (2017) The imperatives of sustainable development. Sustain Dev 25:213–226

Hopwood B, Mellor M, O'Brien G (2005) Sustainable development: mapping different approaches. Sustain Dev 13:38–52

Lockie S, Rockloff S (2005) Decision frameworks: assessment of the social aspects of decision frameworks and development of a conceptual model. Coastal CRC discussions paper, Central Queensland University, Norman Gardens, Australia

Sala S, Ciuffo B, Nijkamp P (2015) A systemic framework for sustainability assessment. Ecol Econ 119:314–325

Sonnemann G, Gemechu ED, Adibi N, De Bruille V, Bulle C (2015) From a critical review to a conceptual framework for integrating the criticality of resources into Life Cycle Sustainability Assessment. J Clean Prod 94:20–34

Stafford-Smith M (2014) UN Sustainability goals need quantified targets. Nature 513:281

Stokstad E (2015) Sustainable goals from UN under fire. Science 347:702–703

United Nations (2015) Transforming our world: the 2030 agenda for sustainable development. Resolution adopted by the General Assembly on 25 September 2015, A/RES/70/1. UN General Assembly, New York

United Nations Department of Economic and Social Affairs (UN-DESA) (2015a) United Nations Conference on the Human Environment (Stockholm Conference). https://sustainabledevelopment.un.org/milestones/humanenvironment. Accessed 27 Apr 2018

United Nations Department of Economic and Social Affairs (UN-DESA) (2015b) Agenda 21-UNCED, 1992. https://sustainabledevelopment.un.org/outcomedocuments/agenda21. Accessed 27 Apr 2018

References

United Nations Department of Economic and Social Affairs (UN-DESA) (2018) Future we want – outcome document. https://sustainabledevelopment.un.org/futurewewant.html. Accessed 24 Apr 2018

United Nations Environment Programme (UNEP), United Nations Institute for Training and Research (UNITAR) (2013) Guidelines for national waste management strategies. Moving from challenges to opportunities. UNEP, Nairobi

United Nations Environment Programme, World Wild Fund for nature, International Union for Conservation of Nature and Natural Resources (UNEP/WWF/IUCCNR) (1980) World conservation strategy-living resource conservation for sustainable development. IUCN/UNEP/WWF, Gland

Verheem R (2002) Recommendations for sustainability assessment in the Netherlands. In commission for EIA. Environmental impact assessment in the Netherlands. Views from the Commission for EIA in 2002, The Netherlands

World Commission on Environment and Development (WCED) (1987) Our common future. Oxford University Press, Oxford

Chapter 2
Prevention and Reuse: Waste Hierarchy Steps Before Waste Collection

Abstract The way how policy instruments and actions can impose measures before products became waste depends on policies based on the waste prevention, reduction, and reuse. A brief review on the concepts in the light of the waste hierarchy principle is discussed, considering the view of European countries and when possible from other countries in the world.

Keywords WHP · Waste Framework Directive · Products reuse · Minimization · Design

2.1 Waste Hierarchy Principle: Saving Materials Before Becoming Waste

Waste is a generic and large concept, which requires definition and, from there, define the strategies to avoid or minimize its generation. Authorities (national and international) defined waste differently:

- Waste Framework Directive (2008/98/EC): Any substance or object which the holder discards or intends to discard or is required to discard.
- US Resource Conservation and Recovery Act (USEPA 2017): any garbage or refuse, sludge from wastewater plant, water supply treatment plant or air pollution control facility, and other discarded materials, resulting from industrial, commercial, mining, and agricultural operations, and from community services.
- Inter-American Development Bank, definition applied at Caribbean and Latin countries (Espinoza et al. 2010): Solid or semisolid waste produced through the general activities of a population center. It includes waste from households, commercial businesses, services, and institutions, as well as common (nonhazardous) hospital waste, waste from industrial offices, waste collected through street sweeping, and the trimmings of plants and trees along streets and in plazas and public green spaces.
- Environment Protection Act (EPASA 2018), for Australia: any discarded, rejected, abandoned, unwanted, or surplus matter, whether or not intended for sale or for recycling, reprocessing, recovery, or purification by a separate

© Springer International Publishing AG, part of Springer Nature 2019 13
A. Pires et al., *Sustainable Solid Waste Collection and Management*,
https://doi.org/10.1007/978-3-319-93200-2_2

operation from that which produced the matter, or anything declared by regulation or by an environment protection policy to be a waste, whether of value or not.
- Act on Waste Management at South Korea (Chung 2011): A material that is unnecessary for human life and business activities such as garbage, combustible ashes, sludge, waste oil, waste acid, waste alkali, carcass, etc. and some waste are defined as waste at courts.
- Law n.12.305 (WIEGO 2018) in Brazil: any material, substance, object, or disposed good resulting from human activities in society, whose final destination proposes to proceed or is obliged to proceed in solid or semisolid states, as well as gases and liquids within containers unfeasible to be released into the public sewage system or water bodies, or that require technically or economically unviable solutions in view of the best available technology.

The way how waste should be managed has been, until now, defined by the waste hierarchy principle (WHP). This principle establishes the preferable order in which the solid waste should be managed and treated, being, firstly, preferred the prevention, reuse, recycling, and recovery over landfill (Hultman and Corvellec 2012). The first time that WHP were introduced in European legislation was at 1975 Directive on waste (European Council 1975) and EU's Second Environment Action Program in 1977 (European Commission 1977) and finally defined at the Community Strategy for Waste Management in 1989 (European Commission (1989). Typically the WHP is presented as an inverted pyramid, where the preferred option is on the top and in bigger proportion than the subsequent management options, like in the case of WHP from European Waste Framework Directive (Fig. 2.1).

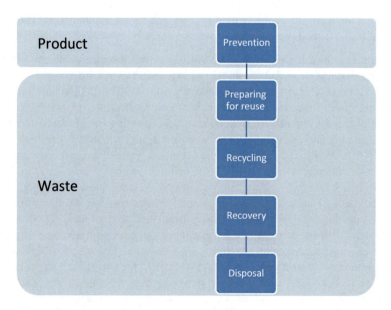

Fig. 2.1 Waste hierarchy principle according to Waste Framework Directive of European Union

2.1 Waste Hierarchy Principle: Saving Materials Before Becoming Waste

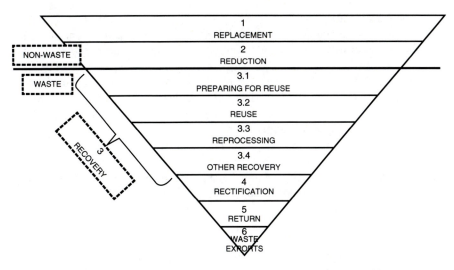

Fig. 2.2 Six stages of the hierarchy of resources use. (Source: Gharfalkar et al. (2015))

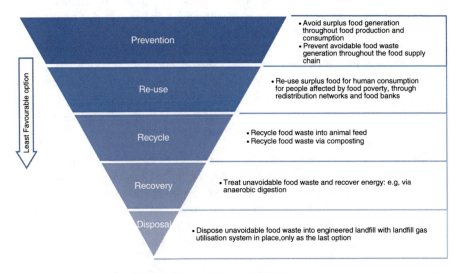

Fig. 2.3 Food WHP. (Source: Papargyropoulou et al. (2014))

An improvement of the Waste Framework Directive has been proposed by Gharfalkar et al. (2015), named "hierarchy of resource use" where a more detailed type of operations can be conducted and prioritized. This new hierarchy intends to help policymakers to provide the adequate incentives on the right waste management operations (Gharfalkar et al. 2015) (Fig. 2.2).

More recently, elaborations of WHP for specific waste streams are occurring. Papargyropoulou et al. (2014) developed a WHP specific for food waste (Fig. 2.3), being to define measures on food waste management but also at the food supply, where waste generation also occur. Knauf (2015) revised the European WHP for waste wood considering the European Union energy policy and European market and life cycle assessment studies on wood waste management, proposing that recycling or other recoveries such as energy recovery have the same level of priority. Richa et al. (2017) proposed and analyzed a WHP combined with circular economy to manage lithium-ion batteries, being highlighted that operations of reuse (direct or cascaded) followed by recycling can be better in terms of ecotoxicity burden that banning such batteries from landfill.

Definitions on WHP all over the world have similarities but also differences that make it difficult to conduct a standard view of all measures to minimize or avoid waste generation. To the better acknowledgment of the concepts, waste prevention and reuse will be characterized by the European view.

2.2 Waste Prevention

In the European view, prevention includes reduction. Reduction includes waste amount reduction, adverse impacts on health and environment, and reduction of harmful content (European Parliament and Council (2008)). A clear message from Waste Framework Directive is that prevention is for products and goods, not for waste. Waste reduction is, sometimes, only seen has the reduction of waste amount landfilled, or send for incineration, not the reduction of waste generated. Even with this narrowed view of waste reduction in terms of its destination, waste prevention concept can be considered, because the adverse impacts from waste management are being prevented (less waste going to landfill or to incineration, lesser environmental adverse impacts).

According to Hutner et al. (2017), types of waste prevention are reduction at source, substitution, and intensification, although intensification is more related to reuse (see Sect. 2.3). Reduction at source occurs during design and production, by applying ecodesign, which is an approach to "design out" waste and other environmental problems but keeping products quality and cost-effectiveness (Bârsan and Bârsan 2014). To prevent waste, ecodesign can focus on the type of materials to be used (environmentally friendly materials, recycled materials), reduction of material input, avoid waste during manufacturing, reduction of packaging, optimization of the product's functionality (which includes multiple functions), prolongation of product lifetime, waste prevention at use stage, and facilitation of maintenance (Wimmer and Züst 2003). Substitution intends to change the materials used in manufacturing to reduce hazardous component (already considered by ecodesign) or to substitute the product or service itself in the sale point by one that generates less waste (durable, reparable) (Hutner et al. 2017).

Waste prevention practices can be implemented by different policy measures. Regulatory, voluntary, and information instruments are possible strategies to implement (Table 2.1). In the study of Kling et al. (2016), the comparison of several economic instruments for waste prevention showed that PAYT is the preferable one

2.2 Waste Prevention

Table 2.1 Policy instruments on waste prevention

Policy instruments	Waste prevention instruments
Regulatory	Landfill ban, incineration bans, plastic bag bans, disposable cutlery bans, to-go or single-use products ban
Market-based	PAYT, landfill tax, incineration taxes and fees, extended producer responsibility principle, precycling insurance, recycling insurance, taxes on products (packaging, plastic bags)
Information	Awareness campaigns, school campaigns, procurement guidelines, information exchange platforms
Voluntary	Home composting, ecodesign of products, designing out waste, bottleless water, nappy laundry services, planning food meals

concerning utility, together with landfill tax. More nonconventional instruments are insurances, for recycling and precycling. Precycling means the "actions taken now to prepare for current resources to become future resources, rather than wastes accumulating in the biosphere" (Greyson 2007). Insurances would serve as a guarantee that future recycling costs or future waste management costs of the product are paid. A recent area where waste prevention is getting further steps is festivals and events. In the study conducted by Martinho et al. (2018), a festival applied mugs to avoid the acquisition of bottled drinks, reusable cutlery at canteen, sugar bowls, proper portion of food, and drinking fountains. The festival is known by the reduced amount of waste generated comparatively to other festivals (Martinho et al. 2018), showing how those measures can be important to promote waste prevention. Another effort to promote waste prevention in Portugal has been the plastic bag tax (Martinho et al. 2017a). The tax was capable to force a change at inquiries, shifting from single-use plastic bags acquisition to reusable bags but also to garbage bags, since single-use plastic bags were used as garbage bags. Plastic bags fee or tax has a considerable positive impact in the reduction on its acquisition in several other European countries (Table 2.2).

The design of instruments requires a profound knowledge of behaviors of the stakeholders which is intended to change the behavior. Without knowing the factors, the instruments to be applied may fail, just because instruments were not transferred considering those factor implications. The study of factors influencing the behavior of waste prevention has been made in the recent years. Cecere et al. (2014) found that prevention behavior is influenced by seldom socially oriented, seldom exposed to peer pressure, and very reliant on purely "altruistic" attitudes. Bortoleto et al. (2012) affirm that clear instructions are needed to citizens prevent waste, where information should emphasize that waste prevention is economically an alternative and has no inconvenient to the citizen. This approach puts in practice the factors of prevention behavior found by Bortoleto et al. (2012): that environmental concern, moral obligation, and inconvenience.

Table 2.2 Policy instruments applied in some European countries for plastic bags

Country	Policy instruments	Outcomes	References
Belgium (2007)	Tax or levy with voluntary agreement	60–80% of reduction	Bio Intelligence Service (2011)
Denmark (1994)	Tax or levy (also for paper bags)	A reduction of 50% on the amount of plastic bags	OECD (2001), The Danish Ecological Council (2015)
Ireland (2002)	Tax or levy	Reduce use by more than 90% and raised revenues around €12–14 million for an environment fund	Convery et al. (2007)
Luxembourg (2004, 2007)	Voluntary agreement to sale "Eco-sac" carrier bag in 2004. Bags started to be charged in 2007, including single-use bags	Saved about 560 million single-use shopping bags until 2013	Valorlux (2014)
Malta (2009)	Tax or levy	Saved around 25 million plastic bags (i.e., more than 50%, corresponding to roughly 150 tons of plastic) in the first 2 years after introducing the tax	Hermann et al. (2011)
Portugal (2017)	Tax	Reduction 20–30% on plastic bags sale in the first 8 months	Silva (2015)
Spain (2009)	Voluntary agreements in Catalonia	A reduction of 40% was achieved in 2010	Bio Intelligence Service (2011)
Romania (2009)	Tax or levy	An increase of plastic bags was verified between 2009 (27 million bags) and 2010 (60 million of bags)	Pre-waste (2011)
UK (Wales in 2011, Northern Ireland in 2013, Scotland in 2014, England in 2015)	Tax or levy with awareness campaigns and voluntary agreement	Wales: 71% reduction in 2015 Northern Ireland: 72% in 2014 Scotland: around 80% in 2015 England: 85% in 2016	BBC (2015), Bio Intelligence Service (2011), DAERA (2016), Howell (2016), Poortinga et al. (2013), The Guardian (2015, 2016)

Source: Adapted from Martinho et al. (2017a)

2.3 Products and Goods Reuse

Besides the definition of prevention, Waste Framework Directive 2008/98/EC also defines specific operation on products and goods before they become waste. The one defined is reuse, as being "any operation by which products or components that are not waste are used again for the same purpose for which they were conceived." It differs from "preparation for reuse" because, in this case, it consists of "checking, cleaning or repairing recovery operations, by which products or components of products that have become waste are prepared so that they can be re-used without any other pre-processing." In practical terms, reuse occurs inside the factory, or a company, or inside homes, when products are repaired or subjected to any operation that makes them usable again. The prevention mentioned here is before the good is put on waste collection system, because if such happens, it means that the users intended to get rid of it. All the reuse at a prevention perspective intends the notion that the users intend to keep using the product or good.

In this reuse concept, the operations that could implement reuse as preventing waste are remanufacturing, refurbishing, where the owner of the product does not change, and the owner itself can make the changes needed to the product keep performing in the same. Intensification considered by Hutner et al. (2017) fits better in reuse definition. The product used by the owner can be increased by sharing the product with other users or by prolonging the use phase of the product through repair. Reuse can be promoted by ecodesign, through improvement of reparability and by design for reuse of parts of the product (Wimmer and Züst 2003).

Several policy instruments can be implemented to promote products or goods reuse. Green public procurement applied can require products or goods with high levels of durability and easiness of repair, making them adequate to be used until exhaustion. Awareness campaigns about sustainable consumption in the way that could promote the acquisition of goods that can be used for such a long time are another type of instrument. The awareness concerning sustainable consumption is relevant. In the study conducted by Martinho et al. (2017b), men are keen to change their smartphone and tablets because they are obsolete and new models appear on the market. The notion of obsolesce is dependent of the appearing of new devices, with questionable new features, being a marketing strategy from devices producers. A better information of citizens concerning the environmental consequences of their consumption pattern is needed to break such pattern.

Voluntary actions such as product service systems can be useful to promote product reuse. Product service systems are "a marketable set of products and services capable of jointly fulfilling a user's need. The product/service ratio in this set can vary, either in terms of function fulfilment or economic value" (Goedkoop et al. 1999). The way how PSS promotes reuse is through the sale of the use (the service) of the product and not the product keeping its ownership (like laundry services), by implementing the leasing, which induces changes in consumer from product acquisition into service acquisition (Mont 2002; Roy 2000). According to Roy (2000), product service system can be divided into four types: service products (or demand services or result services), shared utilization services, product-life extension services, and demand-side

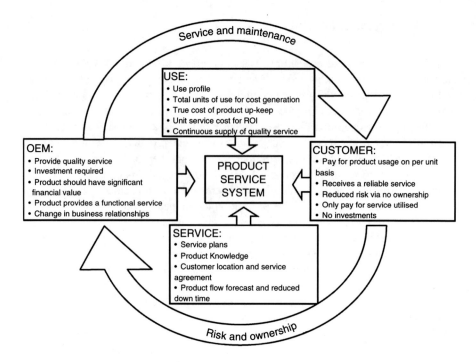

Fig. 2.4 Main features of a product service system. (Source: Bindel et al. (2012))

management (although this last one is applied to energy systems only). Service products intend to sell the result or service that the device/machinery can provide, like in the case of laundry service, which sells clean clothes rather than washing machine (Roy 2000). One of the main examples of service products is Xerox. The modular design strategy of their products allows the remanufacturing of their products, converting 160,000 machines from European customers (in 1997) by reprocessing them, making US$ 80 million of net savings (Maslennikova and Foley 2000). Shared utilization services (also called product-use services or community products) intend to increase the use of products by sharing, like in the case of community wash center instead of individual households washing machines (Roy 2000). Car sharing and bicycle pooling can also be cases of shared utilization service of transportation and can reduce the car use and the need to manufacture and park fewer vehicles (Roy 2000). Product-life extension services aim to increase the useful lifetime of products or goods through a spiral-loop system that minimizes matter, energy, and environmental deterioration, without compromising economic growth and progress (Giarini and Stahel 1993). This is a service being provided to electric and electronic equipment in countries like Portugal, after the 2-year guarantee provided by European regulation. Also, car stations are promoting product-life extension services, once cars are durable and expensive goods that compensate their owner to repair and provide maintenance to avoid the acquisition of a new one.

In a graphic representation of how a product service system works (Fig. 2.4), is notorious that the changing of the ownership from customer to the manufacturer will make the goal is not sale the product, but instead, to increase its operation time, leading more durable products, design for disassembly products, extending as far as technically and economically possible the use phase of the product. Issues such as customer location, product return planning, and service/maintenance plans are additional tasks of the manufacturer that in a product-oriented approach they do not occur (Bindel et al. 2012).

2.4 Final Remarks

The waste hierarchy was useful to include in the policy of the drivers of a better management of resources. Although it is a questionable performance in terms of the overall environmental benefits, its relevance cannot be forgotten in terms of the impact on the way waste should be managed. With the diffusion of circular economy concept all over the world, the waste hierarchy changed for "preservation stages of resource value" by Reike et al. (2018), where the hierarchy "R-ladders" or imperatives constitute a principle to operationalize waste as resources in the economy (Reike et al. 2018).

The definitions and boundaries of the concepts of prevention and reuse and the activities which can contribute to them are difficult to establish. Although the European legislation intended to separate products or goods reuse from waste getting prepared for reuse, in practical terms, the most relevant concern should be to know how can waste prevention be measured, which are the factors that are making prevention and reuse difficult, and what if there is no shifting of waste and environmental impacts. Waste prevention and reuse measures in specific phase of a product may create more waste in subsequent life stages, or the avoidance of toxic materials that are used to increase durability that, in its absence, may force a more frequent replacement (Roy 2000). Life cycle assessment focused on waste is a possible method to assess if waste prevention has occurred effectively.

References

Bârsan L, Bârsan A (2014) Ecodesign education – a necessity towards sustainable products. In: Visa I (ed) Sustainable energy in the built environment – steps towards nZEB proceedings of the conference for sustainable energy (CSE) 2014. Springer, Cham, pp 495–502

Bindel A, Rosamond E, Conway P, West A (2012) Product life cycle information management in the electronics supply chain. Proc Inst Mech Eng Part B 226:1388–1400

Bio Intelligence Service (2011) Assessment of impacts of options to reduce the use of single-use plastic carrier bags. Final report prepared for the European Commission – DG Environment

Bortoleto AP, Kurisu KH, Hanaki K (2012) Model development for household waste prevention behavior. Waste Manag 32:2195–2207

British Broadcasting Corporation (BBC) (2015) Plastic bag use down 71% since 5p charge was introduced. http://www.bbc.com/news/uk-wales-politics-34138414. Accessed 9 Dec 2016

Cecere G, Mancinelli S, Mazzanti M (2014) Waste prevention and social preferences: the role of intrinsic and extrinsic motivations. Ecol Econ 107:163–176

Chung S (2011) Import and export of waste in Korea: regulation and actual practice. In: Kojima M, Michida E (eds) Integration and recycling in Asia: an interim report. Institute of Developing Economies, Chosakenkyu-Hokokushu, pp 45–64

Convery F, McDonnell S, Ferreira S (2007) The most popular tax in Europe? Lessons from the Irish plastic bags levy. Environ Resour Econ 38:1–11

Department of Agriculture and Environment and Rural Affairs (DAERA) (2016) Northern Ireland carrier bag levy statistics. https://www.daera-ni.gov.uk/articles/northern-ireland-carrier-bag-levy-statistics. Accessed 7 Dec 2016

Environmental Protection Agency at South Australia (EPASA) (2018) Regulation waste management. https://www.sciencedirect.com/science/article/pii/S0956053X15000902. Accessed 15 Mar 2018

Espinoza PT, Arce EM, Daza D, Faure MS, Terraza H (2010) Regional evaluation on urban solid waste management in Latin America and the Caribbean – 2010 report. Pan American Health Organization, Inter-American Association of Sanitary and Environmental Engineering, Inter-American Development Bank, Washington, DC

European Commission (1977) 2nd environmental action programme 1977–1981. Off J C 139:1–46

European Commission (1989) A community strategy for waste management, SEC/89/934 (final). Brussels

European Council (1975) Council directive of 15 July 1975 on waste 75/442/EEC. Off J L 194:39–41

European Parliament, Council (2008) Directive 2008/98/EC of the European Parliament and of the Council of 19 November 2008 on waste and repealing certain directive. Off J L312:3–30

Gharfalkar M, Court R, Campbell C, Ali Z, Hillier G (2015) Analysis of waste hierarchy in the European waste directive 2008/98/EC. Waste Manag 39:305–313

Giarini O, Stahel WR (1993) The limits to certainty, 2nd edn. Springer Science+Business Media, Dordrecht

Goedkoop MJ, van Halen CJG, te Riele HRM, Rommens PJM (1999) Product service systems, ecological and economic basis. Report commissioned by Dutch ministries of environment and economic affairs, Pricewaterhouse Coopers NV/Pi!MC, Storrm CS, pre consultants

Greyson J (2007) An economic instrument for zero waste, economic growth and sustainability. J Clean Prod 15:1382–1390

Hermann B, Carus M, Pael M, Blok K (2011) Current policies affecting the market penetration of biomaterials. Biofuels Bioprod Biorefin 5:708–719

Howell D (2016) The 5p plastic bag charge: all you need to know. http://www.bbc.com/news/uk-34346309. Accessed 10 Aug 2016

Hultman J, Corvellec H (2012) The European waste hierarchy: from the sociomateriality of waste to a politics of consumption. Environ Plan-Part A 44:2413–2427

Hutner P, Thorenz A, Tuma A (2017) Waste prevention in communities: a comprehensive survey analyzing status quo, potentials, barriers and measures. J Clean Prod 141:837–851

Kling M, Seyring N, Tzanova P (2016) Assessment of economic instruments for countries with low municipal waste management performance: an approach based on the analytic hierarchy process. Waste Manag Res 34:912–922

Knauf M (2015) Waste hierarchy revisited—an evaluation of waste wood recycling in the context of EU energy policy and the European market. For Policy Econ 54:58–60

Martinho G, Balaia N, Pires A (2017a) The Portuguese plastic carrier bag tax: the effects on consumers' behavior. Waste Manag 61:3–12

Martinho G, Magalhães D, Pires A (2017b) Consumer behavior with respect to the consumption and recycling of smartphones and tablets: an exploratory study in Portugal. J Clean Prod 156:147–158

References

Martinho G, Gomes A, Ramos M, Santos P, Gonçalves G, Fonseca M, Pires A (2018) Solid waste prevention and management at green festivals: a case study of the Andanças festival, Portugal. Waste Manag 71:10–18

Maslennikova I, Foley D (2000) Xerox's approach to sustainability. Interfaces 30:226–233

Mont OK (2002) Clarifying the concept of product–service system. J Clean Prod 10:237–245

Organisation for Economic Co-operation and Development (OECD) (2001) Environmentally related taxes in OECD countries: issues and strategies. OECD, Paris

Papargyropoulou E, Lozano R, Steinberger J, Wright N, bin Ujang Z (2014) The food waste hierarchy as a framework for the management of food surplus and food waste. J Clean Prod 76:106–115

Poortinga W, Whitmarsh L, Suffolk C (2013) The introduction of a single-use carrier bag charge in Wales: attitude change and behavior spillover effect. J Environ Psychol 36:240–247

Pre-waste (2011) Eco-tax on plastic bags in Romania (Pre-waste factsheet 94). http://www.prewaste.eu/index.php?option=com_k2&view=item&id=249&Itemid=94. Accessed 9 Dec 2016

Reike D, Vermeulen WJV, Witjes S (2018) The circular economy: new or refurbished as CE 30? – exploring controversies in the conceptualization of the circular economy through a focus on history and resource value retention options. Resour Conserv Recycl. https://doi.org/10.1016/j.resconrec.2017.08.027 135:246–264

Richa K, Babbitt CW, Gaustad G (2017) Eco-efficiency analysis of a lithium-ion battery waste hierarchy inspired by circular economy. J Ind Ecol 21:715–730

Roy R (2000) Sustainable product-service systems. Futures 32:289–299

Silva AR (2015) Industry reduces personal and cannot compensate for losses caused by environmental tax (in Portuguese: Indústria reduz pessoal e não consegue compensar as perdas provocadas pela taxa ambiental). https://www.publico.pt/economia/noticia/industria-reduz-trabalhadores-e-nao-consegue-compensar-as-perdas-provocadas-pela-taxa-ambiental-1710669. Accessed 10 Aug 2016

The Danish Ecological Council (2015) Fact sheet: Tax on plastic bags. http://www.ecocouncil.dk/en/documents/temasider/1776-150812-tax-on-plastic-bags. Accessed 9 Dec 2016

The Guardian (2015) Scotland's plastic bag usage down 80% since 5p charge introduced. https://www.theguardian.com/environment/2015/apr/17/scotland-plastic-bag-usage-falls-after-5p-charge-introduced. Accessed 9 Dec 2016

The Guardian (2016) England's plastic bag usage drops 85% since 5p charge introduced. https://www.theguardian.com/environment/2016/jul/30/england-plastic-bag-usage-drops-85-per-cent-since-5p-charged-introduced. Accessed 9 Dec 2016

United States Environmental Protection Agency (USEPA) (2017) Criteria for the definition of solid waste and solid and hazardous waste exclusions. https://www.epa.gov/hw/criteria-definition-solid-waste-and-solid-and-hazardous-waste-exclusions. Accessed 1 Mar 2018

Valorlux (2014) PPP-initiative "eco-bag." http://valorlux.lu/sites/valorlux/files/files/VALLO01-4009_FactSheet_GB-4Web.pdf. Accessed 9 Dec 2016

Wimmer W, Züst R (2003) Ecodesign pilot: product investigation, learning and optimization tool for sustainable product development. Kluwer Academic Publishers, Dordrecht

Women in Informal Employment: Globalization and Organizing (WIEGO) (2018) National solid waste policy. http://www.wiego.org/sites/default/files/resources/files/Pereira-Brazilian-Waste-Policy.pdf. Accessed 1 Mar 2018

Chapter 3
Technology Status of Waste Collection Systems

Abstract The increasing rate of waste production per capita, the technological advances in packaging products, and the new waste policy and the legal provisions adopted in developed countries created a constant change in the set of parameters that determine the design of solutions for integrated waste management, where waste collection plays a fundamental role. A vast spectrum of technologies for source-separated waste collection and devices was developed, making the evaluation and selection of the one to be applied a difficult task. The purpose of this chapter is to reduce the complexity of identifying, selecting, and benchmarking waste collection systems, presenting a taxonomic classification for the different technical solutions, related to the relevant parts of collection activities and critical equipment characteristics.

Keywords Containers · Vehicles · Classification · Underground · Surface · Lift · Crane · Compaction · Manual · Assisted

3.1 Waste Collection

3.1.1 Waste Collection Role

Collecting waste is one of the most critical phases of the cycle of waste generation-transformation-elimination (Bautista and Pereira 2006), playing a central but often underestimated role in the waste management system (Bilitewski et al. 2010). Waste collection is a highly visible municipal service that involves large expenditures and operational problems; plus it is expensive to operate regarding investment and operational and environmental costs (Faccio et al. 2011). In fact, due to the massive fuel consumption and labor involved, municipal solid waste (MSW) collection is usually the most polluting and costly component of MSW management (MSWM), representing 50–75% of the total costs (Bilitewski et al. 2010; Tchobanoglous et al. 1993). Waste collection is the contact point between waste generators (citizens) and waste management system and can be associated with different kinds of problems such as littering, overfull containers, low recovery rates, and contamination. A lot of

---

© Springer International Publishing AG, part of Springer Nature 2019
A. Pires et al., *Sustainable Solid Waste Collection and Management*,
https://doi.org/10.1007/978-3-319-93200-2_3

these problems can be solved by the proper implementation of a system when it is new and by increasing facilitators through adequate information and feedback to the public and a good, well-planned collection system (Petersen and Berg 2004).

Although MSW collection has the primary role of providing public health to citizens, several waste streams are source separated to obtain quality waste materials that can be recovered and recycled. Nowadays, other roles have been given to MSW collection, making it more sustainable:

- Technical role: the way how waste is collected can influence its properties and, consequently, the waste treatment technologies. If waste is collected commingled and compacted, its destination can be in landfill, mechanical-biological treatment, or incineration units; however, if specific waste streams are source separated, they have a better quality to be recycled than mixed waste.
- Environmental role: besides recycling, MSW collection has been conducted to reduce fuel consumption or even replace fossil fuels by non-fossil fuels like biogas, with the intention to reduce emissions of greenhouse gases. Also, due to the low average speed of collection vehicles, and numerous stops during collection, the effect they have on congestion, air pollution, and noise is higher than that of other types of freight transportation in cities (Johansson 2006).
- Social role: MSW collection is the WMS identity or municipality identity. Without an appropriate communication, all the effort in promoting waste source separation can fail. The recovery rate depends on the participation activity and separation efficiency of the waste producers (Tanskanen and Melanen 1999). Also, collection operation can create problems of the occupation of public space, noise, odors, and traffic and industrial accidents (Poulsen et al. 1995) that, if not minimized, contribute to a negative image of the entire WMS. It should also highlight the job creation promoted by MSW collection, being the ISWM component in which more jobs are created.
- Economic role: MSW collection is an expensive component of ISWM, regarding investment costs (i.e., vehicles fleet) and operational costs (i.e., fuel, maintenances) (Faccio et al. 2011). It should be regarded that MSW collection is a public good, due to the public health driver, so it has to be available to everyone. When waste streams belong to an extended producer responsibility management system, collection costs should be ensured by the fee paid. However, this is not adequately addressed in practice. The residual fraction collection has to be optimized, to not cumbersome citizens.
- Legal role: in order to fulfill policy and legal provisions adopted in the European Union on waste, a broad spectrum of measures and technical solutions for different types of problems and wastes was developed during the last decades (Bilitewski et al. 2010), promoting a wide range of separate collection systems and giving rise to a number of studies assessing and comparing management strategies (Gallardo et al. 2012; Iriarte et al. 2009).

This evolution of the applied roles has been possible due to technological development, especially in the last decades. MSW collection has evolved from trash cans to robust high-tech material and attractive container design and, at the same time, from dedicated and straightforward collection vehicles to trucks with a global positioning system and radio-frequency identification sensors to identify containers and optimization models to increase efficiency.

Being capable of considering all these roles and taking sustainable decisions in choosing and managing an MSW collection system is not an easy task. It is even more difficult when national legislation implements collection targets to be reached because it will influence MSW collection activities (Pieber 2004; Kogler 2007). At a micro- or local scale, any improvement in MSW collection organization – type, size, and receptacle combination – and the collection frequency will influence the composition of MSW as well as the quality and quantity of the separately collected recyclables and thus demands and costs for the subsequent treatment (Bilitewski et al. 1997; Tchobanoglous et al. 1993). At a macroscale, recovery rate targets are in demand and increase the complexity and total costs of MSW management – dividing the total waste mass into separate waste streams results in an increased number of waste flow paths, functional elements, and interdependence in the waste management systems, increasing the number of containers and the amount of collection work (Kogler 2007; Pieber 2004; Tanskanen and Melanen 1999).

3.1.2 Waste Collection Systems

The process of the waste collection begins when the generated waste is thrown into appropriate receptacles and ends when these receptacles are picked up and emptied by collection vehicles. However, the functional element, referred to as "collection," includes not only the removal of waste but also the transport to the place where the collection vehicle is emptied, including this last operation (Tchobanoglous et al. 1993). Collection and transport must include (Bilitewski et al. 1997):

- Recovery and collection of all household, industrial, and commercial waste, including separate collection of recyclables, removing them from the place where they are produced
- Transport of the collected waste into the processing and disposal facilities

Local governments are usually charged with the responsibility for waste collection and transportation to the disposal facilities, but they may choose to hire private contractors. The functional elements that MSWM involves are waste generation, separation and storage at source, collection, sorting, processing and transformation, and disposal (Tchobanoglous et al. 1993). According to Tchobanoglous et al. (1993), the collection can be decomposed into three operations:

- Deposition, which consists of the set of operations after waste generation, involving waste storage and placement in containers to be removed
- The transfer operation carried out by appropriate personnel and equipment for this purpose, by transferring the waste to the collection vehicles
- The transport, which corresponds to the distance that the collection vehicle makes between the last point of collection and the place of its destination

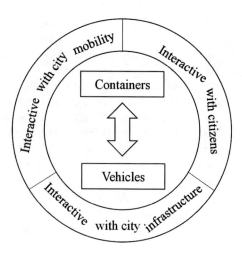

Fig. 3.1 Waste collection system diagram

Upstream of the collection is the waste generation and separation at the source and downstream is transfer and transport, treatment or deposition. The selective collection is not independent of the method of treatment. In fact, in MSW management process, the collection of recyclable materials is of critical importance as the way in which the materials are collected determines the possible options for their recovery and the need for more or less investment in the sorting processes.

Based on the system perspective, a waste collection system (WCS) is composed of the component containers and vehicles, which are interdependent and where interactions occur, forming a relative complex whole. The way how elements interact with each other and with the background system, composed of the waste producers (citizens) and city's infrastructure, will dictate its efficiency and the interaction with city mobility. A WCS has to be attractive, available, near, and safe for citizens to use it. Because WCS involves traffic movement, their schedule needs to be planned to promote its rapid collection and avoid periods of high traffic. Also, the place where to locate containers is influenced by the existing city infrastructure and sidewalk and street inclination, just to name a few. The diagram presented in Fig. 3.1 intends to highlight the complexity.

3.2 Waste Collection System Classification

During the last decades, a broad spectrum of suitable measures and different types of technical solutions for different types of problems and wastes could be developed and realized in technically leading countries (Bilitewski et al. 2010). The need to ensure mandatory recycling and recovery levels for different waste fractions introduced much pressure on waste municipalities systems, forcing them to optimize technical solutions for collection. High recovery rate targets have been set for MSW that municipalities aim to reach mainly by source separation, but the extent to which policies are based on scientific knowledge has been questioned (Dahlén and Lager-kvist 2010).

WCS is a relevant component of a waste management system, being implemented all over the world, but has been classified in a disorganized and dissimilar way. Container and vehicle diversity is quite vast, almost tailor-made for all situations and requirements, so keeping track on their development has become hard. The complexity of equipment, devices, and vehicles increased the difficulty in making a decision on which MSW collection should be implemented to be technically competent, economically affordable, and socially accepted, at the same time, complying with all legal targets and environmental challenges.

The evaluation of collection systems depends on the system boundaries and will always, to some degree, be site-specific. It may not even be desirable to control the factors that cannot be controlled using waste management, but these factors should still be understood as they may offer explanations of variations. One of the factors that can be controlled using waste management is the equipment and technical solutions adopted, but the complexity is high and needs to be reduced to critical factors when searching for causes and effects.

3.2.1 State of the Art

WCS classification has been promoted since the 1990s. Several aspects which could characterize the complex system depend on its components (container and vehicles), how both are interrelated (the collection method), how waste is to be treated and recovered (waste streams), and how WCS is located in the city (i.e., the type of service). According to Bilitewski et al. (2010), a WCS can be defined as a combination of technology and human activities and characterized by (i) the receptacles used for collection, (ii) the applied method of setting them out and picking them up, and (iii) the collection vehicles. The main approaches on WCS classification are going to be presented in this section, divided into container type, vehicle type, collection method, waste streams, and type of service.

(a) Container Type

Container type is referent to the receptacle where waste is disposed temporarily. The variety of containers is quite huge. However, existing classifications found in the literature are characterized mainly by the type (bags, containers, barrels, wheeled, underground), material (plastic, metal), and size (small, medium, large). An instinctive relation exists between container type and its size, being bags and containers without wheels the small-sized containers and wheeled and underground containers the ones with larger dimension. For example, EN840 and EN 12574 family of norms (CEN 2014) classifies containers as two-wheeled with capacity up to 400 L, four-wheeled with a capacity up to 1300 L with flat or dome lid(s), and four-wheeled with a capacity up to 1700 L.

(b) Vehicle Type

The vehicle has the function to discharge the waste container into the vehicle where waste will be transported (Diaz et al. 2005). It can be characterized by the type, which considers collection method, compaction, loading mechanization container lifting device, and by the loading site. Possible vehicle types by collection method are hauled and stationary. By compaction a vehicle can be classified as a compactor (compartmented or not) when waste is compressed, or non-compactor (open or closed truck), and by lifting device vehicles can be classified has hoist truck or lift-off, tilt frame or roll-off, hook lift, crane trucks, trucks with loader up/over, or side loader. Loading site identifies vehicles as front/top, side, and rear.

(c) Collection Method

The collection method is related to the process of emptying the container and its mechanization. Concerning the emptying process, the collection can have different designations. The recipient can be emptied in the same place, being named simple emptying or stationary, can be exchanged by another emptied container, being named as exchange, or can be hauled into the destination, being named hauled or one-way. Only the case of the stationary collection is possible to consider a manual loading system; all the rest is mechanized. Concerning mechanization designations, manual, mechanized, semiautomatic, or automatic special collection systems are all used in literature.

(d) Waste Source and Source Separation

This criterion is related to the source of waste (the place where it is produced) and the source separation considered in the area. Waste source is divided into residential/household, commercial/household-like commercial, and institutional and industrial, being residential divided into dwellings and apartments, mostly. When source separation exists, WCS can be defined by the waste collected, as commingled, residual waste, dry recyclables, and recyclables, just to name a few.

(e) Type of Service: Drop-Off or Pickup Systems

Concerning the service type, different designations exist which are related to the type of waste collected (commingled or separate waste streams) and how the citizen interacts with the WCS. For mixed/commingled waste, the service can be classified as curbside, backyard, alley, setout, and setback and for waste stream designations as a drop-off/bring centers, buyback centers, pickup systems, neighborhood containers, zone containers, green points, and multi-container and special collection (Table 3.1).

Although there is diversity of designations, two main approaches can be adopted:

- (i) Pickup system or curbside collection, where the receptacles are installed/set up for collection close to the houses of the waste generators.
- (ii) Drop-off or bring systems, where accumulated waste amounts are taken by the waste generator to a central location, being dropped into containers specially set up for this purpose. Contrary to the pickup arrangement, the collection vehicles must go to central sites only and not pick up the waste from the curbside in front of each house (Fig. 3.2).

3.2 Waste Collection System Classification

Table 3.1 WCS service type categories and definitions

Service type	Definition
Door-to-door, full-service collection, curbside, alley pickup, or household containers	Containers like bins, racks, sacks, and bags are allocated to individual families, very near to the source of waste generation, where the homeowner is responsible for placing the containers to be emptied at the curb on collection day and for returning the empty containers to their storage location (Dahlén and Lagerkvist 2010; Gonzalez-Torre et al. 2003; Tchobanoglous et al. 1993)
Setout-setback	Containers are set out on the homeowner's property and set back after being emptied by additional crews working in conjunction with the collection crew responsible for loading the collection vehicle (Tchobanoglous et al. 1993)
Backyard carry	The collection crews enter the property to collect refuse. Containers may be transported to the truck, emptied, and returned to their original storage location or emptied into a tub or cart and transported to the vehicle so that only one trip is required (O'Leary 1999)
"Just-in-time" collection	Residents bring out their wastes at the time the collection vehicle reaches a particular spot and rings a bell, a system that works in middle- and upper-class housing of many developing countries (Uriarte 2008)
Drop-off systems or bring systems	It provides containers of different sizes and shapes, and residents are required to deliver recyclables (Dahlén and Lagerkvist 2010; Rhyner et al. 1995)
Multi-container	Citizens dispose each fraction in specific containers located in two areas of the street: organic and residual fraction containers are located on the curb at a maximum distance of 50 m from the dwellings; containers for glass, paper, and packaging are located in areas with groups of containers located at a maximum distance of 300 m from the dwellings (Iriarte et al. 2009)
Neighborhood containers	Individual families are responsible for delivering their waste to a typical container or neighborhood garbage bin near the source of waste generation (Gonzalez-Torre et al. 2003)
Zone containers	Large bins for different waste types are located in central areas that serve one or multiple neighborhoods (Gonzalez-Torre et al. 2003)
Green points	Specifically designed to collect not only separated items from the particular catchment areas and curbside bins but also to selectively collect materials not covered by the other systems, such as hazardous waste, household electrical appliances, and clothes (Gonzalez-Torre et al. 2003)
Buyback centers	Establishments where participants can deliver materials in return for cash payment, such as for recyclable collection (Rhyner et al. 1995)

Fig. 3.2 Schematic drawing of pickup and drop-off arrangement for waste collection. (Source: Bilitewski et al. (2018))

Excluding the container and vehicle type, these classifications have a quite low contribution to distinguishing the several WCS, being unable to promote a robust classification. Taxonomy to classify WCS should show the similarities and differences between WCS and its components, and users should be able to systematically fill and recall information efficiently and effectively to facilitate the use of the taxonomy by diverse scientific and research fields.

The technical details (the features) have implications for planning and operating WCS. Once known and adequately addressed, the features can mitigate WCS costs and environmental impacts. In modeling WCS, parameters such as time per stop (Groot et al. 2014; Sonesson 2000), unload time of a bin (Faccio et al. 2011), and the number of workers (Groot et al. 2014) are all needed.

A taxonomic classification based on the technological features relevant to classify WCS is proposed in the next section. The features highlighted in the taxonomy, such as the container's vehicle coupling, mobility, emplacement, container access for container and body mechanization, lifting mechanization, and loading location for vehicles, influence those variables present in WCS models.

3.2.2 Waste Collection System Types

This taxonomy is divided into three components, container, vehicle, and collection method, and classes and subclasses, which are capable of characterizing the container-vehicle system presented in Fig. 3.3. Trees are used to describe the classes and subclasses of each component. Identification of a feature can reach up to five levels,

3.2 Waste Collection System Classification

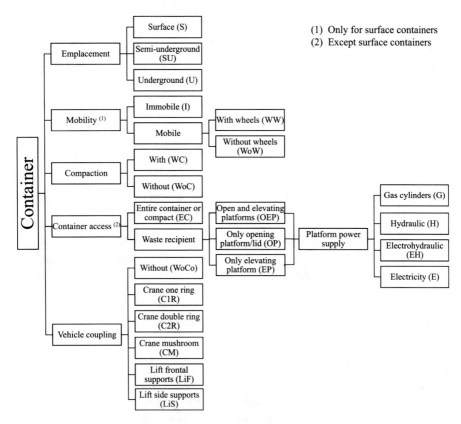

Fig. 3.3 Container classification diagram. (Source: Rodrigues et al. (2016))

and key components will be described to support the taxonomy. The three components result in a nomenclature, to be applied to characterize WCS.

Container Component

Class 1 is the container, and the first-level branch of its classification tree (Fig. 3.4) is divided into relevant technical aspects used to identify container component, identified by subclasses: emplacement (1.1), mobility (1.2), compaction (1.3), container access (1.4), and vehicle coupling (1.5). Container emplacement refers to location related to ground level. Containers can be positioned at the surface (100% of the recipient's capacity is at ground level), entirely underground, or semiunderground. A specific property of surface containers is mobility. Underground and semiunderground are static and must be accessed by the vehicle for waste collection, whereas surface containers can be located and replaced on the street without specific construction work and are easily carried to the collection vehicle. Mobility can be

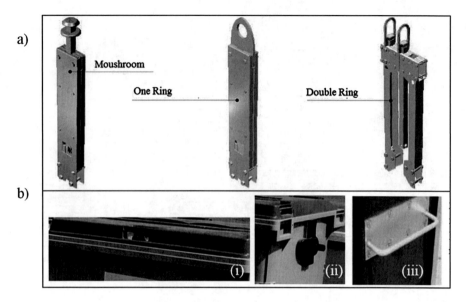

Fig. 3.4 Coupling systems' schematic illustrations. (**a**) Crane coupling systems (OVO Solutions 2012a) and (**b**) lift coupling systems in (i) frontal comb, (ii) HDPE lifting trunnions, and (iii) metallic wings. (Source: Rodrigues et al. (2016))

divided into immobile and mobile containers, with or without wheels. Containers can also be designed to compact waste.

Semiunderground and underground containers must be accessible to the waste collection vehicle. It is not mandatory that all container elements such as platforms, deposition columns, and waste recipients are removed as a unit to dispose waste into the vehicle (compact container). Sometimes only the waste recipient element is removed to be discharged, and vehicle access can be through an open platform, elevated platform, or both. An open platform corresponds to the opening of the surface pull-down lid to access the container; the vehicle pulls the container from its underground location, and the elevating platform raises the container to surface level. When there is no elevating platform, the vehicle itself pulls the container from the underground receptacle. An additional feature characterizing existing platforms is the platform power supply, which can be gas cylinders, hydraulic, or electrohydraulic.

Vehicle coupling defines how the container interacts with the vehicle to promote container discharging. The existing options are absence of coupling system or by the type of coupling system: crane rings, crane mushroom, and crane supports. Rings and mushroom refer to a crane option in the vehicle, and supports are related to the lift option in the vehicle. Crane coupling can be one ring, double ring, or mushroom (Contenur 2014; OVO Solutions 2012b) (Fig. 3.4a). One ring coupling is suitable for truck cranes with a simple forklift, known as simple hook, where the ring is secured on the frame support and detaches the lower lid, which is automatically opened when the pedal (also named *palpeur* system) touches the

3.2 Waste Collection System Classification

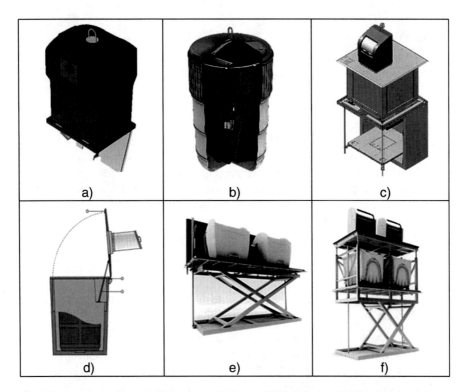

Fig. 3.5 Container schematic illustrations. (**a**) Case 4 (OVO Solutions 2012b), (**b**) case 5 (Sopsa 2012), (**c**) case 6 (OVO Solutions 2012a), (**d**) case 7 (Sotkon 2007), (**e**) case 8 (TNL 2014b), and (**f**) case 9 (TNL 2014c). (Source: Rodrigues et al. (2016))

bottom of the loading truck (OVO Solutions 2012b). The double-ring operation is secured by two sliding rods, and the lifting arm has a double command; one raises the container and opens the lower lid, and the other one keeps the container at the desired height (OVO Solutions 2012b). Mushroom containers, also known as double disc and by the trademark *Kinshofer*, consist of a half sphere or "disc" on the top; the hoisting cable is also equipped with a double command similar to double ring, and the operation is ensured by two tubes sliding one inside the other. This system requires that collection vehicles are equipped with controlled and high-precision positioning and coupling devices, eliminating the need for manual engagement (Kinshofer 2014).

The containers for lift coupling supports have handles or handgrips built into the container body (according to EN 840 (CEN 2014)) with different designations, depending on the lift side (Fig. 3.4b). For frontal handles, a ventral system consists of a frontal comb integrated with the upper body of the container. For side supports, lifting trunnions or points are secured to the upper sides of the container body by two high-density polyethylene (HDPE) lateral pivots and *Ochsner* handles composed of two metal lateral wings (Sulo 2014; Weber 2006). Crane-compatible containers are bottom discharge containers with a trapdoor(s) or cable opening bags; lift-

36   3 Technology Status of Waste Collection Systems

compatible containers have superior discharge capabilities by lid opening and overturning.

Based on the developed classification, ten possible key container component cases describe how the taxonomy works and exemplifies the components of the container to be analyzed (Fig. 3.5):

Case 1. Surface, without wheels, without compaction, vehicle coupling
Case 2. Surface, with wheels, without compaction, lift vehicle coupling
Case 3. Surface, immobile, without compaction, lift vehicle coupling
Case 4. Surface, immobile, without compaction, crane vehicle coupling
Case 5. Semiunderground, without compaction, compact, crane vehicle coupling
Case 6. Underground, without compaction, compact, crane vehicle coupling
Case 7. Underground, without compaction, with opening platform, crane vehicle coupling
Case 8. Underground, without compaction, with open and elevating platform container access, lift vehicle coupling
Case 9. Underground, without compaction, with elevating platform container access, lift vehicle coupling
Case 10. Underground, with compaction, with the open and elevating platform, hook lift vehicle coupling

Case 1 containers are characterized by semitransparent plastic or paper bags or non-wheeled bins, usually with two handles, a cover, and no vehicle lifting handles and with a wide range of capacities, from 0.035 up to 0.11 m^3 (Bilitewski et al. 1997; ISWAWGCTT 2004). Ordinary grocery bags or biodegradable bags for organic waste collection can also be used. Because they have no coupling system with the collection vehicle, all the effort in lifting and disposing is by manual workers.

Case 2 containers include mobile garbage containers with two or four wheels. The generic capacity of these containers goes from 0.12 to 1.1 m^3 (Bilitewski et al. 1997; Kogler 2007) although two-wheeled can start at 0.06 and go up to 0.36 and four-wheeled between 0.66 and 1.1 m^3 (Sulo 2014; Weber 2006). Lift vehicle coupling containers have side and frontal handles and a flat or tilt-curved lid and may have a lid opening system with a pedal or deposition opening adapted to the waste stream (Contenur 2014; Sulo 2014; Weber 2006).

Case 3 steel or HDPE containers were developed for side-loading automated lifts, with a vertical alignment crosshair and four Teflon roller supports at the base of the body instead of wheels. They have an opening lid, or deposition opening adapted to the waste stream. Capacities range between 1.8 and 3.2 m^3 (Contenur 2014; Ros Roca 2014).

Case 4 containers were designed for source-separated collection with a crane and had two main designs: igloo and prismatic (Contenur 2014). Container openings are located on the top, with specific designs for packaging waste type. A container frame attaches the securing system directly to the metal base and to the lower lid using support arms, rods, or clevis fasteners (OVO Solutions 2012b). Capacities range between 2.5 and 3.2 m^3 (Contenur 2014; OVO Solutions 2012b).

Case 5 refers to semiunderground, compact, one-ring crane coupling containers composed of two parts: (1) the outer shell in HDPE and (2) an interior polypropylene

3.2 Waste Collection System Classification

bag where the waste is placed, fixed at the top of the container using a metal ring, and opened by the action of a cable to discharge. Other possible options are a rigid plastic container instead of the flexible bag (Molok 2009; Sopsa 2012) or a concrete monobloc wheel in place of the HPDE outer shell (Sopsa 2012). These cylindrical-shaped containers have a capacity range from 0.3 to 5 m^3, being 3 and 5 m^3 as the most common for municipal waste (Molok 2009).

Case 6 refers to the underground, entire/compact containers for crane vehicle coupling, installed inside an underground watertight concrete bunker with a fixed pedestrian platform in galvanized steel (Contenur 2014; OVO Solutions 2012b). At surface level, only the inlet structure (column) and pedestrian platform are visible. These containers are called compact containers because the column, container, and pedestrian platform are a unit removed together. The stainless steel container is emptied by one or more opening flaps underneath, designed to collect liquid. Capacity ranges from 1 to 5 m^3 (OVO Solutions 2012b).

Case 7 consists of underground containers with opening platform access, with one ring crane vehicle coupling. Case 7 containers are distinguished from case 6 by container access because the only element hoisted is the waste recipient, not the compact container. Access to the waste recipient is ensured by the pedestrian platform, which opens (in contrast to case 6) and has a manual hook engagement to the ring container (Resolur 2013; Sotkon 2007). The platform power supply can be hydraulic, electrohydraulic, or gas cylinders (Sotkon 2007; TNL 2014a). Containers' capacity ranges from 1 to 5 m^3 (Resolur 2013; Sotkon 2007; TNL 2014a), which can be bottom discharged using a trapdoor located at the base or by overturning using both vehicle coupling options, crane and adapted rear lift.

Case 8 consists of underground containers with open and elevating platforms for container access and lift vehicle coupling, which stands on the platform and is elevated to the surface level rather than discharged by automated lifting and side-loading vehicles. Both platforms are powered by an electrohydraulic unit, activated inside the vehicle cabin using a remote control (Contenur 2014; Equinord 2009; TNL 2014b). Containers' capacity ranges from 3.2 to 4 m^3 (Contenur 2014; TNL 2014b).

Case 9 differs from case 8 in container access, in which case 9 is by an elevating platform only. With a capacity range from 0.8 to 1 m^3, the container is emptied by semiautomated lifting rear-end loading vehicles (Contenur 2014; TNL 2014c). The elevating platform is operated either by remote control console or independent central electrohydraulic or collection vehicle (Contenur 2014; Equinord 2009; TNL 2014c).

Case 10 consists of underground compaction containers with a top-loading chamber, with openings and elevating platforms and hook lift vehicle coupling. The elevating platform lifts the compacting container box up to the street level, and the opening platform rotates on the back axle to facilitate container access (TNL 2014d). Both platforms are powered by an electrohydraulic power station (TNL 2014d). A system on the compaction plate controls the container's filling rate, with capacities between 12 and 25 m^3 (Equinord 2009; Villiguer 2014).

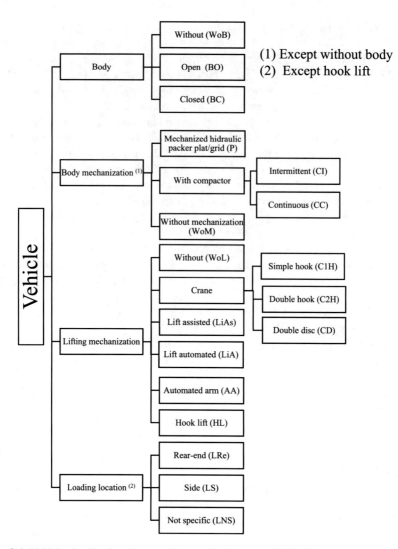

Fig. 3.6 Vehicle classification diagram. (Source: Rodrigues et al. (2016))

Vehicle Component

Vehicle class 2, presented in Fig. 3.6, is divided into subclass body (2.1), body mechanization (2.2), lifting mechanization (2.3), and loading location (2.4). The location waste determines if the body is discharged, which can be open, closed, or nonexistent, in which case the whole container is put into the vehicle, as occurs in the hauled collection. The open and closed body can be non-compartmented or compartmented to separately collect two or more types of waste at the same time (e.g., vertically split body, dual compartment), classified as multi-compartment or

3.2 Waste Collection System Classification

single compartment body, respectively. The body can be mechanized with different structures (body mechanization), such as a sweep plate, grid, or a compactor, which work continuously or intermittently. Also, lifting can be variously mechanized (lifting mechanization) with hooks, lifts (forks, bars, or both), hook lifts, and automated arms. Different lifting devices can be used in the same vehicle, classified using the corresponding taxonomic characteristics. All crane-based lifting devices can be interchangeable because they are non-fixed elements; using all in the same vehicle is possible. Available options for loading location include rear-end, side, and even nonspecific, as in the case of manual loading where the body is opened to dispose waste bags. The proposed taxonomy is presented in Fig. 3.6.

The literature analysis found ten key vehicle components representing all possible taxonomic components:

1. Body open, non-mechanized, crane lifting, not specific loading site
2. Body closed, mechanized packer plate/grid, lift assisted, rear-end loading site
3. Body closed, with intermittent compaction, lift assisted, rear-end loading site
4. Body closed, with intermittent compaction, crane lifting (and lift assisted), rear-end loading site
5. Without body, hook lift
6. Body closed, with intermittent compaction, crane lifting, not specific loading site
7. Body closed, with intermittent compaction, lift assisted, side-loading site
8. Body closed, with continuous compaction, lift assisted, rear-end loading site
9. Body closed, with intermittent compaction, lift automated, side-loading site
10. Body closed, with intermittent compaction, arm automated, side-loading site

Case 1 vehicles (Fig. 3.7a) are composed of an open box body and a hydraulic crane, which can be manually operated from the crane footboard, on the floor, or remotely. Different coupling systems can be installed on the crane, depending on compatibility with different container crane vehicle coupling types.

Case 2 vehicles, also called as satellite units, are composed of a rear-loading forklift mechanism and a simple hydraulic sweep plate or grid that clears the rear of the hopper to provide load security and distribution inside the load box but provides no compaction or semi-compaction (Heil Farid 2014; Ros Roca 2014).

Case 3 vehicles are composed of a hydraulically powered compression/ejection plate, a load box, an articulated sweep plate, and a rear tailgate with a large hopper capacity and a lifting mechanism (Ecofar 2013). Front or lateral support coupling containers are raised by a loading fork that hooks onto the front of the container or by retractable lift bars (Bilitewski et al. 1997), respectively. A moving plate scoops the waste out from the loading hopper and compresses it against a moving wall (intermittent compaction), with a leachate tank at the bottom of the body. With the body full of waste, the compaction wall moves and ejects waste through an open tailgate.

Case 4 vehicles (Fig. 3.7b) are similar to case 3 but have a telescopic crane, an enlarged loading hopper, and a tailgate with a higher load volume to receive big underground waste containers (Ros Roca 2014; Soma 2014) or discharge from

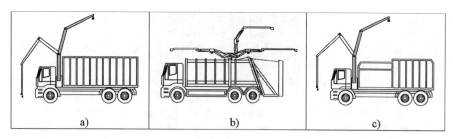

Fig. 3.7 Vehicle schematic illustrations. (**a**) Case 1 (Sotkon 2007), (**b**) case 4 (Sotkon 2007), and (**c**) case 6 (Sotkon 2007). (Source: Rodrigues et al. (2016))

satellite vehicles. Containers are collected with the crane or with both crane and adapted rear lift (Soma 2014).

Case 5 vehicles are designated as hook lift or container vehicles, mostly used to collect high-volume containers. Underground compactor containers (container case 10) are lifted over the collection vehicle chassis with a hook lift system. These demountable body handling technology vehicles are known by trademarks such as "Ampliroll" and "Multilift."

Case 6 vehicles (Fig. 3.7c) are composed of a self-contained waste compaction mobile unit with a top-loading compacting chamber, where waste is unloaded and compacted. The container body is fed by a longitudinal sliding drawer in the compacting chamber through bottom tabs and unloaded by the tailgate, hydraulic, or gravity-opened doors (Mofil 2014). A hydraulic crane collects containers.

Case 7 vehicles are side-loading vehicles with ejection plates, also called satellite vehicles because a transfer system transfers the payload to a full-size rear loader (Ecofar 2013; Heil 2014). These vehicles are a one-piece body construction in which the waste processing and unloading are carried out by the hydraulic ejection panel (Ecofar 2013). These vehicles may have single- or dual-side hopper doors for manual loading operation or a side lift with a loading fork (Ecofar 2013; Heil 2014).

Case 8 vehicles are for continuous compaction, differentiated from case 3 by the compaction system, which consists of a fixed compacting screw system in the rear and a spiral screw conveyor inside the cylindrical body drum that continuously mixes and compacts the entire load during collection (FAUN 2014).

In cases 9 and 10, the vehicles are automatic side lift or arm grabber, operated by the driver inside the vehicle, using a joystick and a video system (Heil 2014; Heil Farid 2014). The vehicle stops alongside the container, and the arm (single or double) grabs the container, empties it, and replaces it automatically (Kogler 2007). A continuously reciprocating metal pusher plate at the loading hopper forces the waste through an aperture into the main body, compacting against the material already loaded.

3.2 Waste Collection System Classification

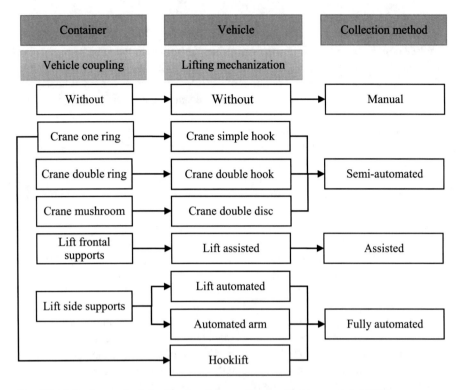

Fig. 3.8 Collection method classification diagram. (Source: Rodrigues et al. (2016))

Collection Method Component

Collection method categories describe how the container interacts with the vehicle and workforce. The collection method can be manual, assisted, semiautomated, and fully automated. Manual collection occurs when the worker carries, lifts, and unloads waste bins or bags into the vehicle. Assisted collection is a mix of manual and mechanical processes in which the container displacement near the vehicle is manual and mechanization occurs only in the lifting and emptying by the vehicle. Semiautomated collection is a mechanized process of all steps involved in collection procedure, but the worker must be outside the vehicle to control the coupling and provide manual assistance on vehicle-container coupling and uncoupling. Fully automated collection involves no direct intervention of workers, and the container-vehicle interaction is controlled by a single operator inside the vehicle cabin.

A relation between container and vehicle components is needed to classify WCS by collection methods. Because collection methods are related to mechanization and provide a link between container and vehicle, the features to be addressed are container-vehicle coupling and vehicle lifting mechanization (Fig. 3.8).

Key Collection Methods

The key container and vehicle components can characterize the four key collection methods. Manual collection occurs for containers classified as surface, without wheels, and without vehicle coupling (e.g., bags and bins without wheels) and collected by vehicles classified as open body, non-mechanized body, without lifting or specific loading tools, or closed body, with intermittent compaction and lifting mechanization, which is not used.

In the assisted collection method, wheeled mobile surface containers with lift vehicle coupling frontal or lateral supports are rolled by the workers to the collection vehicle, which can have an intermittent or continuous compactor, semiautomated lifting, and a rear, frontal, or side-loading location. Three or more workers are usually needed for assisted and manual collection.

Semiautomated collection methods can use underground containers with platform access and crane vehicle coupling, collected by vehicles with the closed or open body or an intermittent compactor or non-mechanized body with a hook lifting or double-disc rear loading or nonspecific location. Two workers (driver and crane operator) are usually sufficient, although a single operator can control the double-disc system.

Fully automated collection methods have no direct intervention of workers because the driver inside the vehicle operates all collection processes. An example is a surface container without wheels, with side supports, collected by a closed-body vehicle with intermittent compaction and automated lifting and side loading. Fully or semiautomated collection methods are also characterized by a relatively higher container capacity than assisted or manual collection methods, which are workforce dependent.

References

Bautista J, Pereira J (2006) Modeling the problem of locating collection areas for urban waste management an application to the metropolitan area of Barcelona. Omega 34:617–629

Bilitewski B, Hardtle G, Marek K (1997) Waste management. Springer, Berlin

Bilitewski B, Wagner J Reichenbach J (2010) Best practice municipal waste management. Federal Environmental Agency, Intecus

Bilitewski B, Wagner J, Reichenbach J (2018) Best practice municipal waste management. Umweltbundesamt, pdf publication: https://www.umweltbundesamt.de/publikationen

Contenur (2014) Catalogue contenur. http://www.contenur.com/en/. Accessed 15 Feb 2014

Dahlén L, Lagerkvist A (2010) Evaluation of recycling programmes in household waste collection systems. Waste Manag Res 28:577–586

Diaz LF, Savage GM, Eggerth LL (2005) Solid waste management. United Nations Environment Programme, Paris

Ecofar (2013) Products Ecofar. http://www.ecofar.it/getcontent.aspx?nID=3&l=en. Accessed 1 Mar 2014

Equinord (2009) Products Equinord. http://www.equinord.com/web/Default.aspx. Accessed 16 Feb 2014

References

European Committee for Standardization (CEN) (2014) CEN/TC 183/WG 1 – waste containers. http://standards.cen.eu/dyn/www/f?p=204:32:0::::FSP_ORG_ID,FSP_LANG_ID:7293,25& cs=1963AB0E62521CCFF7015C670244E3A73. Accessed 10 Feb 2014

Faccio M, Persona A, Zanin G (2011) Waste collection multi objective model with real time traceability data. Waste Manag 31:2391–2405

FAUN (2014) Rear loaders. https://www.faun.com/en/home/refuse-collection-vehicles/rear-loaders.html. Accessed 5 Mar 2014

Gallardo A, Bovea MD, Colomer FJ, Prades M (2012) Analysis of collection systems for sorted household waste in Spain. Waste Manag 32:1623–1633

Gonzalez-Torre P L, Adenso-Dıaz B, Ruiz-Torres A (2003). Some comparative factos regarding recycling collection systems in regions of the USA and Europe. J Environ Manag 69:129–138

Groot J, Bing X, Bos-Brouwers H, Bloemhof-Ruwaard J (2014) A comprehensive waste collection cost model applied to post-consumer plastic packaging waste. Resour Conserv Recycl 85:79–87

Heil (2014) Rear loaders. http://www.heil.com/#. Accessed 5 Mar 2014

Heil Farid (2014) Product groups. http://www.heilfarid.eu.com/heil-farid-products/product-groups. Accessed 1 Mar 2014

Iriarte A, Gabarrell X, Rieradevall J (2009) LCA of selective waste collection systems in dense urban areas. Waste Manag 29:903–914

ISWA Working Group on Collection and Transportation Technology (ISWAWGCTT) (2004) Overview of household collection systems in different cities and regions. Report for the International Solid Waste Association

Johansson OM (2006) The effect of dynamic scheduling and routing in a solid waste management system. Waste Manag 26:875–885

Kinshofer (2014) Container discharge units. Kinshofer. http://www.kinshofer.com/eng/index.php/en/crane-3/domestic-waste/container-discharge-units. Accessed 1 Mar 2014

Kogler T (2007) Waste collection – a report with support from ISWA Working Group on Collection and Transportation Technology. Report for the International Solid Waste Association

Mofil (2014) Mofil monobloc compactors. http://www.mofilpt/fich_up/compactador%2015m3.pdf. Accessed 17 Feb 2014

Molok (2009) Products Molok. http://www.molok.com/main.php?loc_id=8. Accessed 20 Feb 2014

O'Leary PR (1999) Decision maker's guide to solid waste management, 2nd edn. United States Environmental Protection Agency, Office of Solid Waste, RCRA Information Center, Washington, DC

OVO Solutions (2012a) Underground systems. http://www.ovosolutions.com/363/underground-systems.htm. Accessed 10 Mar 2014

OVO Solutions (2012b) Cyclea. http://www.ovosolutions.com/47/cyclea.htm. Accessed 15 Feb 2014

Petersen CHM, Berg PEO (2004) Use of recycling stations in Borlänge, Sweden – volume weights and attitudes. Waste Manag 24:911–918

Pieber MK (2004) Waste collection from urban households in Europe and Australia. Waste Manag World: July-August 2004, 111–124

Poulsen OM, Niels OB, Niels E, Ase MH, Ulla II, Lelieveld D, Malmros P, Matthiasen L, Nielsen BH, Nielsen EM, Schibye B, Skov T, Stenbaek EI, Wilkins CK (1995) Collection of domestic waste – review of occupational health problems and their possible causes. Sci Total Environ 170:1–19

Resolur (2013) Advantages (in Spanish: Ventajas). http://www.soterrado.es/index.html. Accessed 1 Mar 2014

Rhyner CR, Schwartz LJ, Wenger RB, Kohrell MG (1995) Waste management and resource recovery. CRC Press/Lewis Publishers, Boca Raton

Rodrigues S, Martinho G, Pires A (2016) Waste collection systems part a: a taxonomy. J Clean Prod 113:374–387

Ros Roca (2014) Products Ros Roca. http://www.rosroca.com/en/products/integral-waste-collec tion.html. Accessed 15 Feb 2014

Soma (2014) USW containers. http://www.soma.pt/old/english/products/residuossolidos/caixasrsu/ hallerx2c.html. Accessed 5 Mar 2014

Sonesson U (2000) Modelling of waste collection – a general approach to calculate fuel consumption and time. Waste Manag Res 18:115–123

Sopsa (2012) Products semi underground. http://www.sopsa.pt/en/node/250. Accessed 20 Feb 2014

Sotkon (2007) Mbe Sotkon – Underground containers for MSW – Technical description (in Spanish: Mbe Sotkon – Contenedores Subterráneos para RSU – Memória Técnica). http://www.construnario.com/catalogo/mbe-sotkon-sl/catalogos. Accessed 15 Oct 2010

Sulo (2014) Products. http://www.sulo.com/index.php/en. Accessed 15 Feb 2014

Tanskanen J, Melanen M (1999) Modelling separation strategies of municipal solid waste in Finland. Waste Manag Res 17:80–92

Tchobanoglous G, Theisen H, Vigil S (1993) Integrated solid waste management: engineering principles and management issues. McGraw-Hill, New York

TNL (2014b) Waste system Sidetainer. TNL. http://static.lvengine.net/tnl/Imgs/articles/article_70/ sideTAINER_TNL_EN_v3.pdf. Accessed 1 Mar 2014

TNL (2014c) Waste system ecotainer. TNL. http://static.lvengine.net/tnl/Imgs/articles/article_69/ ecoTAINER_TNL_EN_v3.pdf. Accessed 1 Mar 2014

TNL (2014d) Waste system bigtainer. TNL. http://static.lvengine.net/tnl/Imgs/articles/article_78/ 590bigTAINER_TNL_EN_v3_1.pdf. Accessed 1 Mar 2014

Uriarte FA (2008) Solid waste management: principles and practices. The University of the Philippines Press, Quezon City

Villiger (2014). Sub-Vil: the Idea to Make Waste Disappear Underground. Villiger. http://www.villiger.com/sub-vil-en.html. Accessed 20 Feb 2014

Weber (2006) Products Weber. http://www.w-weber.com/engl/start.html. Accessed 15 Feb 2014

Chapter 4
Preparation for Reusing, Recycling, Recovering, and Landfilling: Waste Hierarchy Steps After Waste Collection

Abstract The less useful operations in accordance to waste hierarchy principle will be driven in this section. Reusing, recycling, treating, and landfilling are all operation options for waste, which need to be considered regarding its impact on the environment and how their management can potentiate a better use of resources. A brief review on the concepts is presented, in the light of European waste management definitions and existing technologies.

Keywords WHP · MSW · Upcycling · Downcycling · Incineration · Biological treatment · End-of-waste criteria

4.1 Waste Hierarchy Principle: All After Becoming Waste

The management options when products become waste are vast, although the hierarchy is quite similar between them. In the "hierarchy of resource use" of Gharfalkar et al. (2015), waste should be managed following the preference order:

- Preparing for reuse is referent to options of cleaning, checking, repairing after the product has become waste, and making the object be used again as for the same purpose (like in definition of Waste Framework Directive 2008/98/EC).
- Reuse via resale of used, repaired, refurbished, reconditioned, or remanufactured products; reuse via renting, leasing, or servitization of products; and reuse without any further operation (secondhand, thirdhand, always with owners changing).
- Reprocessing: upcycling, recycling, and downcycling.
- Other recovery: recovery of energy and recovery of other substances or materials to be used as fuels or for backfilling.
- Rectification: considered for treatment before disposal.
- Return: disposal of waste.
- Waste exports: waste exports are seen as waste trafficking, considered by Bartl (2014) where waste exports are not in light with the global system with finite resources and where countries may divert waste from their landfills and send them to less developed countries.

© Springer International Publishing AG, part of Springer Nature 2019
A. Pires et al., *Sustainable Solid Waste Collection and Management*,
https://doi.org/10.1007/978-3-319-93200-2_4

45

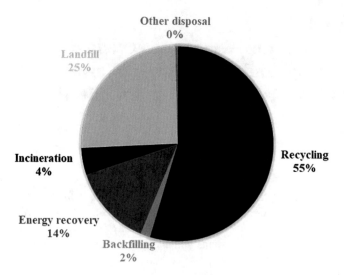

Fig. 4.1 Municipal solid waste generated and type of treatment, in 2014. (Source of data: Eurostat (2017))

For food waste management, Papargyropoulou et al. (2014) defined a waste hierarchy to help on its sustainable management, regarding avoiding food waste and food loss in the life cycle of food production. The waste hierarchy proposed presents the following order of operations (Papargyropoulou et al. 2014):

- Reuse: which includes food for human consumption, for people affected by food poverty through redistribution networks and food banks.
- Recycling: recycle food waste into animal feed and via composting.
- Recovery: treat unavoidable food waste and recovery energy, including anaerobic digestion.
- Disposal: dispose unavoidable food waste into the sanitary landfill with landfill gas extraction and recovery.

The diversity of waste hierarchy options shows how essential and discussible is the definition of the most sustainable correct direction to follow when managing waste. To better present the treatment options possible to waste, the defintions on waste hierarchy at Waste Framework Directive will be followed in the next sections.

Waste Framework Directive divides waste hierarchy into management for waste in preparing for reuse, recycling, and disposal. The management of waste generated at European Union is presented in Fig. 4.1. The analysis shows that there is no statistical information on waste send for "preparing for reuse," being mixed with the recycling operation, which will be the dominant operation. Also, 54.6% of municipal waste (excluding imports and export of waste) for the 28 countries of European Union is sending for recycling, 25.3% for landfilling, 13.6% for energy recovery, 4.4% for incineration (without energy recovery), 1.7% for backfilling, and 0.4% for other disposal operation (Eurostat 2017). Recycling is, in fact, the leading solution for waste, but landfill has a relevant role in the integrated waste management. More than half of the countries are preferring recycling operation in opposition to the other waste management operation options, being the recycling leaders the countries Belgium, the Netherlands, and Slovenia, with recycling rates above 70%. Landfilling is still the preferred destination for countries like Bulgaria, Estonia, Greece, Spain, Cyprus, Hungary, Malta, Romania, and Slovakia.

4.2 Preparing for Reuse

The definition of preparing for reuse from Waste Framework Directive includes the "checking, cleaning or repairing recovery operations, by which products or components of products that have become waste are prepared so that they can be re-used without any other pre-processing." The definition considered in European Union legislation requires that the product has become waste, i.e., it has entered a collection system to be discarded or delivers it to another entity to get rid of it. The frontier of the owner defines the difference between being a waste prevention measure and preparing for reuse measure.

There have been different approaches to promote preparing for reuse. European legislation (and subsequent transpose to the national law of Member States) includes targets of preparation for reuse together with recycling for several waste materials, plastic, paper, glass, metal from and households, and for construction and demolition waste. Market-based instruments applied to preparing for reuse are deposit-refund systems and extended producer responsibility instrument. For several years in Portugal, before the entrance of compliance management for packaging waste, glass bottles were subjected to deposit-refund systems, to be collected and refilled again. Under the responsibility inherent at extended producer responsibility, the manufacturers can develop their products under design for disassembly, making products adequate to be, at waste phase, reparable to others to use them, at second-hand market or donations. Information campaigns on preparing for reuse also occur through the elaboration of indicators and awareness campaigns. Voluntary instruments such as norms, standards, and guidelines to conduct verification and guarantee for the electric and electronic waste are also being applied in European countries. In the UK, the PAS 141:2011 standard sets out the requirements for preparing waste electrical and electronic equipment (WEEE) for reuse, including suggestions for handling, tracking, segregating, storing, and protecting the appliances and its components for the preparation for reuse (Lu et al. 2018). In Flanders region of Belgium exists the standard for reuse of WEEE from Public Waste Agency of Flanders (OVAM), where environmental criteria are also considered, namely, the energy labeling to improve the environmental performance of reused appliance (Lu et al. 2018). In Germany, the standard *VDI 2343 – recycling of electrical and electronic equipment* – also allows promoting the benefits of reuse. Bovea et al. (2016) have developed a protocol specific for small WEEE from households, classifying appliances by potential reuse and the tests to be conducted, being based in other protocols

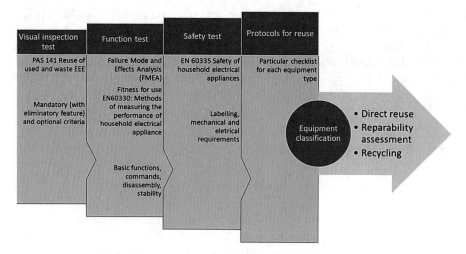

Fig. 4.2 Proposed methodology for the preparation for reuse of small WEEE. (Source: Adapted from Bovea et al. (2016))

already existing. In Fig. 4.2 the protocol is presented. The first step is the visual inspection, which should be done following PAS 141; the function test verifies if the appliance is operating according to its functions; the safety test verifies the aspects related to electrical, mechanical, and thermal risks; and the reuse protocols will define the reuse potential and which operations to be made to the appliances are to be reused (Bovea et al. 2016).

4.3 Recycling

Recycling means "any recovery operation by which waste materials are reprocessed into products, materials or substances whether for the original or other purposes. It includes the reprocessing of organic material but does not include energy recovery and the reprocessing into materials that are to be used as fuels or for backfilling operations" (European Waste Framework 2008/98/EC). Looking at Fig. 4.3, European countries most devoted to recycling (i.e., where waste generated is mostly sent for recycling) are Belgium, Czech Republic, the Netherlands, and Slovenia, just to name a few, and countries with low waste recycling are, for example, Bulgaria, Estonia, and Romania.

Three types of recycling can be described: upcycling, recycling, and downcycling. The main difference of those definitions will be addressed in the next subsections.

4.3 Recycling

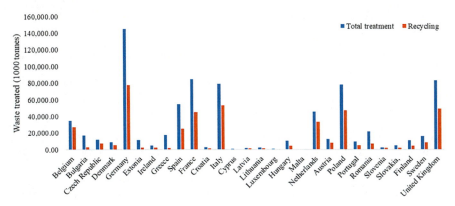

Fig. 4.3 Recycling of municipal solid waste in European countries. (Source of data: Eurostat (2017))

4.3.1 Upcycling

According to Cohen and Robbins (2011), upcycling was firstly introduced by William McDonough and Michael Braungart on the book "Cradle to Cradle: Remaking the Way We Make Things" as "the practice of taking something that is disposable and transforming it into something of greater use and value (McDonought and Braungart (2002))." Other definitions also go on the same concept, increasing value. Upcycling is referent to processes that can increase the value of the recycled material over time, where the recycled material is reemployed for a more significant use or with a higher environmental value (Chandler and Werther 2014). Another view of upcycling is the one brought by Huysman et al. (2017), occurring when, for example, the plastic is of good quality but is used to replace another material that presents a higher environmental burden when compared to the virgin plastic.

To achieve upcycling concept, the industry needs to avoid the use of harmful substances and materials, recycle and upcycle for the continuous life of the products manufactured, decrease the consumption of energy and water, and also pay fair wages to employees (DeLong et al. 2017). Cases of upcycling are making purses out of used tires or used spare parts from end-of-life vehicles (McKenna et al. 2013), turning curtains into garments, or making old pair of jeans into a bag (Hjelmgren et al. 2015). In this cases, upcycling imposes the conversion of the product into other more valuable products. However, upcycling may also occur inside the same product. In the case study presented by Niero et al. (2017), a methodology to promote eco-efficiency and eco-effectiveness for aluminum cans of Carlsberg intends to upcycle the can continuously, in which every time that the can is recycled, it improved.

One of the areas of upcycling is being discussed in the textile sector. Waste textiles have been considered a waste stream needing better-dedicated management. Hjelmgren et al. (2015) identified the barriers to a large-scale upcycling of clothing in Swedish clothing sector as the shortage of suitable production facilities which are located outside Sweden and the need for significant amount of waste materials to make production and transportation efficient, just to name a few (Table 4.1).

Table 4.1 Barriers to upcycling textile waste

Products	Production facilities
Costs due to capital tied in inventory	Transportation costs and lost/reduced value that is created from an environmental perspective due to the transportation of finished products, or high manufacturing costs due to small-scale
Costs for producing clothes which are not in demand	Costs of inventories
Reduced perceived value of products using the same brand name as the product made of waste material	When using a highly specialized production facility, production of products made of waste material has a significant negative impact on the utilization of the facility. The need for distributors to change their purchasing routines. Concerns about sanitation of used clothes
Fibers origin	The lack of transparency and traceability in the supply chain concerning the fibers origin (the input material), to deal with potential perceived risks that traces of hazardous substances

Sources: Hjelmgren et al. (2015); Meyers (2014); Watson et al. (2017)

4.3.2 Recycling

In recycling, the process used to recycle the waste maintains its value over time. Recycling cases occur when the waste materials are recycled again into the initial products, i.e., in the cases of closed-loop recycling. Cases of recycling or also of closed-loop recycling are glass recycling, where the glass can be recycled several amounts of times without losing its properties. Herat (2008) compared a recycling (closed loop) and a downcycling (open loop into a lower-value product) of cathode-ray tube (CRT) glass. The closed-loop solution for CRT glass was glass-to-glass recycling, where the process allowed to obtain leaded and unleaded CRT glass. The open cycling tested was glass-to-lead recycling, where CRT glass was subjected to a smelting process, recovering lead and copper. Glass-to-glass recycling has barriers such as the difference in CRT glass composition due to different producers, high labor cost of dismantling, cheap and ready availability of other recycled glass, and high collection costs from significant barriers (Herat 2008).

4.3.3 Downcycling

Downcycling is a recycling process where the value of the recycled material decreases over time, being used in less valued processes, with lesser quality material and with changes in inherent properties, when compared to its original use (Ashby et al. 2007; Chandler and Werther 2014; Geyer et al. 2015). Cases of downcycling are recycling of printing paper into toilet paper (McKenna et al. 2013). Most of the time, the actual recycling of municipal waste streams (e.g., paper/cardboard, plastic) is considered more like a downcycling and not recycling. Such is related to the poor design of products, which are not conceived to be recycled and disassembled, and end-of-life management of products and materials, getting contaminated with other substances or materials, leading to recycled materials with low quality, limiting the applications of those materials (de Aguiar et al. 2017; Reuter et al. 2013).

A particular case of downcycling is the one related with recycled aggregates from construction and demolition waste. Recycled aggregates are results from concrete crushing, sieving, and decontamination (if needed), being adequate for use as bulk fill, fill in drainage, sub-base or base material in road construction, and also aggregate for a new concrete (Florea and Brouwers 2013; Hansen 2002). The first three operations use downcycling, being the most applied operation to recycled construction and demolition waste in Europe (Florea and Brouwers 2013; Hansen and Lauritzen 2004). Countries like Belgium and the Netherlands are facing the problem of aggregate market saturation, where the use of such recycled material is no longer applicable, due to its low quality, and the applications of such low quality material is ceasing (viz. road construction) (Di Maria et al. 2018; Hu et al. 2013). The only way to move from downcycling into recycling is by improving the quality of recycled aggregates, by removing impurities by advanced recycling techniques, or by selective demolition of buildings, which includes the progressive dismantling of the buildings, although the high costs of such procedure are not promoting it (Di Maria et al. 2018).

4.3.4 Recycling Challenges

One of the main drawbacks of the recycling is the difficulty in promoting a homogenous market for recyclates and other products made of waste (including other recovery at Sect. 4.4), in such a way that the industry could have trust on the waste products and where the bureaucracy related to waste transportation and management could be softer. Waste Framework Directive intended to promote the introduction of waste products in the economy by defining the end-of-waste criteria for specific waste, where waste products could respect specific requirements to ensure that they are secondary raw material for the industry. These requirements are (European Parliament and Council 2008):

- "the substance or object is commonly used for specific purposes;
- a market or demand exists for such a substance or object;
- the substance or object fulfills the technical requirements for the specific purposes and meets the existing legislation and standards applicable to products; and
- the use of the substance or object will not lead to overall adverse environmental or human health impacts."

End-of-waste criteria are to be applied to specific waste streams, being the Joint Research Centre responsible for its selection. The ones selected so far are (Villanueva et al. 2010):

1. Streams used as feedstock in industrial processes, a pathway that most often controls the risks of health and environmental damage via industrial permits. The streams identified in this subcategory are:

   - Metal scrap of iron and steel, aluminum, copper
   - Plastics
   - Paper
   - Textiles
   - Glass
   - Metal scrap of zinc, lead, and tin
   - Other metals

2. Streams used in applications that imply direct exposure to the environment. In these cases, the EoW criteria to be developed in the further assessment shall probably include limit values for pollutant content or leaching, taking into account any possible adverse environmental and health effects. The streams in this subcategory are:

   - C&D waste aggregates
   - Ashes and slag
   - Biodegradable waste materials stabilized for recycling

3. Streams that may be in line with the EoW principles. However, it is not clear in all cases that (a) their current management in the EU takes place via recycling or (b) that recycling is a priority compared to controlled energy recovery or landfilling in suitable facilities. More detailed information is needed about their subfractions and their available outlets before they opt for selection. By the results collected, the waste streams proposed for this category are solid waste fuels, wood, waste oil, tires, and solvents.

4.3.5 Remarks

An aspect that should be highlighted when identifying those recycling measures is the missing concept of value. What is a more significant value than the initial one? Is the market value of final products made with recycled materials? Alternatively, are regarding environmental impacts? Alternatively, in the destination regarding market, but not regarding market value but regarding demand – a more valuable product can be made of recycled materials, but the demand for it can be too low, not allowing an adequate avoidance of virgin resources by replacing them with the recycled material. For that reason, maybe it is better to mention quality and not value.

Identifying which is the route of the waste being managed can be hard. Even end-of-waste criteria defined by the Waste Framework Directive only want the waste to be a product and define its features, although it is not helpful in this area. Again, the hierarchy of recycling options probably requires other methodologies to help to understand the more sustainable ones and such recycling process compared to the other waste operation from the waste hierarchy.

4.4 Other Recovery

Other recovery management option means (European Parliament and Council 2008):

> Any operation the principal result of which is waste serving a useful purpose by replacing other materials which would otherwise have been used to fulfill a particular function, or waste being prepared to fulfill that function, in the plant or the wider economy.

In other recovery, the most spread technology is waste-to-energy (WtE). Depending on which side of the world, the type of technologies included in WtE varies. In Asian countries, WtE includes physical, thermal, chemical, and biological techniques (Pan et al. 2015). Concerning municipal solid waste, the most devoted WtE techniques are co-combustion, co-digestion, and fermentation/compost, being generated by biogas, heat- and refuse-derived fuel, presented in Fig. 4.4.

There are particular situations on recovery technologies at European Union, in the light of waste hierarchy. One case is defining when energy recovery vs. incineration is occurring, and the second case is when biological treatment (in this case by anaerobic digestion) is recovery or recycling. Those situations are particularly relevant when targets need to be fulfilled by European countries, this way, respecting the European legislation.

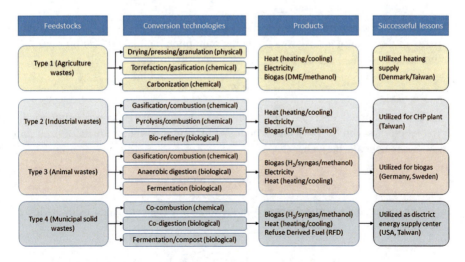

Fig. 4.4 Technology tree for WtE techniques. (Source: Pan et al. (2015))

Table 4.2 Average energy recovery efficiency, according to R1 formula, by type of plant in Europe

Type of plants	Average energy recovery efficiency (R1 formula)	Average waste flow (t/y)
CHP plants	0.71	230,000
Mainly electricity-producing plants	0.49	150,000
Mainly heat-producing plants	0.64	90,000

Source: Grosso et al. (2010)

Looking at incineration, the Directive 2008/987EC defines that it is a case of energy recovery through the definition of efficiency by R1 formula (Eq. 4.1). The units in operation before 1 January 2009 capable of reaching 0.60 of energy efficiency (equal or above) and units permitted after 31 December 2008 capable to reach 0.65 of energy efficiency are units where energy recovery occurs. The rest of the units not capable of doing it are making incineration (a disposal operation). This situation has made several incineration plants that do not recover the heat directly but for electric energy use without being capable of meeting the energy efficiency, which is the case of incineration plants in Portugal. In 2014, the energy recovery of municipal solid waste was 471 thousand tons, when incineration was 998 thousand tons (Eurostat 2017), showing the difficulty in increasing efficiency of electricity-producing plants in reaching the required efficiency. According to Table 4.2, the average energy recovery of efficiency by the R1 formula for electric energy production units is 49%. The R1 formula is given by Grosso et al. (2010):

$$\text{Efficiency} = \frac{E_p - (E_f - E_i)}{0.97 \times (E_w - E_f)} \qquad (4.1)$$

"where E_p is the annual energy produced as heat or electricity. It is calculated with energy in the form of electricity (E_{el}) being multiplied by 2.6 and heat produced for commercial use (E_{th}) multiplied by 1.1 (GJ/year). In formula it results:

$$E_p = 1.1 \times E_{th} + 2.6 \times E_{el}$$

E_f is the annual energy input to the system from fuels, contributing to the production of steam (GJ/year); it is obtained by summing the products of each fuel flow by its net calorific value (NCV):

$$E_f = \sum m_{fuel,i} \times NCV_{fuel,i}$$

E_w is the annual energy contained in the treated waste calculated using its lower net calorific value (GJ/year):

$$E_w = m_{waste} \times NCV_{waste}$$

E_i is the annual energy imported, excluding E_w and E_f (GJ/year); 0.97 is a factor accounting for energy losses due to bottom ash and radiation."

The issue of anaerobic digestion is concerning the capability of digestate to meet the recycling definition, i.e., in producing a product. Here, also the composting is included in the discussion. The products of composting and anaerobic digestions are compost or digestate which, according to the Commission Decision 2011/753/EU (Commission 2011), is used as recycled product, material, or substance for land treatment resulting in a benefit to agriculture or ecological improvement (European Commission 2017). The issue here is on when the anaerobic digestion or composting processes are included in mechanical-biological treatment units, which are treating residual waste or mixed waste, i.e., municipal waste which has not been source separated. In those units, only if the owner of the unit can prove that the produced compost or digestate brings a benefit to agriculture or ecological improvement can it be seen as a product and, in that case, a recycling operation (European Commission 2017). The way to prove such benefit is made through the compliance with national norms and standard for compost and digestate, which is defined by each European country.

Eurostat is focusing on the presentation of recovery as the main incineration including energy recovery. The recovery rates vary from 1% from Bulgaria and Greece to more than 30% for countries like Denmark, Germany, Luxembourg, Finland, and Sweden (Eurostat 2017).

4.5 Disposal

Disposal definition considered is "any operation which is not recovery even where the operation has as a secondary consequence the reclamation of substances or energy." The disposal is the last option for waste, in the light of waste hierarchy but also of the circular economy, because the waste material will get lost to the economy but also the environment, not being available to replace virgin materials. Ways defined to avoid the disposal management option of waste have been defined by policy instruments, like bans of materials from landfill, higher landfill and incineration fees for recyclable materials, and the use of policy instrument to promote the other waste management hierarchy options.

The two most known and spread disposal techniques are engineering (also sanitary) landfills and incineration (without energy recovery). Figure 4.5 shows the countries Spain, the UK, Poland, France, Italy, Germany, and Bulgaria with considerable annual amounts of municipal waste sent to landfill and incineration without energy recovery in 2014. On the other hand, other countries like Luxembourg have no landfilling, no incineration without energy recovery, and no other disposal.

The dependence of landfilling has made countries to divert waste from this operation, namely, by landfill taxes and taxes for specific waste features going to landfill. Another perspective to reduce the environmental impact from landfills is its mining. Landfill mining has been used all over the world in the last 62 years; it started in 1953 in Israel and rapidly spread to the USA, Canada, India, and several

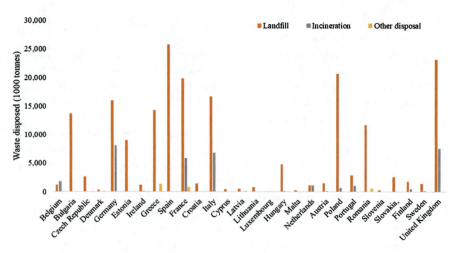

Fig. 4.5 Waste disposal destinations in European countries. (Source of data: Eurostat (2017))

countries in Europe like Germany, Sweden, Belgium, and Italy (Hogland et al. 2004; Kurian et al. 1999; Jones et al. 2013). Landfill mining is being addressed as the new source of raw materials for Europe. However, landfill mining can have other drivers that have justified it. The needs to recover the land for added value activities like construction, to remove waste and stabilize hazardous fractions, to extend landfill capacity, to generate revenues from materials obtained and fuel for energy production, and to reduce landfill closure costs are drivers to the landfill mining (Collivignarelli et al. 1997; USEPA 1997).

Although all the drivers promoting landfill mining, most of the time, this operation is not economically feasible, leading to the concept of "temporary stage," where materials without value to be mined are conditionally stored (Breure et al. 2018; Jones et al. 2013). Besides the pragmatic economic affordable issue, other barriers such as misleading and missing legislation, shortage of environmental standards for the materials to be explored, shortage of best available techniques that support the technical operation of a landfill mining activity, lacking of standardization of safety and health, public skepticism, the missing of studies and life cycle assessment showing the environmental benefit of landfill mining, and the decreasing of recoverable waste in landfills are to be solved to enable landfill mining to be a reality (Pires et al. 2016).

4.6 Final Remarks

Using the waste hierarchy ordination of waste operations to manage municipal solid waste (or another type of waste) may be a challenge and can be costly, and the environmental benefit can be questioned. Aspects related to infrastructure location, features of material to be recycled, and quality of recycled material as well as of waste-derived fuel can dictate different destinations that may impose different impacts on the environment, different financial resources, and different revenues that have made researchers doubt the waste hierarchy. One thing is sure: waste hierarchy helps to save resources. Although waste hierarchy seems static, the concepts of the waste operations prioritized are not closed and in continuing update, as long as technology also evolves.

Research on waste hierarchy and how the waste collection can contribute to the hierarchy is needed. European regulations are based on the scientific evidence that source separation of waste is a requirement to ensure recycling, being this aspect more determinant of the biodegradable municipal waste. If biodegradable municipal waste is not source separated, the quality of compost of digestate is questioned, not ensuring the occurred recycling but recovery only. For other materials, the mechanical processing of mixed waste is capable of bringing high amounts of recyclable waste that citizens are not source separating, making more material available for recycling, although with a loss of quality. An analysis of the entire life cycle of the waste – from the source of the waste as a product until the last destination – is required to ensure that the best destination is given regarding sustainability.

References

Ashby M, Shercliff H, Cebon D (2007) Materials: engineering science processing and design. Elsevier, Amsterdam

Bartl A (2014) Ways and entanglements of the waste hierarchy. Waste Manag 34:1–2

Bovea MD, Ibáñez-Forés V, Pérez-Belis V, Quemades-Beltrán P (2016) Potential reuse of small household waste electrical and electronic equipment: methodology and case study. Waste Manag 53:204–217

Breure AM, Lijzen JPA, Maring L (2018) Soil and land management in a circular economy. Sci Total Environ 624:1125–1130

Chandler D, Werther WB (2014) Strategic corporate social responsibility – stakeholders, globalization, and sustainable value creation. SAGE Publications, Thousand Oaks

Cohen N, Robbins P (eds) (2011) Green business – an A-to-Z guide. SAGE Publications, Los Angeles

Collivignarelli C, Baldi M, Bertanza G, Bina S, Conti F (1997) Characterisation of waste from landfill mining: case studies. In: Christensen TH, Cossu R, Stegmann R (eds) Proceedings Sardinia 97, sixth international landfill symposium. CISA, Environmental Sanitary Engeneering Centre, Cagliari

Commission (2011) Commission decision of 18 November 2011 establishing rules and calculation methods for verifying compliance with the targets set in Article 11(2) of Directive 2008/98/EC of the European Parliament and of the Council. Off J Eur Union L310:11–16

de Aguiar J, de Oliveira L, da Silva JO, Bond D, Scalice RK, Becker D (2017) A design tool to diagnose product recyclability during product design phase. J Clean Prod 141:219–229

DeLong M, Casto MA, Min S, Goncu-Berk G (2017) Exploring an up-cycling design process for apparel design education. Fash Pract 9:48–68

Di Maria A, Eyckman J, van Acker K (2018) Downcycling versus recycling of construction and demolition waste: combining LCA and LCC to support sustainable policy making. Waste Manag https://doi.org/10.1016/j.wasman.2018.01.028 75:3–21

European Parliament, Council (2008) Directive 2008/98/EC of the European Parliament and of the Council of 19 November 2008 on waste and repealing certain directive. Off J L312:3–30

European Commission (2017) Guidance on municipal waste data collection – Eurostat Unit E2 Environmental statistics and accounts, sustainable development. http://ec.europa.eu/eurostat/documents/342366/351811/Municipal+Waste+guidance/bd38a449-7d30-44b6-a39f-8a20a9e67af2. Accessed 15 Feb 2018

Eurostat (2017) Waste management indicators. http://ec.europa.eu/eurostat/statistics-explained/index.php/Waste_management_indicators. Accessed 10 Apr 2018

Florea MVA, Brouwers HJH (2013) Properties of various size fractions of crushed concrete related to process conditions and re-use. Cem Concr Res 52:11–21

Geyer R, Kucenski B, Zink T, Henderson A (2015) Common misconceptions about recycling. J Ind Ecol 20:101–1017

Gharfalkar M, Court R, Campbell C, Ali Z, Hillier G (2015) Analysis of waste hierarchy in the European waste directive 2008/98/EC. Waste Manag 39:305–313

Grosso M, Motta A, Rigamonti L (2010) Efficiency of energy recovery from waste incineration, in the light of the new Waste Framework Directive. Waste Manag 30:1238–1243

Hansen TC (2002) Products, codes, standards, and testing methods for recycled aggregate concrete. In: Hansen TC (ed) Recycling of demolished concrete and masonry. Taylor & Francis, Oxon, pp 109–122

Hansen TC, Lauritzen E (2004) Concrete waste in a global perspective. In: Liu TC, Meyer C (eds) Recycling concrete and other materials for sustainable development. American Concrete Institute, Farmington Hills, pp 35–45

Herat S (2008) Recycling of cathode ray tubes (CRTs) in electronic waste. Clean (Weinh) 36:19–24

Hjelmgren D, Salomonson N, Ekström KM (2015) Upcycling of pre-consumer waste – opportunities and barriers in the furniture and clothing industries. In: Ekström KM (ed) Waste management and sustainable consumption – reflections on consumer waste. Routledge, Oxon

Hogland W, Marques M, Nimmermark S (2004) Landfill mining and waste characterization: a strategy for remediation of contaminated areas. J Mater Cycles Waste Manag 6:119–124

Hu M, Kleijn R, Bozhilova-Kisheva K, Di Maio F (2013) An approach to LCSA: the case of concrete recycling. Int J Life Cycle Assess 18:1793–1803

Huysman S, De Schaepmeester J, Ragaert K, Dewulf J, De Meester S (2017) Performance indicators for a circular economy: a case study on post-industrial plastic waste. Resour Conserv Recycl 120:46–54

Jones PT, Geysena D, Tielemans Y, van Passel S, Pontikes Y, Blanpain B, Quaghebeur M, Hoekstra N (2013) Enhanced landfill mining in view of multiple resource recovery: a critical review. J Clean Prod 55:45–55

Kurian J, Esakku S, Palanivelu K, Selvam A (1999) Studies on landfill mining at solid waste dumpsites in India. In: Christensen TH, Cossu R, Stegmann R (eds) Proceedings of the Sardinia 99, seventh international waste management and landfill symposium. CISA, Environmental Sanitary Engineering Centre, Cagliari

Lu B, Yang J, Ijomah W, Wu W, Zlamparet G (2018) Perspectives on reuse of WEEE in China: lessons from the EU. Resour Conserv Recycl 135:83–92 https://doi.org/10.1016/j.resconrec.2017.07.012

McDonough W, Braungart M (2002) Cradle to cradle: remaking the way we make thinks. North Point Press, New York, USA

McKenna R, Reith S, Cail S, Kessler A, Fichtner W (2013) Energy savings through direct secondary reuse: an exemplary analysis of the German automotive sector. J Clean Prod 52:103–112

Meyers GJ (2014) Designing and selling recycled fashion: Acceptance of upcycled secondhand clothes by female consumers, age 25–65. Dissertation. North Dakota State University

Niero M, Hauschild MZ, Hoffmeyer SB, Olsen SI (2017) Combining eco-efficiency and eco-effectiveness for continuous loop beverage packaging systems: lessons from the Carlsberg circular community. Appl Implement 21:742–753

References

Pan SY, Du MA, Huang IT, Liu IH, Chang EE, Chiang PC (2015) Strategies on implementation of waste-to-energy (WTE) supply chain for circular economy system: a review. J Clean Prod 108:409–421

Papargyropoulou E, Lozano R, Steinberger J, Wright N, bin Ujang Z (2014) The food waste hierarchy as a framework for the management of food surplus and food waste. J Clean Prod 76:106–115

Pires A, Martinho G, Silveira A, Gomes A, Cardoso J, Lapa N (2016) The urgent need of policy and regulation drivers to promote landfill mining in Portugal. In: Pereira MJ, Carvalho MT, Neves PF (eds) Proceedings of the international symposium on enhanced landfill mining, Lisbon, February 2016. Instituto Superior Técnico, Lisbon, pp 67–78

Reuter M, Hudson C, van Schaik A, Heiskanen K, Meskers C, Hagelüken C (2013) Assessing mineral resources in society: metal recycling opportunities, limits, infrastructures. United Nations Environment Programme, Paris

United States Environmental Protection Agency (USEPA) (1997) Landfill reclamation, solid waste and emergency response (5306W) EPA 530-F-97-001

Villanueva A, Delgado L, Luo Z, Eder P, Catarino AS, Litten D (2010) Study on the selection of waste streams for end-of-waste assessment. Office of the European Union, Luxembourg, Luxembourg

Watson D, Elander M, Gylling A, Andersson T, Heikklilä P (2017) Stimulating Textile-to-Textile Recycling. TemaNord 2017:569. Nordic Council of Ministers, Copenhagen

Chapter 5
Economic Perspective

Abstract Solid waste management requires specific rules to ensure the collection and management respecting the citizens and the environment. A brief review on how solid waste management is defined by international and national regulations is provided. The intention is to cover broader waste streams from municipal solid waste, including batteries and accumulators, end-of-life vehicles, packaging waste, waste from electric and electronic equipment, waste oils, biodegradable municipal waste, and waste tires.

Keywords Waste framework directive · Basel convention · Transboundary shipment · National legislation · International legislation · Packaging waste · Waste oils · WEEE · ELV

5.1 International Legislation on Waste

Although waste is a domestic and local issue to be managed by municipalities, the globalization of markets obligated the definition of international rules to regulate how waste can be managed between countries and to drive the guidelines to be implemented for each country. Concerning the first goal, the most known international waste regulations applicable to municipal solid waste is Basel Convention; for the second goal, the most known international guiding regulation is the European Union waste policies.

5.1.1 Basel Convention

The Basel Convention on the Control of Transboundary Movements of Hazardous Wastes and their Disposal is in force since 1992 and intends to protect human health and the environment against the adverse effects of hazardous waste (Basel Convention 2011), with the goals being:

© Springer International Publishing AG, part of Springer Nature 2019
A. Pires et al., *Sustainable Solid Waste Collection and Management*,
https://doi.org/10.1007/978-3-319-93200-2_5

- To reduce hazardous waste generation and promotion adequate management in environmental terms
- To restrict its import/export movements except when it follows principles of environmentally sound management, and
- To comply with the regulatory system of transboundary movements

In practical terms, the Basel Convention intends to prohibit the shipment of hazardous waste from Annex VII countries (Organization for Economic Cooperation and Development (OECD), the European Community (EC), and Lichtenstein, but the United States has not ratified) to non-Annex VII countries (the rest of signatories countries) (Lepawsky 2014; Wang et al. 2016). The assumption presents in the Basel Convention was that non-Annex VII countries have no hazardous waste (probably due to the missing industry level that Annex VII countries possess). The reality shows that the protected (developing countries) countries have hazardous waste and they can trade and ship hazardous waste between them or from developing countries to developed countries and no environmentally sound management is ensured (Lepawsky 2014, 2015). Nowadays, the trade of e-waste (or waste electrical and electronic equipment), a hazardous waste due to the presence of substances from List A of Annex VIII of the Basel Convention, between developing countries have increased, with consequences to public health and the environment. Disposal plants release toxic chemicals, volatile organic chemicals, and heavy metals because recycling is made by heating and manual removal of components from circuit boards, open burning to reduce volumes and recover metals, and open acid digestion to recover precious metals (Robinson 2009; Wang et al. 2016). This reveals the inadequacy of Basel Convention pointed out by Lepawsky (2014) in facing the world trade of e-waste by defining trade bans focused on flows North to South and not on ensuring environmentally sound recycling processes all over the world, once that e-waste exists everywhere (Fig. 5.1).

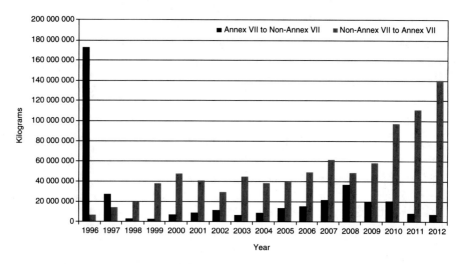

Fig. 5.1 International trade of e-waste, 1996–2012. (Source: Lepawsky (2015))

5.1.2 European Union Waste Policies

European countries entering into the European Union need to implement policies included in the European Community Treaty (Hedemann-Robinson 2007). The policies include but are not limited to directives, regulation, and decisions (Chang et al. 2013). Directives must be implemented, but the country may choose how to implement; regulations enter directly into the force of all Member States; strategies are guidelines to be considered in the future waste legislation, helping Member States to get prepared for the future changes. From all forms of legal documents, the directives are the ones most promoted by European Commission (Fischer and Davidsen 2010).

The main framework for waste management is the Waste Framework Directive 2008/98/EC (European Parliament and Council 2008). The Waste Framework intends to define and implement the following (Chang et al. 2013):

- Waste hierarchy principle promotion
- Authorities responsibilities, including waste planning
- "Polluter-pays" principle, where the cost of disposing must be borne by the waste producer, who request waste be properly handled by a waste collection subcontractor or disposer and/or by the upstream polluter or the producer of the product becoming waste
- Hazardous waste management mandatory aspects such as labeling, record keeping, monitoring and control obligations, and banning of mixing hazardous waste

There are also directive regulation waste operations, namely, incineration and landfill directives. Incineration Directive 2000/76/EC introduces measures to prevent environmental impacts and human health issues from incineration and co-incineration. Landfill Directive 1999/31/EC focuses on landfill activities such as landfill gas, waste acceptance criteria, meteorological data collection, monitoring, after-closure care, and technical requirements.

Specific waste streams directives have also been elaborated by European Union: batteries and accumulators (B&A), waste from electric and electronic equipment (WEEE), end-of-life vehicles (ELV), mining, packaging, polychlorinated biphenyl/terphenyls (PCB/PCT), sewage sludge, ships, titanium dioxide, and waste oils. PCB/PCT, mining, sewage sludge, ships, and titanium dioxide will not be considered because they are related to industry.

5.2 National Waste Regulation in European Union Countries

When materials and objects are classified as waste, technical, functional, environmental, and health protection aspects to manage it are regulated at the European Union, with direct and indirect consequences on the waste management infrastructure, as well as on the waste markets and responsibilities on its management. According to Chang et al. (2013), the transposition of the European Union waste directives involves several issues, where two situations may occur.

> Either the Member State does not possess specific legislation concerning waste (as happens for most waste streams), or the country already has legislation to manage waste. In this situation, the legislation can be in accordance with the directive; however, it is possible that they are not in accordance, creating infractions on transpositions of directives. These cases are discussed in the European Court of Justice.

After legislation is transposed, national environmental agencies are normally the ones that verifies how is legislation being implemented in the field. Chang et al. (2013) describes succinctly the responsibilities of the several players in the municipal waste management.

> Central/federal governments are responsible for the conception of waste management plans, in which the goals of the legislation and the strategies to ensure its success are defined. Provincial/regional entities have the duty to apply national policies (...) Municipalities/local authorities are responsible for solid waste generated by municipalities/households and similar waste, excluding industrial waste, which is the responsibility of the owner. Municipalities can act together through regionalization like an association or a task force, which allows improved waste management efficiency. Waste management operations under the municipality's scope of work are collection, energy recovery, and landfill disposal; and treatment. (...) Waste management plans are also proposed for legislation via a single municipality or municipalities' associations.

Although the role of municipalities in the waste management is to control this public health issue and to comply with the European Union's requirements, a new paradigm is raised with the appearance of extended producer responsibility. This principle entered in the end of 1990s for packaging waste, helping municipalities in supporting the recycling activity. The extended producer responsibility (EPR) has shifted the municipalities' role to the manufacturers and importers of products. The original purpose for the application of EPR to waste management was twofold: to relieve municipalities of some of the financial burden of waste management and to provide incentives to producers to reduce the use of primary resources, promote the use of more secondary materials, and undertake product design changes to reduce waste (OECD 2001). In practical terms, the EPR involves the payment to the producer responsibility organization (PRO) to transfer its responsibility to manage their product when reach waste stage including its collection. The local authorities/municipalities and distributors are accountable for the waste collection defined by the PRO, and the PRO will finance local authorities and private operators to send waste for recycling, energy recovery facilities, or other destinations, depending of the goals of the PRO (Chang et al. 2013; Pires et al. 2015). The monetary and material flow involved in a EPR system for packaging waste is presented in Fig. 5.2.

The way how EPR is being regulated and how EPR manages specific waste streams will be addressed in the next subsections. Waste streams regulated by EPR chosen are waste batteries and accumulators, end-of-life tires, end-of-life vehicles, packaging waste, waste electrical, and electronic equipment (WEEE).

5.2 National Waste Regulation in European Union Countries

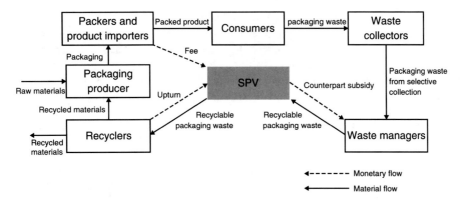

Fig. 5.2 The extended producer responsibility scheme for packaging in Portugal managed by the Portuguese Green Dot system. SPV – Sociedade Ponto Verde (the PRO corresponding to Green Dot System in Portugal). (Source: Pires et al. (2015))

5.2.1 Batteries and Accumulators

In the European Union, batteries and accumulators are regulated by Directive 2006/66/EC, where it established the rules to put on market these products and how they should be managed when reach their end-of-life stage. The waste management requirements are the following (European Parliament and Council 2006):

- A 25% collection rate for waste portable batteries to be met by September 2012, rising to 45% by September 2016
- A prohibition on the disposal by landfill or incineration of waste industrial and automotive batteries in effect, setting a 100% collection and recycling target
- The setting of recycling efficiencies to ensure that a high proportion of the weight of waste batteries is recycled (65% of lead acid batteries, 75% of nickel-cadmium batteries, and 50% of other waste batteries)

Although the targets are defined, there is no formal information concerning the targets accomplish by European Union Member States. Some information exists at Tsiarta et al. (2015), where 20 Members States reported collection rates around 25% by 2012, complying with the target. The existing collection schemes in Member States reported by Tsiarta et al. (2015) show the existence of community collection, free of charge, or collection points near the final distributor of portable batteries without the obligation of a new purchase. For automotive batteries, there are collection schemes like in Portugal, near the car shops, but there are countries where car batteries are collected and processed within a free market.

In Portugal, batteries are managed by five PROs: one is dedicated specifically to managing portable and industrial batteries and accumulators; two are licensed to manage portable and industrial batteries and accumulators and WEEE; another manages end-of-life vehicles beyond industrial and vehicle batteries and accumulators; and one system manages vehicle and some industrial batteries. Waste portable batteries and accumulators have selective network structures based on municipal systems, distributors, and other collection points, but also selective collection network at distributors and others receive the waste batteries and accumulators by end user without any charge (APA 2018; Tsiarta et al. 2015). The impact of the directive has resulted in the achieving and/or exceeding in 2012 all three recycling efficiency rates outlined by the directive by 19 countries (for all batteries or for lead-acid only) (Tsiarta et al. 2015).

5.2.2 End-of-Life Vehicles (ELV)

The Directive 2000/53/EC and its amendments intend to set measures to prevent the amount of waste from ELV and their components and promote their reuse, recycled, or recovery when possible. The directive stipulates that Member States shall take the necessary measures to ensure that the following targets are attained by economic operators (European Parliament and Council 2000):

- No later than 1 January 2006, for all end-of-life vehicles, the reuse and recovery shall be increased to a minimum of 85% by an average weight per vehicle a year. Within the same time limit, the reuse and recycling shall be increased to a minimum of 80% by an average weight per vehicle a year.
- No later than 1 January 2015, for all end-of-life vehicles, the reuse and recovery shall be increased to a minimum of 95% by an average weight per vehicle a year. Within the same time limit, the reuse and recycling shall be increased to a minimum of 85% by an average weight per vehicle a year.

The new vehicle manufacturers (together with importers and distributors) should promote the absence of hazardous substances such as lead, mercury, cadmium, and hexavalent chromium, must provide systems to collect ELVs, and, where technically feasible, used parts from repaired passenger cars. The regulation also demands for a certification of destruction when the vehicle reaches ELV stage. The delivery of the vehicle should be made with no expense for the vehicle's owner, being the treatment to be supported by the manufacturer.

Directive 2005/64/EC are defined as the technical rules that vehicle's parts and materials may be reuses, recycled, and recovered, ensuring safety and no environmental risks. According to the directive, the new vehicles to be sold in European Union may be reused and/or recycled to a minimum of 85% by mass or reused and/or recovered to a minimum of 95% by mass, excluding airbags, seat belts, and steering locks (European Union Law 2015).

The performance of the ELV management schemes have allowed to comply with the targets of 2006, like is presented in Fig. 5.3. However, the targets of 2015 are still in clearance, but the reuse/recycling target of 85% was reached in 2014.

5.2 National Waste Regulation in European Union Countries

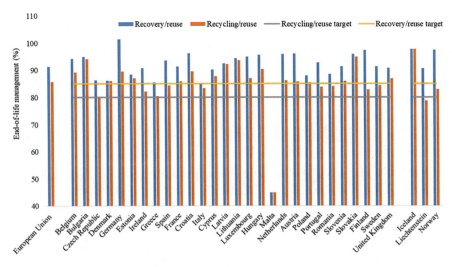

Fig. 5.3 Recovery and recycling rate for end-of-life vehicles in 2014 at European Union countries. (Source of data: Eurostat (2017))

5.2.3 Packaging Waste

Packaging waste is regulated mainly by Directive 94/62/EC and its amendments, where there are defined measures that could be implemented to manage packaging waste, to develop packaging reuse systems, and to source separate collection. The targets have been evolving since 2001, being the most recent ones (European Union Law 2014).

- By no later than 31 December 2008, at least 60% by weight of packaging waste to be recovered or incinerated at waste incineration plants with energy recovery.
- By no later than 31 December 2008, between 55% and 80% by weight of packaging waste to be recycled.
- No later than 31 December 2008, the following targets for materials contained in packaging waste must be attained:
  - 60% for glass, paper, and board
  - 50% for metals
  - 22.5% for plastics and
  - 15% for wood

The results reached for European Union Member States are presented in Fig. 5.4, for recycling and recovery targets. In general, the targets have been reached for most countries. The application of source separation schemes, namely, drop-off systems and door-to-door collection schemes, together with policy instruments, namely, voluntary programs, information and awareness campaigns, funding programs, standards, and eco-labels (Chang et al. 2013).

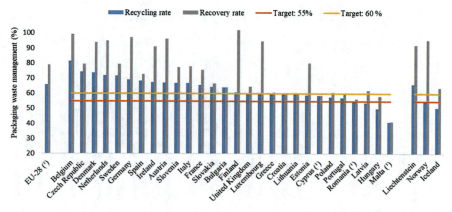

Fig. 5.4 Recycling and recovery rates for all packaging, 2015. (Source of data: Eurostat (2018a))

5.2.4 Waste Electrical and Electronic Equipment

The European Union has pointed out that Member States should encourage the development of devices, components, or materials that could be dismantled and recovered, reused, and recycled. First directive on WEEE was Directive 2002/96/EC, which has been substituted by Directive 2012/19/EU. In this directive, the established targets are the following (European Parliament and Council 2012):

> From 2016, the minimum collection rate shall be 45% calculated on the basis of the total weight of WEEE collected (...) From 2019, the minimum collection rate to be achieved annually shall be 65% of the average weight of EEE placed on the market in the three preceding years in the Member State concerned, or alternatively 85% of WEEE generated on the territory of that Member State.

The WEEE Directive has been capable to demand for several collection schemes to ensure that WEEE producers could deliver them free of charge. There are several ways to deliver WEEE (European Parliament and Council, 2012):

- Deliver at least free of charge at a waste collection system.
- Near distributors: to deliver free of charge or one-to-one basis for the same type of device or equivalent.
- Distributors with sales are of EEE at least 400 m^2 to receive very small WEEE free of charge.

The results reached in 2015, according to Eurostat (2018a, b) and presented in Fig. 5.5, point out that the target for 2016 was reached by 13 of 28 European Union countries, being households the main source of WEEE in all countries. The most collected WEEE belongs to large household appliances (52%), IT, and telecommunication equipment (16%), followed by consumer equipment (15%), small appliances (10%), and the rest of WEEE which corresponds to 7%.

5.3 Final Remarks

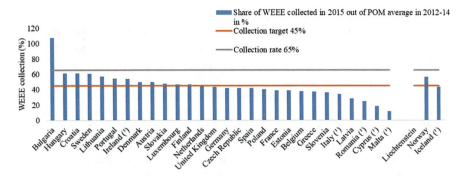

Fig. 5.5 Total collection rate for WEEE, 2015 (in accordance with calculation of Directive 2012/19/EC). (Source of data: Eurostat (2018b))

5.2.5 Waste Oils

The success of EPR to manage waste streams have led the application of EPR to waste lube oils in countries like Portugal, Spain, Belgium, and Italy. Other countries prefer to let the market works, which likely occurs with the management of industrial waste. One of the issues in managing waste oils by EPR is necessary to have in mind the disappearance of lube oil during use, which constraint the separate collection of this waste. According to Lohof (1991), the amount of waste oils generated during its use as lubricant is not 100% but only around 50%. This consumption or leakage of lube oils during use has posed constraints in the use of EPR concept itself in Portugal. The solution was to establish the separate waste collection rates feasible with this reality.

Although there is no official European Union statistics on waste oils, the Groupement Européen de l'Industrie de la Régénération (GEIR) mentions that there is around 5.7 Mtonnes of waste oils in European countries, having the potential to be collected around 2.7 M tonnes (GEIR, 2015). From the total generated, 65% are collected to be recovered (Botas et al. 2017).

5.3 Final Remarks

The regulation on waste management at international and national levels intends to ensure the environmentally sound management, where the local authorities and citizens have an important role to make regulation work. The elaboration of regulation of waste collection is vital to ensure that waste is minimized, that waste is source separated in such way that it conserves its properties to be recycled, and to define who is responsible for its collection and treatment.

In the study by BiPRO/CRI (2015), they realized that European capital cities where mandatory separate collection for specific municipal waste streams results in higher recycling rates for municipal waste; the technical collection system infrastructure is crucial for the success of the collection system; the door-to-door reaches best quality of recyclables and collection rates; brings system works better for glass collection; comingled collection of recyclables may occur but with reservations due to contamination potential; and introducing separate collection of biodegradable waste in the door-to-door system increases sorting of dry recyclables. In particular case study conducted in Portugal verifies some of those technical aspects of the BiPRO/CRI (2015) but highlights another aspect: the need to optimize continually the source separated collection system, to ensure that they are operating to provide the best service to users at the lowest economic and environmental cost (Martinho et al. 2017; Pires et al. 2017).

A new player is now raising at local level, motivated by the goal of local authorities to promote "zero-waste" initiatives – the grassroots environmental organizations (GEOs). GEOs are a type of nongovernmental organization (NGO) which is characterized by being local, focused on solving environmental problems at a local level. This profile can differ from NGOs which have a more national and international approach to solve environmental problems. In Esporles (Mallorca, Spain), a pay-as-you-throw scheme was implemented based on unit pricing for waste, where the GEO composed of local community was capable to accept the entrance of the unit-pricing scheme, because they were involved in the participatory process, what increased the acceptance of the unit-pricing scheme, and reducing the collateral effects (Weber et al. 2017). The governance of municipal waste requires society, probably by GEOs, to get involved to ensure a bottom-up approach in dealing with the challenges of waste and resource scarcity, to make regulation accepted by the citizens, and to leverage waste management in such way that could contribute to solve waste global problems at a local level.

References

Agência Portuguesa do Ambiente (APA) (2018) Waste batteries and accumulators (in Portuguese: Resíduos de pilhas e baterias). https://www.apambiente.pt/index.php?ref=16&subref=84&sub2ref=197&sub3ref=281. Accessed 27 Apr 2018

Basel Convenion (2011) Convention overview. http://www.basel.int/TheConvention/Overview/tabid/1271/Default.aspx. Accessed 23 Apr 2018

BiPRO/CRI (2015) Assessment of separate collection schemes in the 28 capitals of the EU. Final report. November 2015

Botas JA, Moreno J, Espada JJ, Serrano DP, Dufour J (2017) Recycling of used lubricating oil: evaluation of environmental and energy performance by LCA. Resour Conserv Recycl 125:315–323

Chang NB, Pires A, Martinho G (2013) Environmental legislation: EU countries solid waste management. In: Jørgensen SE (ed) Encyclopedia of environmental management, Four volume set. CRC Press, Boca Raton, pp 892–913

European Parliament, Council (2000) Directive 2000/53/EC of the European Parliament and of the Council of 18 September 2000 on end-of life vehicles. Off J L269:34–43

References

European Parliament, Council (2006) Directive 2006/66/EC of the European Parliament and of the Council of 6 September 2006 on batteries and accumulators and waste batteries and accumulators and repealing directive 91/157/EEC. Off J L266:1–14

European Parliament, Council (2008) Directive 2008/98/EC of the European Parliament and of the Council of 19 November 2008 on waste and repealing certain directive. Off J L312:3–30

European Parliament, Council (2012) Directive 2012/19/EU of the European Parliament and of the Council of 4 July 2012 on waste electrical and electronic equipment (WEEE). Off J L197:38–71

European Union Law (2014) Packaging and packaging waste. http://eur-lex.europa.eu/legal-content/EN/TXT/?uri=LEGISSUM:l21207. Accessed 27 Apr 2018

European Union Law (2015) Re-use, recycling and recovery of vehicle parts and materials. http://eur-lex.europa.eu/legal-content/EN/TXT/?uri=LEGISSUM:n26102. Accessed 27 Apr 2018

Eurostat (2017) End-of-life vehicle statistics. http://ec.europa.eu/eurostat/statistics-explained/index.php?title=End-of-life_vehicle_statistics&oldid=376304. Accessed 20 Apr 2018

Eurostat (2018a) Packaging waste statistics. http://ec.europa.eu/eurostat/statistics-explained/index.php/Packaging_waste_statistics. Accessed 23 Apr 2018

Eurostat (2018b) Waste statistics – electrical and electronic equipment. http://ec.europa.eu/eurostat/statistics-explained/index.php?title=Waste_statistics_-_electrical_and_electronic_equipment. Accessed 23 Apr 2018

Fischer C, Davidsen C (2010) Europe as a recycling society – The European recycling map. ETC/SCP working paper 5/2010, European Topic Centre on Sustainable Consumption and Production, Copnhagen, Denmark, pp 1–22

Groupement Européen de l'Industrie de la Régénération (GEIR) (2015) Waste lube oil management in Europe. In: Green planet association annual conference

Hedemann-Robinson M (2007) Enforcement of European Union environmental law: legal issues and challenges. Routledge-Cavendish, Oxon

Lepawsky J (2014) Are we living in a post-Basel world? Area 47:7–15

Lepawsky J (2015) The changing geography of global trade in electronic discards: time to rethink the e-waste problem. Geogr J 181:147–159

Lohof (1991) Where has all the used oil gone? Where can it go? Int Environ Aff 3:259–281

Martinho G, Gomes A, Santos P, Ramos M, Cardoso J, Silveira A, Pires A (2017) A case study of packaging waste collection systems in Portugal – part I: performance and operation analysis. Waste Manag 61:96–107

Organisation for Economic Co-operation and Development (OECD) (2001) Extended producer responsibility: a guidance manual for governments. OECD, Paris

Pires A, Martinho G, Ribeiro R, Mota M, Teixeira L (2015) Extended producer responsibility: a differential fee model for promoting sustainable packaging. J Clean Prod 108:343–353

Pires A, Sargedas J, Miguel M, Pina J, Martinho G (2017) A case study of packaging waste collection systems in Portugal – part II: environmental and economic analysis. Waste Manag 61:108–116

Robinson BH (2009) E-waste: an assessment of global production and environmental impacts. Sci Total Environ 408:183–191

Tsiarta C, Watson S, Hudson J (2015) Final implementation report for the Directive 2006/66/EC on batteries and accumulators. http://ec.europa.eu/environment/waste/batteries/pdf/batteries_directive_report.pdf. Accessed 20 Apr 2018

Wang Z, Zhang B, Guan D (2016) Take responsibility for electronic-waste disposal. Nature 536:23–25

Weber G, Calaf-Forn M, Puig-Ventosa I, Cabras I, D'Alisa G (2017) The role of environmental organisations on urban transformation: the case of waste management in Esporles (Mallorca). J Clean Prod https://doi.org/10.1016/j.jclepro.2017.08.241 195:1546–1557

Chapter 6
Psychosocial Perspective

Abstract Strategies have been applied to encourage recycling behavior, which is the most participative component of the integrated waste management, where the citizens are the key to ensure the success of a separate waste collection system. A short review of the most studied variables and factors involved in the study of consumption and recycling behavior is made in this chapter. At the end of the chapter are presented several limitations of existing recycling behavior models which need to be solved and which criteria should be used to measure the efficacy and efficiency of interventions used to change behavior. The urgent need to understand the recycling phenomena is notorious to address the environmental problem of waste and to increase recycling, promoting the circular economy.

Keywords Recycling behavior · Household waste · Motivation · Contextual factors · Sociodemographic variables · Convenience factors

6.1 Contributions of Social Psychology to Source Separate Waste Collection

One of the primary prerequisites for the circular economy is to make the waste incorporated again in the economic system. However, recycling industries will only be viable if they have access to waste to process, both in quantity and quality. The access to waste demands the implementation of highly efficient and effective separate waste collection systems, more sustainable in economic and environmental terms, by the entities responsible for waste management.

The waste collection is the component interface between the waste management entity and the users to whom the service is supplied (service image), and it is very vulnerable and dependent on the users' behavior. Once that collection is the starting point to a circular economy, probably the most important one, it is fundamental to pay close attention to the collection systems. The attention needs to focus on research and innovative technological solutions development or in social research that may contribute to increase the collected quantity and, consequently, to increase the sustainability of these systems.

© Springer International Publishing AG, part of Springer Nature 2019
A. Pires et al., *Sustainable Solid Waste Collection and Management*,
https://doi.org/10.1007/978-3-319-93200-2_6

The success of any separate collection system hinges on the participation rate of the waste producers and the low-level contamination of materials. How to motivate individuals to respond positively to recycling programs? In a short- or a long-term period, which motivational techniques have more effect on behavior? These are some of the questions that since the 1970s, many social psychologists, accompanying the implementation of waste collection systems all over the world, have sought to answer, by developing theories and models to be applied to experimental researches which aim to understand and predict the selective separation behaviors.

Research in separation behavior can be divided into two major groups, according to their purposes and methods (Martinho 1998):

- One line of research has been centered in developing behavioral models and identifying the predictive variables of behavior (e.g., contextual, sociodemographic, and psychosocial). These studies usually aim to discover which characteristics differ, for example, recyclers from non-recyclers or which factors are more determinant to recycling behaviors.
- Another line of research has sought to evaluate the effect of several types of intervention on the determinants of waste reduction and recycling behavior and identify which are most useful to change those behaviors.

The research outcomes are not always consensual, which leads, in some cases, to ambiguous results. However, a group of converging aspects seems to exist, either about most determining factors of recycling participation, either concerning the models and theories that explain the behaviors, or concerning the most promising strategies for the success of the recycling programs.

This research is crucial to the separate collection systems' managers because their success is not limited by techno-economical considerations, being much more dependent on the adherence of the services' users. To obtain recyclable materials in quantity and quality, it is necessary that the users separate the waste they produce correctly and place them in the correct places, according to the waste type. For this reason, it is essential to understand the social and psychological aspects that may determine recycling behaviors, predict behaviors when it is intended to implement a separate collection project, and evaluate the most effective strategies to change behaviors to promote the adherence of population to the separate collection.

6.2 Determining Factors of Recycling Behaviors

Until today, research aimed to discover the differences between, for example, recyclers and non-recyclers or which factors are the most determinant of recycling behaviors. The results of this research line show that there are sociodemographic, contextual, and psychosocial factors that may explain these differences and be determinant of recycling behaviors.

In the meta-analysis performed by Miafodzyeva and Brandt (2013), which includes 63 articles about recycling behaviors, published in a period of 20 years (1990–2010), the variables that have been the subject of research are grouped into

6.2 Determining Factors of Recycling Behaviors

three broad categories, namely, sociodemographic, technical/organizational (external factors), and psychosocial variables (e.g., norms, motivations, attitudes, habits). In what concerns to the determinant and most important factors of recycling behaviors, those authors highlight the convenience, information, moral norms, and environmental concern/awareness.

Concerning the most studied sociodemographic variables (e.g., gender, age, income, education, profession), the articles which seek to connect them with recycling behaviors are very ambiguous and a little consensual. Gender is a variable which in some studies is related to recycling behaviors (McDonald and Ball 1998; Barr et al. 2003; Meneses and Palacio 2005; Martinho et al. 2015, 2017a) and, in others, has no correlation (Hornik et al. 1995; Hage et al. 2008; Hage and Söderholm 2008). The same happens with the variable age. For instance, to Hage and Söderholm (2008) and Miafodzyeva et al. (2013), there is no connection between age and recycling behaviors, whereas to De Feo and De Gisi (2010), Saphores et al. (2006), and Domina and Koch (2002), age affects behavior. Some studies show that younger individuals are more likely to recycle, while in others it is older individuals that are linked to recycling behavior. As for the income, education, and profession variables, to some there is no correlation (Hage and Söderholm 2008; Vencatasawmy et al. 2000; Miafodzyeva et al. 2013); to others, there is (Schultz et al. 1995; Owens et al. 2002).

The outcomes from two different studies carried out in different time periods and in different metropolitan areas of Lisbon, one by Martinho (1998), about the bottle bank system, and another by Lima and Batel (2011), about the multi-material system, also reveal non-consensual results. In the study by Martinho (1998), the recycling group, comparatively to the non-recycling group, included older individuals with higher education and socio-professional status; the gender of the individuals did not differentiate the groups. According to Lima and Batel (2011), male individuals between 35 and 44 years old with higher education and social class are more likely to recycle; female individuals over the age of 65, with lower levels of education and belonging to medium/low social class, are the least likely.

The inconsistency of these outcomes is mainly due to methodological differences and deficiencies, namely, the sample dimension, period of observation, the type of waste being separated, the exemption of other factors that might mask casual relations, the measure of intention versus behavior, and the interviewed family member, among other factors. Another problem, identified by Becker (2014), lays on the fact that researchers sought more to evaluate the direct influence of these variables in behavior than the indirect influence that they might have on contextual and psychosocial factors associated with behavior.

Among the contextual factors that could be determinant of recycling behavior, literature indicates also many other factors: the waste management policies and the operational systems' conditions, in which fall the waste's final destination (e.g., landfill, incineration), the type of undifferentiated waste collection system (e.g., close or not to the separate collection), the type of separate collection systems (e.g., door-to-door vs recycling banks, mandatory vs voluntary), the type and number of recycling bins (e.g., bags, box, containers, and their dimensions), the distance and location of the recycling banks, the collection schedule, the aesthetic and sanitation of the containers, and the information about what, when, and where to deposit (Howenstine 1993; Margai 1997; Martinho 1998; Lima and Branco 2016). It also appointed the existence of operational barriers, real or perceived (e.g., busy roads, location of the recycling banks in dangerous areas), and the type of house (e.g., homes vs apartment buildings, or owned vs rented, buildings with vertical pipelines, or waste room), among others (Lansana 1993; Vitor and Martinho 2009; Margai 1997; Rogoff and Williams 1994; Waite 1995; Lindsay and Strathman 1997). From these studies, it can be concluded that a system will produce better outcomes if:

- It includes a more significant number of waste compounds duly collected as a recyclable mixture.
- The deposition equipment is conveniently placed and close (e.g., frequent crossing sites and close to houses).
- The recycling system does not demand radical changes in individuals' habits.
- The system maintenance is visible to users (e.g., appearance, sanitation, safety).
- An excellent source of practical information is provided.
- The strategies for adequate behavioral changes are implemented, e.g., the user charges are proportionate to the amount of waste (systems like pay-as-you-throw (PAYT)).

These contextual and convenience factors may be perceived in different ways by individuals and influence their recycling behaviors (De Young 1993). These factors are also the ones that are most important when it comes to a decision and action by the individuals in charge of the waste management. Therefore a recycling system will be more successful if it applies measures to avoid or reduce recycling's most common barriers, those considered the least convenient for users.

Compared to sociodemographic, psychological variables seem to possess a superior discriminatory power among recyclers and non-recyclers. Most studies confirm that it is in the group of psychosocial variables that lays the main difference between these two groups. Included in this category are specific attitudes, social and personal norms, the attribution of responsibility, awareness of the problem, social influence, perceived behavioral control, and behavior intentions. These variables are part of a behavior model structure primarily applied to recycling behavior, as described below.

6.3 Understanding and Predicting Models of Recycling Behaviors

The conceptual bases, theories, and models developed by behaviorists and social psychologists during the 1970s and 1980s are the starting points of most studies about personal and contextual factors associated with recycling behaviors and about the techniques used to promote these behaviors.

6.3 Understanding and Predicting Models of Recycling Behaviors

Although many theories and models have contributed to the understanding of recycling behaviors-attitudes relation and the predictive power of several variables of recycling behavior, the two most essential models remain: the Schwartz (1970) model of altruistic behavior, which is based on the norm activation model theory, and the model based on the rational choice theory, namely, the theory of reasoned action by Fishbein and Ajzen (1975), which was reviewed by Ajzen (1985) in the theory of planned behavior.

These models have been applied mainly to environmental behaviors, including recycling behaviors, with the goal of identifying the most determinant and predictive behavior variables. For those technicians, who must design a separate collection system, knowing the best model to predict the populations' behaviors is crucial to the dimensioning of the entire system and its proper functioning.

6.3.1 Schwartz Model of Altruistic Behavior

Theoretically, recycling has a great chance of success, because a lot of individuals socially and politically supports it. However, in many communities the low rates of separate collection reveal that there is no real intrinsic meaning; a disparity between the social support and the individuals' recycling behavior can be observed.

Several authors (Hopper and Nielsen 1991; Barr 2007; Abrahamse and Steg 2009) have suggested the possibility of recycling being conceptualized as altruistic behavior since a large normative approval characterizes altruism but most times with limited behavioral participation. Several study outcomes support the hypothesis that normative beliefs guide recycling behavior, as recyclers many times describe moral and altruistic motives as the main reasons to recycle (e.g., Hopper and Nielsen 1991; Oskamp et al. 1991; Vining et al. 1992; Lindsay and Strathman 1997; Vining and Ebreo 1992; Davies et al. 2002; Chaisamrej 2006). In this conceptual framework, the central problem is to understand the process by which altruistic social norms can be translated into concordant and proper individual behaviors.

The psychosocial model of altruistic behavior developed by Schwartz (1970, 1977) has been used mainly as theoretical and experimental support in environmental behavior studies. The distinctive characteristic of this model is that it considers that although most people verbally agree with a norm which influences a specific moral behavior, not all of them will act accordingly. This assumption corresponds to the current recycling situation. As Hopper and Nielsen (1991) refer, it is not necessary to convince people that recycling is good but to persuade them to act by that idea. As per the altruistic behavior model created by Schwartz (1977), the process begins with the social norms concerning behavior, which represent the values and attitudes of "others."

However, these norms are too broad to command behavior. It is necessary that social norms be implemented at the personal level to become personal norms. Although derived from socially shared norms, personal norms are distinct because the consequences of violating or supporting them are tied to a self-concept.

To violate a personal norm generates guilt and to support one originates pride. Meaning, social norms exist at the social structure level, while personal norms are remarkably internalized moral attitudes (Hopper and Nielsen 1991).

The other crucial point of the model is the relation between personal norms and behavior. Individuals can internalize norms and still not act according to them. To Schwartz (1977), personal norms are only activated and influence behavior when the decision-maker is aware of the consequences of the action toward others (or toward the environment) and feels personally responsible for the actions and their consequences. When these two variables, which Schwartz designated "awareness of consequences" and "attribution of responsibility," are present at a high level, personal norms will guide behavior.

By definition, altruistic behavior is normative behavior, and norms are developed by social interaction (Hopper and Nielsen 1991). This presupposes that in a cognitive-normative approach, based on social influence as an interventional strategy factor over norms, the awareness of consequences and the attribution of responsibilities toward recycling can both contribute to the increase of individuals' participation in a recycling program.

The definition and operationalization of norms have been understood in two different ways. To some authors (e.g., McCaul and Koop 1982; Oskamp et al. 1991; Lindsay and Strathman 1997), norms are the individual's perception of others' participation in recycling activities measured, for example, g perception over the number of recycler neighbors. Other authors (e.g., Hopper and Nielsen 1991; Vining and Ebreo 1990) understand it as an implicit community rule, being evaluated regarding perception or pressure applied by others (i.e., such as neighbors, relatives, friends, neighborhood leaders).

Normative components are often significantly related to recycling behaviors. In some countries, a non-recycler is considered an individual with a weak civic sense, and the chance of being called out by a neighbor due to adverse social behavior, such as not recycling, has excellent power over behavior. This social pressure is one of the explanations to the fact that separate door-to-door collection is typically more successful than recycling banks. In door-to-door systems, families' behaviors are more visible to others; it is known who recycles and who does not.

6.3.2 Theory of Reasoned Action and Theory of Planned Behavior

Fishbein and Ajzen's (1975) theory of reasoned action, and later, Ajzen's (1985) theory of planned behavior, is a theoretical cognitive-behavioral model that lays on the assumption that most socially relevant actions (such as recycling) are deliberate (rational).

Fishbein and Ajzen (1975) affirm that the immediate determinant of behavior is an intention, being itself the best predictor of behavior. The stronger the individual's behavioral intention is, the higher the chance that the behavior will occur. In turn, the

6.3 Understanding and Predicting Models of Recycling Behaviors

intention to produce a behavior directly depends on a social factor, the subjective norm, and a personal one, the attitude toward the behavior.

The subjective norm refers to the perception that an individual has of the social pressures applied to produce or not a behavior. The attitude toward behavior refers to the positive or negative evaluation that the individual does of that behavior. The individual's beliefs determine both, but for each case, the type of belief is different: the attitude toward the behavior derives from behavioral beliefs, while the subjective norm derives from normative beliefs (Ajzen and Fishbein 1980).

The model also considers the influence of other variables in each of these determinants of intention. To the attitude toward behavior, it adds the outcomes evaluation that the individual does of the individual's action. To the subjective norm, it adds the will that the individual has to follow the norms imposed by the individual's specific reference groups. The attitude and subjective norms, taken into account due to their relative importance, are considered by the authors as determinants of behavioral intention. The relative importance of these two factors may vary according to the context.

In this model, the authors do not exclude the influence of external variables (e.g., sociodemographic, personality traits, general attitudes associated with objects or targets). Nevertheless, they consider that these variables have only potential importance; their relation to behavior is indirect. These variables only affect behavior as far as they influence their determinants.

The theory of reasoned action has mainly been applied in several branches of social psychology, including the environmental behaviors (e.g., recycling, energy consumption, use of public transportation). However, it has also suffered some revisions and critiques. One of the critiques concerns the fact that two types of belief only determine intentions and behaviors, attitudes toward behavior and subjective norms (Bandura 1977), or the experiences of past behaviors (Fazio and Zanna 1981; Cooper and Croyle 1984; Cialdini et al. 1981; Echabe et al. 1988), or the fact that attitudinal and normative beliefs' effects are considered independent of each other and complementary (Andrews and Kandel 1979; Liska 1984; Grube et al. 1986). The works of Bagozzi et al. (1990) and Schlegel and DiTecco (1982) likewise suggest that attitudes do not perfectly mediate the effect of cognitions in intentions; they might also have an independent effect in intentions. In turn, Grube et al. (1986) defend that behavioral norm (or perceived behavior) must be considered as a different variable from the subjective norm, having an independent contribution in intentions and behaviors. In the model, the effect of attitudes in behavior is unidirectional, but reciprocal effects can be seen between attitudes and behaviors in certain conditions, depending on the strength of the attitudes and the singularity of behavior (Zanna et al. 1982).

Based on work carried out within the framework of behaviors-attitudes relations and the observation that the individual's behaviors are not always under the individual's total control, Ajzen reviewed his initial theory and proposed the theory of planned behavior. Like in the original theory, the primary factor is the intention to produce a behavior. However, in this new model, a third conceptual determinant independent from intentions is added, the perceived behavioral control (Ajzen 1985).

Internal or external factors may stop the transformation of a behavioral intention into the similar behavior to the individual. Internal factors include the locus of control, orientation toward action, information, capabilities, and emotions over which the individual has less control. In the same way, external factors like time, opportunities, convenience, and dependence on others may limit the behavior's performance without changing the determinants of intention (Ajzen 1985; Beale and Manstead 1991). This way, the perceived behavioral control corresponds to the perception of how easy or difficult it is to perform the behavior due to the influence of internal or external factors, being assumed by Ajzen (1991), as capable of reflecting past experiences or anticipating deterrents and obstacles. The meaning of this variable is, in many aspects, similar to the concept of personal efficiency defined by Bandura (1982).

According to the theory of planned behavior, the perceived behavioral control influences behavior, indirectly or directly, through behavioral intentions, being reciprocally influenced by attitudes and subjective norms. Ajzen and Madden (1986) proposed two versions for this approach: the first one that considers perceived behavioral control was correlating to attitude and the subjective norm and exerting an influence that is independent of behavioral intention and a second version that considers the possibility of a direct connection between perceived behavioral control and behavior. In this last one, it is a real control that is expected to exert a direct influence over behavior and not the perceived control.

The perceived behavioral control may have a direct connection to behavior without the mediation of intention. However, this direct effect is only expected in two situations. When the predicted behavior is not entirely under volitional control and when the perceptions of behavioral control reflect real control of the situation with some level of accuracy.

Similar to the theory of reasoned action, the theory of planned behavior has suffered several critiques and benefited from applications developed in different types of behaviors. For example, to Fishbein and Stasson (1990), the concept of intention becomes ambiguous concerning less volitional behavior, since there may exist many measures of intention (e.g., self-prediction, desire), which may lead to less trustworthy outcomes. The extensive application of the theory of planned behavior to several environmental behaviors has shown differences not only in the model structure and the intensity of the connection between variables and behavior but has also shown a more significant predictive power of the model with the addition of other variables.

Some authors have stated that attitudes are a stronger predictor of intention than the subjective norm (e.g., Bentler and Speckhart 1979; Boyd and Wandersman 1991; Farley et al. 1981; Tonglet et al. 2004). To Vallerand et al. (1992), attitudes predict intentions better than the subjective norm because attitudes capture individual attitudes toward an act, while the subjective norm deals with a more foreign concept, the self-perception of what others, significant to the individual, think.

Although always being based on the attitude-behavior relation, several authors have added variables to the original models. With these variables, they have gained a more significant predictive power of behaviors. For example, Martinho (1998) adds social identity, Eagly and Chaiken (1993) add the variable habits, acting either as attitudes' precedents or as direct determinants of behavior, and Stern and Oskamp (1987) introduce external or contextual factors (e.g., government regulations, monetary incentives, constructed environment, publicity and information). Some years later Stern (2000) adds two more variables, habits and.personal capabilities, stating that sociodemographic factors serve well to describe the personal capabilities factor.

The habits, a variable that recently social psychologists are giving more importance, may be characterized by the periodic repetition of specific behaviors, which don't demand high attention or later evaluation by those who carry them out (Henriksson et al. 2010; Webb et al. 2009). The force of habits is determined by the frequency of past behavior (Egmond and Bruel 2007). In what concerns domestic waste management, the oldest and most normal behavior is their mixed deposition in a single container.

For many, recycling is a behavior that still isn't habitual and frequent, and changing habits is not so easy unless it stems from necessity or tangible and relevant goal or a change of attitudes and of the importance attributed to recycling (John et al. 2013). Individuals that do not have recycling habits usually express lower intentions of recycling in the future and feel a weaker normative pressure to recycle. In opposite, recyclers reveal that they have developed recycling habits (Knussen and Yule 2008; Söderholm 2011).

6.4 Strategies to Change Behaviors and Their Evaluation

For professionals and separate collection system managers, it is essential to know which is the best strategy to change behaviors and how to evaluate the efficiency and effectiveness of those strategies to promote and maintain the desired behaviors.

Recycling can be framed in the conceptual framework of the standard dilemma or social dilemmas (Hardin 1968; Dawes 1980; Vlek and Steg 2002; McAndrew 1993). According to Dawes (1980), social dilemmas can be defined as the conflict between the individual interests (competition situation) and the collective interests (cooperation situation). In fact, recycling has costs for individuals; it demands time and efforts to keep, separate, and transport the recyclable materials and to change habits. In most cases, it does not grant immediate or tangible individual rewards, even though it is beneficial for society, especially in the future. The cost/personal convenience relation and the benefits for society/environment will determine the individuals' cooperation behavior (i.e., participating in recycling) or competition (i.e., not participating in recycling).

The opposition between the individual's interest and collective interest may be solved in two ways (De Young 1993; Gardner and Stern 1996; Gifford 2014; Vlek 1996): (1) modifying the objective data of the situation, in a way that ensures that the competitive option stops being the profitable one (e.g., extrinsically incentives, such as social coercion and fines) or (2) psychologically redefining the situation, in a way that the dilemma stops existing (e.g., intrinsically incentives, changing values and attitudes, altering the systems, so it becomes more convenient). The strategies or interventions for the recycling behavior changing may be included in one of these goals.

Geller (1989) and Dwyer et al. (1993) organize and differentiate the strategies meant to change prior and after behaviors. According to these authors, any intervention which has the goal of facilitating or increasing recycling behaviors before the behavior occurs (e.g., separating and depositing recyclable materials) is classified as prior. In this group are included the strategies of engineering and design, or operational strategies, communication through verbal and written messages, modeling and demonstration, target setting and compromise processes, and environmental awareness and education. Within the following strategies are included the ones that show a consequence upon producing the behavior, for instance, positive motivational techniques (e.g., material rewards, information feedback, social recognition, intrinsic satisfaction) or coercive motivational techniques (e.g., penalties, social pressure, intrinsic dissatisfaction).

In turn, for De Young (1993) techniques differ toward the starting point of change and the level of involvement of individuals as active participants in producing behavior. The author refers that a distinction should be made between information and the motivations offered/gained by others (external) and those acquired by the individual (internal) as a result of direct experience, which is generally considered by individuals as less quantifiable and of intangible nature. Using the contribution of Geller's (1989) approach, Dwyer et al. (1993), De Young (1993), and Martinho (1998) propose an organization of the different intervention techniques taking into consideration the moment in which they occur (prior to or after behavior) and the starting point of change (external or internal), according to Table 6.1.

Many studies have evaluated the isolated effect of each one of these strategies in the recycling behavior change, revealing that the combination of different strategies ensures better outcomes regarding the amount of collected materials (Stern and Oskamp 1987). However, for separate collection systems' managers, generalizing or applying the outcomes of these studies to their contexts is difficult. First, nearly all studies present short experimental demonstrations, which vary between 1 day and several months and are applied to very particular situations (e.g., students, office workers). Second, most studies are not evaluating the cost/benefit impacts of their interventions in quantity and quality of participation, and the rare studies where this is done reveal that the experimental intervention costs exceed the benefits, mostly because the period of evaluation is short and the indirect costs/benefits or externalities were not considered Stern and Oskamp 1987; Rousta et al. 2016).

The evaluation of a particular intervention designed for the recycling behavior change must be done in multiple dimensions. De Young et al. (1993) consider the following components in the evaluation of behavioral efficiency: (1) reliability, (2) how quick the change is, (3) particularism, (4) generalization, and (5) durability. Besides these criteria, Martinho (2009) proposes two more: (6) economic evaluation of interventions and (7) users' level of satisfaction (Fig. 6.1). This group of criteria includes the different evaluation dimensions that are important for politics and technicians in charge of separate collection systems.

6.4 Strategies to Change Behaviors and Their Evaluation

Table 6.1 The organizational structure of the strategies applied to the recycling behavior change

Source of change	The position of the strategy toward the moment of performance of the behavior	
	Antecedent	Consequent

External (tangible)	Techniques/operational
	  Waste management policies, type of separate collection system, number of separations to carry out in origin, type and number of recipients, recipients' distance and location, collection frequency, and schedule
		Positive motivation techniques
		  Material rewards (e.g., prizes, raffles)
		  Outcomes feedback
		  Social recognition
		Coercive motivational techniques
		  Punishment and penalties
		  Social coercion

	Communication/program promotion Appeals to neighborhood leaders, goals and target setting, modeling and demonstration, environmental education	
Internal (intangible)	Compromised processes (verbal or written) Modeling with experimentation	Positive and internal coercive motivations (e.g., sense guilt, duty, intrinsic satisfaction)

Source: Martinho (1998)

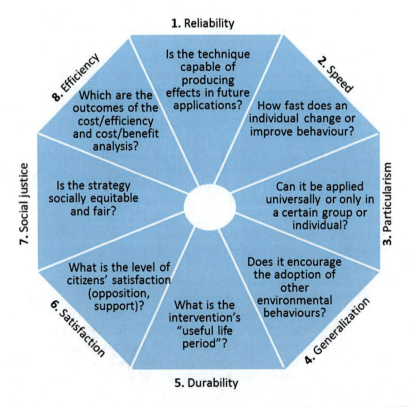

Fig. 6.1 Criteria for the evaluation of behavior change interventions. (Source: Martinho (2009))

Concerning reliability (trust) the point is to evaluate the capacity of a technique to produce effects not only during the first time it is applied to an individual or group but in future applications to the same individual or group. Some material rewards (e.g., prizes, raffles) techniques failed due to the effect of loss of novelty (Needleman and Geller 1992; Luyben and Cummings 1981–82; Katzev and Pardini 1987–88; Schultz et al. 1995; McClelland and Canter 1981).

The speed with which a technique can induce behaviors change, meaning, how fast individual changes or improves the behavior and adopts pro-environmental practices after the first intervention, can be a decisive strategic factor to promote a recycling program, to start a new program, and to capitalize the investment. Some external incentives, such as the elimination of systems' operational barriers (e.g., dumpsters being closer to users) and certain types of material incentives, especially raffles and lotteries, produce fast and immediate changes. However, the first has a more positive effect, because they remain on the field, while the second lead to participation decline when they are removed (Oskamp et al. 1994; Porter et al. 1995; EEB 2005; OCDE 2006).

Particularism intends to evaluate if a technique can be designed to be applied universally or only to a group or, in extreme cases, to an individual. Foa (1971) discussed the particularism of several extrinsic motivators, some of the universal character (e.g., money, information, prizes), others more particular (e.g., social recognition, services, personal attention). The author suggests that money is the less particular motivator, it enjoys universal acceptance, it has the same value for everyone, and the relation between the intervenient and the receptor does not matter. In opposite, personal attention is critical; the person who gives the attention matters, and its efficiency is strictly connected to the source of attention. However, a more particularistic technique is harder to put in practice; it has to be more localized to a specific situation, involving more costs and human resources.

Generalization is connected to the level in which the required behavior frequency increase has repercussions in other environmental behaviors. Considering the distinct current environmental issues, a technique capable of simultaneously encouraging the adoption of many behaviors is precious to the environmental programs' investments and for the environment itself. A study by Berger (1997), based on official Canadian statistics (censuses), reveals that, although the different environmental behaviors lay in different structures, recycling may function as a first step to adopting other environmental behaviors. Recycling was positively correlated to energy conservation, water, composting, usage of own bags to shop, among others.

Another evaluation criterion is the durability of intervention effects, which analyzes the "useful life period" of intervention. To know if behavior change remains without the necessity to repeat interventions is fundamental since, besides economic reasons, environmental problems' scale demands long-term behavioral changes. Many interventions that produce fast effects are less durable; as soon as the intervention technique is removed, behaviors almost always return to the starting point, especially in the case of techniques that use external incentives and neither appeal to social norms' influence nor the development of intrinsic motivation. Nevertheless, some individuals, after being induced to perform a specific behavior, may acquire new habits and discover that barriers or inconveniences that they initially predicted are not significant after all and may discover other indirectly induced motives such as social rewards, monetary savings, intrinsic satisfaction, sense of competence, among others.

Few are the studies that have focused on cost/benefit analysis toward the strategies. Some refer that many techniques are economically inefficient toward the amount of recovered materials. But even in the studies that evaluated the economic component (Springer and Haver 1994; Watts et al. 1994), the revenue generated by the sale of materials is not considered, not even the value the residents give the fact that they have neither a recycling system implemented in their residential area nor the related externalities, such as the avoided costs by depositing in a landfill or CO_2 emissions.

At last, another rarely used indicator is the assessment of the level of satisfaction of the users concerning the service rendered and the intrinsic satisfaction that is offered to them to participate in favor of the environment and to preserve resources. A technique may generate immediate positive effects such as the ones that use the material or coercive incentives, but it can be negatively evaluated by the population. The level of satisfaction of users must be a part of the evaluation criteria of intervention techniques and recycling systems since it is an indicator of great importance to the support of implemented policies.

From the outcomes of several studies about this issue, combining a small group of interventions is required to cover the different evaluation dimensions. Systems which include good periodic procedural information (e.g., what, how, and when), minimize systems' operational barriers (e.g., dumpsters being closer to users, collection schedule, and dumpster sanitation), provide outcome and challenge feedback (e.g., what has been achieved and next goals), and a PAYT-like tariff system (e.g., proportional payment to the quantity of produced waste) are the ones which achieve the best results in terms of evaluation criteria and in terms of the quantity and quality of materials. Also the use of taxes can induce change of behavior, like is the case of plastic bag tax applied in Portugal, where the tax applied reduced the consumption of single-use plastic bags (Martinho et al. 2017b).

A tariff system like PAYT is one of the complete intervention instruments to change behaviors; it is reliable, fast, universal, durable, equative, and fair and can ensure the economic sustainability of waste management systems. Its implementation in different cities has demonstrated a reduction in the production of undifferentiated waste and a considerable increase in recycling rates that may reach values higher than 70% (Reichenbach 2008; Morlok et al. 2017). The effectiveness and efficiency of these different intervention strategies depend on community characteristics. Besides the sociodemographic and cultural characteristics, the contextual characteristics (e.g., urbanistic, type of house and systems' operational conditions) differ from place to place. For these reasons, before any intervention, it is necessary to know the starting situation and divide the population into homogeneous groups to apply adapted interventions to each context (Jesson 2009). The idea that the intervention type must be selected according to the defined goal and according to the target group's characteristics is part of the social marketing concept (Kotler and

Zaltman 1971; Bloom and Novelli 1981) that should also be applied to recycling behavior strategies, as defended by Geller (1989), Schultz et al. (1995), Howenstine (1993), and Rousta et al. 2016.

6.5 Current Limitations and Future Perspectives for Social Psychology

Although they have been important, the outcomes from many studies carried out for approximately 40 years by social psychologists around recycling behaviors, as indicated by Schmuck and Vlek (2003), have not been translated into practical and effective results for the decision-makers, including the separate collection systems' managers. The unsuccess of the results of social studies is mostly due to the following methodological issues:

1. Very restricted studies applied to a small number of individuals or social groups with specific characteristics and in specific contexts, which make it impossible to generalize results and which many times matches more a scientific curiosity than a practical application goal.
2. Reduced level of accuracy in the identification and description of the separate collection operational system in which the study is focused, lacking details to enable the technicians to use and compare the results to their cases; it is not enough to say if the system is the door-to-door or the bring type; it is necessary to define what materials are to be separated and how, the type and dimension of recipients used for the deposition of recyclables, the distance they are from the populations, the relation between the conventional collection system and the separate collection system, the collection schedule, the existence or lack of waste tariffs, how the system is managed and maintained, and the type and frequency of information campaigns and education carried out.
3. The definition of the behavior being studied is vague in many studies; in some cases, they refer to recycling behavior, a somewhat vague concept which cannot have the same meaning for all.
4. Behaviors are measured differently; in some cases behavioral intention is measured, while in others it is the recycling behavior, which is measured indirectly by self-report (verbal answers) and rarely has a direct measure (behavioral answers).
5. The indicators used to measure behaviors rarely are the same as those used by technicians, for instance, for technicians the participation rate is associated to a frequency (1 month); in social researches the time reference is not the same.
6. Sociodemographic variables have been studied as determinant factors of behavior and less as an influencer of those determinants, such as attitudes, habits, and social norms.
7. In behavioral models, the separation of the population into homogeneous groups is not always carried out; the model structure and connections between variables

6.5 Current Limitations and Future Perspectives for Social Psychology

can change according to the population characteristics; it is important to identify those variables which are the most explanatory of a certain group's behavior, for instance, perceived barriers or attitudes, which are fundamental for the selection of intervention techniques best adapted to the population's characteristics.

8. Many studies ignore some sociodemographic, contextual, and psychosocial variables that may be relevant to recycling behavior, namely, social identity and cohesion, building particularisms, existence or lack of waste room, number and type of information and sensitization campaigns, and type of undifferentiated waste collection system.

9. Intervention strategies for behaviors change rarely and are evaluated in their different evaluation dimensions; the behavioral changes monitorization is restricted to a few weeks after the intervention; therefore it's not possible to obtain a long-term evaluation of its effects; it is also necessary to define the type of behavior that has been changed; the interventions can have different effects on different recycling variables (e.g., quantity, participation frequency, reduction of contaminants); additionally, there aren't many studies that evaluate the potential effect of the combination of different change behaviors strategies.

10. The attitudes, the central concept of social psychology, and the most studied variable within the environmental behavior framework aren't always operationalized and measured in the same way; in some cases, general attitudes toward recycling are measured; others measure specific attitudes toward separation and separate deposition behavior of a certain waste flux. Moreover, rare are the studies that present the analysis of the reliability and validity of the attitude scales; the psychometric qualities of these scales are unknown.

11. Some studies about the effect of material incentives already have been carried out (e.g., prizes, raffles, fines), but the concrete effect of the "PAYT" variable and the influence that different types of PAYT exert over determinants of behavior, intention, and behavior itself still need more in-depth study.

12. From the several studies about recycling behavioral models, it can be concluded that a singular universal model that explains and predicts recycling behavior with some level of certainty does not exist; further studies in this area are necessary, based on bigger populational samples and in behavioral indicators measured directly.

With the new information and communication technologies that have recently surfaced in the market, which include a wide range of innovative technical solutions applied to containers, vehicles, and the monitorization and circuit management global system, many of these methodological issues may have short-term solutions.

The development of these technologies is largely attributable to the need to reduce collection costs, to the waste management community policies, to the implementation of PAYT systems (Gallardo et al. 2012; Bilitewski 2008), and to the application of the new circular economy paradigms, of the Internet Of things, of the smart cities, and of the big data analytics to the waste management sector (Medvedev et al. 2015; Kedia 2016; Perchard 2017).

These devices and technologies designed for the intelligent and integrated waste collection systems' management enable an optimized planning of circuits, with the resultant reduction of costs and emissions to the atmosphere, a better knowledge of the producers and the systems, less complaints, transparent and fair tariffs, recycling increase and statistical data retrieval, and much more precise systems' operational effectiveness indicators.

This transitions from the traditional "heavy" work needed for the monitorization of the circuits, quantification and characterization of waste, data treatment and analysis, and fleets and teams management, to a smarter, automatic, continuous, and integrated management, which besides giving better operational, economic, and environmental collection outcomes, by permitting the acquisition of a significant amount of data, it opens up a huge window of opportunities for new research lines and the development of the waste management sector. The outcomes of studies may contribute to the resolution of the different problems, whether in the engineering field or in the social psychology applied to waste production and recycling behaviors. It is anticipated close cooperation between waste management entities and university multidisciplinary research teams for the development of studies focused on, for instance:

- The exploration of spatial and time waste production patterns, translated into production forecast models, the production of real-time dynamic circuits, increase of the number of dumpsters, equipment and human resources planning, and management models.
- More precise estimates of the specific weight of the different waste fluxes inside dumpsters, which may be associated with the data obtained through waste characterization campaigns that are carried out annually by urban waste management system.
- Definition of the circuits' performance indicators, specific to each type of dumpster-vehicle system, as proposed by Rodrigues et al. (2016) and to each type of collection system.
- Creation of systems that enable the traceability of individual-specific waste fluxes (e.g., construction and demolition waste, electrical and electronic equipment waste).
- Design of integrated collection platforms that allow, through georeferentiation of the behavioral variables (e.g., quantity and composition of undifferentiated waste, quantity and quality of differentiated waste), sociodemographic variables (e.g., number of family members, age, gender, socioeconomic status, professions), and contextual variables (e.g., type of house, cultural aspects, type of intervention/ sensitization), knowing the profile of recyclers and non-recyclers and, based on these variables, to develop new explanatory and predictive models of behaviors.
- Identification, through alerts, of the areas where it is necessary to implement behavior change strategies, characterization and separation of the population into more homogeneous groups, and evaluation of the more adequate strategies and the effect of applying those interventions in behaviors change by monitoring the collected and recycled quantities and the increase of the recyclables' quality.

- Possibility that the evaluation and decision support models and systems may be continuously calibrated as more data is acquired, including those that are frequently used in waste management, such as (Pires et al. 2011) cost/benefit analysis models, forecast models, simulation models, optimization models, multicriteria analysis, information management system, development of scenarios, analysis of the material fluxes, analysis of the life cycle, among others.
- New strategies to promote and communicate services to citizens, behavior feedback, and persuasive messages (e.g., information about the last and the next collection date, collected amounts, value to pay, appeals to participation, challenges).

Separate collection systems are strategic for the current paradigm of the circular economy, for the local and regional economies, and the achievement of community and national goals concerning the separate collection, recycling, and landfill diversion. The newly available technologies for waste management will demand considerable financial resources and a straight collaboration between collection operators and universities to carry out research and to develop solutions, which will be more economical and adapted to the reality of each community. Multidisciplinary teams composed of specialists from engineering areas (e.g., environment, electrotechnical, computer science) and social sciences (e.g., economists and social psychologists).

References

Abrahamse W, Steg L (2009) How do socio-demographic and psychological factors relate to households' direct and indirect energy use and savings? J Econ Psychol 30:711–720

Ajzen I (1985) From intentions to actions: a theory of planned behaviour. In: Kuhl J, Beckmann J (eds) Action-control: from cognition to behaviour. Springer, Heidelberg, pp 11–39

Ajzen I (1991) The theory of planned behaviour. Organ Behav Hum Decis Process 50:179–211

Ajzen I, Fishbein M (1980) Understanding attitudes and predicting social behaviour. Prentice-Hall, Inc, Englewood Cliffs

Ajzen I, Madden TJ (1986) Prediction of goal-directed behavior: attitudes, intentions, and perceived behavioral control. J Exp Soc Psychol 22:453–474

Andrews KH, Kandel DB (1979) Attitude and behavior: a specification of the contingent consistency hypothesis. Am Sociol Rev 44:298–310

Bagozzi RP, Yi Y, Baumgartner J (1990) The level of effort required for behavior as a moderator of an attitude-behavior relation. Eur J Soc Psychol 20:45–59

Bandura A (1977) Self-efficacy: toward a unifying theory of behavioral change. Psychol Rev 84:191–215

Bandura A (1982) Self-efficacy mechanism in human agency. Am Psychol 37:122–147

Barr D (2007) Factors influencing environmental attitudes and behaviors: a UK case study of household waste management. Environ Behav 39:435–473

Barr S, Ford NJ, Gilg AW (2003) Attitudes towards recycling household waste in Exeter, Devon: quantitative and qualitative approaches. Local Environ 8:407–421

Beale DA, Manstead ASR (1991) Predicting mothers' intentions to limit frequency of infants' sugar intake: testing the theory of planned behaviour. J Appl Soc Psychol 21:409–431

Becker N (2014) Increasing high recycling rates socio-demographics as an additional layer of information to improve waste management. Dissertation, University of Lund

Bentler PM, Speckhart G (1979) Models of attitude-behavior relations. Psychol Rev 86:452–464

Berger I (1997) The demographics of recycling and the structure of environmental behaviour. Environ Behav 29:515–531

Bilitewski B (2008) From traditional to modern fee systems. Waste Manag 28:2760–2766

Bloom PN, Novelli WD (1981) Problems and challenges in social marketing. J Mark 45:79–88

Boyd B, Wandersman A (1991) Predicting undergraduate condom use with the Fishbein and Ajzen and the Triandis attitude-behavior models: implications for public health interventions. J Appl Soc Psychol 21:1810–1830

Chaisamrej R (2006) The integration of the theory of planned behavior, altruism, and self-construal: implications for designing recycling campaigns in individualistic and collectivistic societies. Dissertation, University of Kentucky

Cialdini RB, Petty RE, Cacioppo JT (1981) Attitude and attitude change. Annu Rev Psychol 32:357–404

Cooper J, Croyle R (1984) Attitude and attitude change. Annu Rev Psychol 35:395–426

Davies J, Foxall GR, Pallister J (2002) Beyond the intention-behavior mythology: an integrated model of recycling. Mark Theory 2:29–113

Dawes RM (1980) Social dilemmas. Annu Rev Psychol 31:169–193

De Feo G, De Gisi S (2010) Domestic separation and collection of municipal solid waste: opinion and awareness of citizens and worker. Sustainability 2:1297–1326

De Young R (1993) Changing behavior and making it stick: the conceptualization and management of conservation behaviour. Environ Behav 25:485–505

De Young R, Duncan A, Frank J, Gill N, Rothman S, Shenot J, Shotkin A, Zweizig M (1993) Promoting source reduction behavior: the role of motivational information. Environ Behav 25:70–85

Domina T, Koch K (2002) Convenience and frequency of recycling: implications for including textiles in curbside recycling programs. Environ Behav 34:216–238

Dwyer W, Leeming F, Cobern M, Porter B, Jackson J (1993) Critical review of behavior interventions to preserve the environment: research science 1980. Environ Behav 25:275–321

Eagly AH, Chaiken S (1993) The psychology of attitudes. Harcourt Brace Jovanovich College Publishers, Orlando

Echabe AE, Rovira DP, Garate JF (1988) Testing Ajzen and Fishbein's attitudes model: the prediction of voting. Eur J Soc Psychol 18:181–189

Egmond RB, Bruel R (2007) Nothing is as practical as a good theory: analysis of theories and a tool for developing interventions to influence energy-related behaviour. http://www.cres.gr/behave/pdf/paper_final_draft_CE1309.pdf. Accessed 10 Mar 2017

European Environmental Bureau (EEB) (2005) EU environmental policy handbook – a critical analysis of EU environmental legislation. EEB, Bruxels

Farley JU, Lehmann DR, Ryan MJ (1981) Generalizing from "imperfect" replication. J Bus 54:597–610

Fazio RH, Zanna MP (1981) Direct experience and attitude behavior consistency. In: Berkowitz L (ed) Advances in experimental social psychology. Academic Press, New York

Fishbein M, Ajzen I (1975) Belief, attitude, intention, and behavior: an introduction to theory and research. Addison-Wesley, Reading

Fishbein M, Stasson M (1990) The role of desires, self-predictions, and perceived control in the prediction of training session attendance. J Appl Soc Psychol 20:173–198

Foa UG (1971) Interpersonal and economic resources. Science 171:345–351

Gallardo A, Bovea MD, Colomer FJ, Prades M (2012) Analysis of collection systems for sorted household waste in Spain. Waste Manag 32:1623–1633

Gardner GT, Stern PC (1996) Environmental problems and human behaviour. Allyn & Bacon, Boston

Geller ES (1989) Applied behavior analysis and social marketing: an integration to preserve the environment. J Soc Issues 45:17–36

Gifford R (2014) Environmental psychology: principles and practice, 5th edn. Optimal Books, Colville

References

Grube JW, Morgan M, McGree ST (1986) Attitudes and normative beliefs as predictors of smoking intentions and behaviors: a test of three models. Br J Soc Psychol 25:81–93

Hage O, Söderholm P (2008) An econometric analysis of regional differences in household waste collection: the case of plastic packaging waste in Sweden. Waste Manag 28:1720–1731

Hage O, Söderholm P, Ch B (2008) Norms and economic motivation in household recycling: evidence from Sweden. Resour Conserv Recycl 53:155–165

Hardin G (1968) The tragedy of the commons. Science 162:1243–1248

Henriksson G, Åkesson L, Ewert S (2010) Uncertainty regarding waste handling in everyday life. Sustainability 2:2799–2813

Hopper JR, Nielsen JN (1991) Recycling as altruistic behaviour: normative and behavioral strategies to expand participation in a community recycling program. Environ Behav 23:195–220

Hornik J, Cherian J, Madansky M, Narayana C (1995) Determinants of recycling behavior: a synthesis of research results. J Socio Econ 24:105–127

Howenstine E (1993) Market segmentation for recycling. Environ Behav 25:86–102

Jesson JK (2009) Household waste recycling behavior: a market segmentation model. Soc Mark Q 15:25–38

John P, Cotterill S, Richardson L, Moseley A, Stoker G, Wales C, Smith G (2013) Nudge, nudge, think, think: experimenting with ways to change civic behavior. Bloomsbury Academic Publishing, London

Katzev RD, Pardini AU (1987–88) The comparative effectiveness of reward and commitment approaches in motivating community recycling. J Environ Syst 17:93–111

Kedia P (2016) Big data analytics for efficient waste management. Int J Res Eng Technol 5:208–211

Knussen C, Yule F (2008) "I'm not in the habit of recycling" the role of habitual behavior in the disposal of household waste. Environ Behav 40:683–702

Kotler P, Zaltman G (1971) Social marketing: an approach to planned social change. J Mark 35:3–12

Lansana FM (1993) A comparative analysis of curbside recycling behavior in urban and suburban communities. Prof Geogr 45:169–179

Lima ML, Batel S (2011) Psychosocial study of predictors of recycling behaviors in different waste disposal systems (in Portuguese: Estudo psicossocial dos preditores de comportamentos de reciclagem em distintos sistemas de deposição de resíduos). ISCTE-IUL, Lisbon

Lima ML, Branco C (2016) Implementation, evaluation and validation of social indicators predictors of participation associated with the recycling of urban waste (in Portuguese: Implementação, avaliação e validação de indicadores sociais preditores da participação associados à reciclagem de resíduos urbanos (Relatório 1: indicadores sociais de gestão de resíduos). CIS-IUL/ISCTE-IUL, Lisbon

Lindsay J, Strathman A (1997) Predictors of recycling behavior: an application of a modified health belief model. J Appl Soc Psychol 27:1799–1823

Liska AE (1984) A critical examination of the causal structure of the Fishbein/Ajzen attitude-behavior model. Soc Psychol Q 47:61–74

Luyben PD, Cummings S (1981–82) Motivating beverage container recycling on a college campus. J Environ Syst 11:235–245

Margai FL (1997) Analyzing changes in waste reduction behavior in a low-income urban community following a public outreach program. Environ Behav 29:769–792

Martinho G (1998) Determining factors for recycling behaviors. Case study: glass system (in Portuguese: Factores determinantes para os comportamentos de reciclagem caso de estudo: sistema de vidrões). Dissertation, NOVA University of Lisbon

Martinho MG (2009) Recycling psychology. An eco-different (in Portuguese: Psicologia da reciclagem. Um eco.diferente) honorable mention at communicating environment prize by Calouste Gulbenkian Foundation

Martinho G, Pires A, Portela G, Fonseca M (2015) Factors affecting consumers' choices concerning sustainable packaging during product purchase and recycling. Resour Conserv Recycl 103:58–68

Martinho G, Magalhães D, Pires A (2017a) Consumer behavior with respect to the consumption and recycling of smartphones and tablets: an exploratory study in Portugal. J Clean Prod 156:147–158

Martinho G, Balaia N, Pires A (2017b) The Portuguese plastic carrier bag tax: the effects on consumers' behavior. Waste Manag 61:3–12

McAndrew FT (1993) Environmental psychology. Wadsworth, Belmont

McCaul KD, Koop JT (1982) Effects of goal setting and commitment on increasing metal recycling. J Appl Psychol 67:377–379

McClelland L, Canter R (1981) Psychological research on energy conservation: context, approaches, method. In: Baum A, Singer J (eds) Advances in environmental psychology: energy conservation, psychological perspectives, vol 3. Lawrence Erlbaum, Hillsdale, pp 1–25

McDonald S, Ball R (1998) Public participation in plastics recycling schemes. Resour Conserv Recycl 22:123–141

Medvedev A, Fedchenkov P, Zaslavsky A, Anagnostopoulos T, Khoruzhnikov S (2015) Waste management as an IoT-enabled service in smart cities. In: Balandin S, Andreev S, Koucheryavy Y (eds) Internet of things, smart spaces, and next generation networks and systems. ruSMART 2015. Lecture notes in computer science, vol 9247. Springer, Cham, pp 104–115

Meneses GD, Palacio AB (2005) Recycling behavior: a multidi-mensional approach. Environ Behav 37:837–860

Miafodzyeva S, Brandt N (2013) Recycling behavior among households: synthesizing determinants via a meta-analysis. Waste Biomass Valorization 4:221–235

Miafodzyeva S, Brandt N, Anderson M (2013) Recycling behavior of householders living in multicultural urban area: a case study of Järva, Stockholm, Sweden. Waste Manag Res 31:447–457

Morlok J, Schoenberger H, Styles D, Galvez-Martos J-L, Zeschmar-Lahl B (2017) The impact of Pay-As-You-Throw schemes on municipal solid waste management: the exemplar case of the county of Aschaffenburg, Germany. Resources 6:8

Needleman LD, Geller ES (1992) Comparing interventions to motivate work-site collection of home-generated recyclables. Am J Community Psychol 20:775–787

Organisation for Economic Co-operation and Development (OCDE) (2006) Impacts of unit-based waste collection charges. http://www1.oecd.org/officialdocuments/publicdisplaydocumentpdf/?cote=ENV/EPOC/WGWPR(2005)10/FINAL&docLanguage=En. Accessed 3 May 2018

Oskamp S, Harrington M, Edwards T, Sherwood D, Okuda S, Swanson D (1991) Factors influencing household recycling behaviour. Environ Behav 23:494–519

Oskamp S, Williams R, Unipan J, Steers N, Mainieri T, Kurland G (1994) Psychological factors affecting paper recycling by businesses. Environ Behav 26:477–503

Owens J, Dickerson S, Macintosh DL (2002) Demographic covariates of residential recycling efficiency. Environ Behav 32:637–650

Perchard E (2017) Smart streets: how the internet of things is revolutionising waste. Resource sharing knowledge to promote waste as a resource. https://resource.co/article/smart-streets-how-internet-things-revolutionising-waste-11721. Accessed 10 Mar 2017

Pires A, Martinho G, Chang NB (2011) Solid waste management in European countries: a review of systems analysis techniques. J Environ Manag 92:1033–1050

Porter BE, Leeming FC, Dwyer WO (1995) Solid waste recovery: a review of behavioral programs to increase recycling. Environ Behav 27:122–152

Reichenbach J (2008) Status and prospects of pay-as-you-throw in Europe—a review of pilot research and implementation studies. Waste Manag 28:2809–2814

Rodrigues S, Martinho G, Pires A (2016) Waste collection systems part B: benchmarking indicators. Benchmarking of the Great Lisbon Area, Portugal. J Clean Prod 139:230–241

Rogoff MJ, Williams JF (1994) Approaches to implementing solid waste recycling facilities. Noyes Publications, Park Ridges

Rousta K, Bolton K, Danhén L (2016) A procedure to transform recycling behavior for source separation of household waste. Recycling 1:147–165

Saphores JD, Nixon H, Ogunseitan O, Shapiro A (2006) Household willingness to recycle electronic waste: an application to California. Environ Behav 38:183–208

References 93

Schlegel RP, DiTecco D (1982) Attitudinal structures and the attitude-behavior relation. In: Zanna MP, Higgins ET, Herman CP (eds) Consistency in social behavior: the Ontario symposium, vol 2. Erlbaum, Hillsdale, pp 283–301

Schmuck P, Vlek C (2003) Psychologists can do much to support sustainable development. Eur Psychol 8:66–76

Schultz PW, Oskamp S, Mainieri T (1995) Who recycles and when? A review of personal and situational factors. J Environ Psychol 15:105–121

Schwartz SH (1970) Moral decision making and behavior. In: Macauley M, Berkowitz L (eds) Altruism and helping behavior. Academic Press, New York, pp 127–141

Schwartz SH (1977) Normative influences on altruism. In: Berkowitz L (ed) Advances in experimental social psychology, vol 10. Academic Press, New York, pp 221–279

Söderholm P (2011) Environmental policy and household behavior: sustainability and everyday life London. Earthscan, Washington, DC

Springer T, Haver R (1994) Preventing waste at the source: educating the public. Resour Recycl 13:95–100

Stern PC (2000) Toward a coherent theory of environmentally significant behavior. J Soc Issues 56:65–84

Stern PC, Oskamp S (1987) Managing scarce environmental resources. In: Stokols D, Altman I (eds) Handbook of environmental psychology, vol 2. Wiley, New York, pp 1043–1088

Tonglet M, Phillips P, Bates M (2004) Determining the drivers for householder pro-environmental behavior: waste minimization compared to recycling. Resour Conserv Recycl 42:27–48

Vallerand RJ, Pelletier LG, Blais MR, Brière NM, Senècal C, Vallières EF (1992) The academic motivation scale: a measure of intrinsic, extrinsic and amotivation in education. Educ Psychol Meas 52:1003–1017

Vencatasawmy CP, Ohman M, Brannstrom TA (2000) A survey of recycling behavior in households in Kiruna, Sweden. Waste Manag 18:545–556

Vining J, Ebreo A (1990) What makes a recycler? A comparison of recyclers and nonrecyclers. Environ Behav 22:55–73

Vining J, Ebreo A (1992) Predicting recycling behavior from global and specific environmental attitudes and changes in recycling opportunities. J Appl Soc Psychol 22:1580–1607

Vining J, Linn N, Burdge R (1992) Why recycle? A comparison of recycling motivations in four communities. Environ Manag 16:785–797

Vitor F, Martinho MG (2009) Factors influencing households' participation in organic waste separation. In: Proceedings of the ISWA 2009 World Congress, Lisbon, 12–15 Oct 2009

Vlek C (1996) Collective risk generation and risk management the unexploited potential of the social dilemas paradigm. In: Liebrand WBG, Messick DM (eds) Frontiers in social dilemmas research. Springer, Berlin, pp 11–38

Vlek C, Steg L (2002) The commons dilemma as a practical model for research and policy making about environmental risks. In G. Bartels & W. Nelissen (Eds.), Marketing for sustainability. Towards transactional policy-making (pp. 205–303). Amsterdam, Berlin, Oxford: IOS Press

Waite R (1995) Household waste recycling. Earthscan Publications Ldt, London

Watts A, Fuszek R, Arndt M (1994) In the multi-family way: testing collection strategies and examining costs. Resour Recycl 13:51–56

Webb T, Sheeran P, Luszczynska A (2009) Planning to break unwanted habits: habit strength moderates implementation intention effects on behavior change. Br J Soc Psychol 48:507–523

Zanna MP, Higgins ET, Herman CP (1982) Consistency in social behaviour. Erlbaum, Hillsdale

Chapter 7
Economic Perspective

Abstract In a sustainable waste collection and management system, the economic component is traditionally an imperative factor for the decisions and justifications of all operations. The calculation of the cost and revenue components associated with waste collection systems must be taken into account and analyzed from an integrated waste management perspective. Additionally, the willingness and the affordability of citizens to finance the waste collection and management system are also vital to the success of any collection and management system. In this chapter presented and discussed are the waste collection costs, the financial concerns of waste management systems and instruments of waste policy, and the public and private sector financing.

Keywords Benchmarking · Cost functions · Financing · Fuel consumption · PAYT · Public-private partnership · Recycling · Solid waste

7.1 Waste Collection Costs

Waste collection and transportation can represent the most expensive component of an integrated waste management system (IWMS), as it involves intensive labor and many vehicles with high fuel consumption. Several authors indicate costs above 50% on municipal solid waste (MSW) collection, although, in some cases, it may represent 70–80% of the total costs (Tchobanoglous et al. 1993; Bilitewski et al. 1994; Sonesson 2000; Johansson 2006; Faccio 2011; Sora and González 2014; Greco et al. 2015).

Moreover, the mentioned costs will increase with the growing implementation of separate collection for different materials (e.g., package and packaging waste, organic) and with the development and application of information and communication technologies (ICT), as well as with the progress of platforms and programs for integrated collection management (smart waste collection).

Waste system planning and management consist of a trade-off between a set of goals that can be opposite, such as reduction of collection costs or increasing amounts of waste for recycling and recovery. It requires a holistic and interactive approach regarding costs and benefits associated with various components of an

© Springer International Publishing AG, part of Springer Nature 2019
A. Pires et al., *Sustainable Solid Waste Collection and Management*,
https://doi.org/10.1007/978-3-319-93200-2_7

integrated waste management system (Shekdar and Mistry 2001). In this context, it is essential to know the costs and the primary drivers of waste collection systems and identify optimization opportunities and benchmarking analysis for the most cost-effective waste collection system decisions (Jacobsen et al. 2013; Huang et al. 2011; Greco et al. 2015).

However, given the vast diversity and complexity of waste collection systems, cost-benefit analysis is not a simple task since it requires collection and integration of big amount of data. Benchmarking analysis between different MSW collection systems is a complex or even impossible exercise and must always be done with due caution and reference to the local context, since economic efficiency depends on many specific and contextual factors, which cannot be strictly comparable between them (D'Onza et al. 2016; Hage 2008). As Hage (2008) points out, there is no simple way to understand waste collection system costs without considering the local context in which they apply. There are countless contextual variables that may influence the waste collection costs, namely, amount, composition, and specific weight of waste streams to collect (e.g., undifferentiated waste or material separate collection); type, number, and location of waste containers (e.g., bags, bins/containers, pneumatic systems, fill rate); system collection type (e.g., drop-off systems, door-to-door collection); vehicle type (e.g., capacity, compaction rate, mechanization, fuel consumption, maintenance costs); smart equipments and technologies used; route optimization efficiency; collection frequency; demographical characteristics (e.g., population served, mean household size, population density); urban agglomeration type and structure; terrain altitude and topography; traffic conditions and restrictions; climatic conditions; garage and treatment or disposal facility location; team size, number of working hours, and staff average salaries; market structure (e.g., public/private monopoly or competitive); waste policies (e.g., recycling targets, landfill diversion); subsidies and tax; market price for recyclable waste; behavior of collection service users; commuting and tourism; and methodology used to calculate costs.

The empirical results of studies carried out in municipalities of several countries show that the main drivers of solid waste collection are amount of waste, collection type and target material (e.g., undifferentiated, paper/paperboard, multi-material, organic), population density and municipality size (economies of scale), collection frequency, number of collection points/pickup per area, and salaries (Stevens 1978; Callan and Thomas 2001; Dijkgraaf and Gradus 2004; Dubin and Navarro 1988; Ohlsson 2003; Greco et al. 2015; Karadimas et al. 2007; Greco et al. 2015).

Besides the diversity and complexity of contextual and operational/technological variables, costs of waste collection systems benchmarking analysis also face the difficulty to obtain comparable data, such as lack of MSW definition harmonization, container-vehicle systems diversity, and methods used to calculate current and foreseen costs (Rodrigues et al. 2016a). Dahlén et al. (2009) identified and characterized different error sources from waste collection data, which make comparison rough, into these groups: (i) "general data problems," such as the incorrect MSW definition; (ii) "data uncertainties related to specific waste categories," such as number and frequency of collection, for example, paper/cardboard collected in commercial sector (it does not depend on collection system or population participation); (iii) "unreliable data from recycling centers," because they receive waste from commercial sector further household MSW; and finally, (iv) "household waste component analysis data not comparable" because there are no standardized methods.

As geographical and demographic data (e.g., routes distances, number of inhabitants), geographic information systems (GIS), and optimization software are more accessible nowadays, they allow better scaling and planning of the waste collection service (Rada et al. 2013; Sanjeevi and Shahabudeen 2016). Since 2000, GIS and optimization software have been used to reduce the distances traveled by the vehicles (Sanjeevi and Shahabudeen 2016) and to optimize the container location and coverage (López et al. 2007). However, the software requires detailed data for functioning correctly, such as the number and capacity of the containers, the collection vehicle's capacity, the specific weight of waste inside the containers, the filling rates, the waste collection times, and the number of working hours, among others (Bosch et al. 2001; Sonesson 2000; Ghose 2006; González-Torre et al. 2003; Johansson 2006). These data are not always available and often rely on statistics and empirical data, such as the specific weight of the waste inside the container and waste collection times for each type of system and waste stream. Because gathering this data is challenging, standard costs and calculation models are often obtained in other contexts, and because of that, the results can be conditioned.

These difficulties may justify the fact that waste collection systems have not been further studied in integrated waste management systems. This statement is supported by Allesch and Brunner (2014), who reviewed 151 studies on waste management assessment methods, analyzing their objectives, methodologies, investigated systems, and the results about the economic, environmental, and social issues. The authors concluded that although economic aspects are considered in approximately 50% of the studies, only 9% of them analyze the waste collection subsystem.

Many of the publications are based on comparative analysis between different collection systems: Hage (2008) used cost indicators to compare the separate collection of household plastic packaging waste in Sweden; Larsen et al. (2010) used the impact assessment categories of the life cycle analysis, as well as the collection costs, recycling rates, sorting efficiency, and waste quantities; Vidal (2001) developed evaluation indexes focused on legal, social, environmental, and economic performance to evaluate four presorting practices in Spain; Gamberini et al. (2013) applied demand profiles and annual cost indexes for waste management, expressed as €/year and €/year/inhabitant in several Italian communities.

The disadvantage of the use of models is that cost data must be accurate, and representative of the specific case study since the use of standard data is not always considered a safe approach. Besides that, cost data may not be readily accessible, and, occasionally, several assumptions must be analyzed (Komilis 2008). More recently, the development and implementation of ICT (i.e., volumetric sensors, RFID systems, GPRS and GPS technologies) allowed real-time monitoring of the routes, and these tasks are executed in a more precise way.

The waste collection and transport costs can be expressed as a total annual value (€/year) or a total cost per ton of waste collected (€/t), per inhabitant served (€/inhab/year), or per area (€/m^2). The most common indicator is the cost per ton collected (Sora and González 2014). These indicators can also be disaggregated by type of waste collected (e.g., undifferentiated or specific stream), by type of collection system (e.g., drop-off systems, door-to-door collection), or by cost components (e.g., human resources, fleet, administrative costs and revenue). The cost indicator to be used and the level of disaggregation will depend on the intended objectives for the cost analysis and the available information.

However, there is no common basis for the indicator's calculation, defining which cost components should be included (or excluded) or how they should be evaluated, or the boundaries of the subsystem to be considered for the analysis. Some analysis only determines the absolute costs related to the waste collection and transport, and others, in the specific case of waste selective collection systems, determine the relative costs by accounting, for example, the avoided costs of landfilling or the related environmental and economic benefits of natural resource savings, the sale of recyclables income, or reduction of CO_2 emissions, among others. These approaches require the use of proper methodologies, such as the economic life cycle analysis of materials, or the calculation of the equivalent CO_2 emissions, which also have their limitations.

The lack of a conventional standard approach often results in incomplete or inaccurate costs data, leading to the risk of faulty conclusions or decisions, or making it impossible to compare and analyze most cost-effective waste collection systems. Table 7.1 presents some examples of cost values obtained for different urban waste collection and transport systems and for different countries or cities, whose differences reflect this context, type of variables considered, and methods or models used.

In theory, all direct and indirect costs and revenues of the waste collection and transport system must be considered for the costs calculation, namely, the amortization of equipment (e.g., investments in vehicles and containers and their realistic depreciation and amortization rates); the operating and maintenance costs of containers and vehicles (e.g., insurance, maintenance pieces, fuel and other fluids, materials and labor for repairing works, tires and washing expenses); labor costs directly related with the collection activity (salaries and social costs for vehicle drivers and other employees) and for staff who do not work directly with the MSW; consumable, clothing, and miscellaneous material costs; administrative costs, which include the costs of installations, namely, the amortization and the maintenance costs related to buildings, administrative and IT equipment, hygiene and cleaning, insurances, security, energy and water consumption, communications, interest, and tax charges; and the costs with information and awareness of citizens and economic agents.

However, because it is not always easy to gather all that information, some authors only account the most relevant direct costs related to the human resources (labor force) and the fleet, and disregard, for example, the costs related with the infrastructures, administrative services, and others. In fact, those are the components

7.1 Waste Collection Costs

Table 7.1 Cost indicators of different urban waste collection and transport systems

Waste type/ stream	System type	Cost/waste collected (ton)	Observations
Mixed waste	n/a	$77.82	Data from cost analysis (Callan and Thomas 2001)
		$40–90	Costs including pick-up, transfer, and transport to the final disposal site (for developed countries) (Hoornweg and Bhada-Tata 2012)
		$30–75	Costs including pick-up, transfer, and transport to the final disposal site (for developing countries) (Hoornweg and Bhada-Tata 2012)
		$85–250	Costs including pick-up, transfer, and transport to the final disposal site (for G8 countries) (Hoornweg and Bhada-Tata 2012)
		79.34 €	Data from full cost accounting (D'Onza et al. 2016)
	Road containers, door-to-door	56–126 €	Costs for municipal waste management in the EU – average values (Hogg 2002)
		30–67 €	Costs for municipal waste management in the EU – urban values (Hogg 2002)
		55–71 €	Costs for municipal waste management in the EU – rural values (Hogg 2002)
		37.50 €	Values for Portuguese municipalities (Levy 2004)
		$12.19– 111.40	Solid waste collected in 1996 at US municipalities (Bohm et al. 2010)
	Road containers	123.59 €	Based on real data on Catalan municipalities (Sora and González 2014)
		19.60 €	Results from a cost calculation method proposed by Boskovic et al. (2016). Mixed waste with separate collection of recyclables
		16.40 €	Results from a cost calculation method proposed by Boskovic et al. (2016). Mixed waste without separate collection of recyclables
	Door-to-door	154.39 €	Based on real data on Catalan municipalities (Sora and González 2014)
Recyclable materials (mix)	n/a	224.38 €	Data from full cost accounting (D'Onza et al. 2016)
	Road containers, door-to-door	$72.57– 342.80	Recyclable materials collected in 1996 (Bohm et al. 2010)
	Road containers	59.00 €	Costs for municipal recyclable waste in the EU cities (Hogg 2002)
		19.70 €	Results from a cost calculation method proposed by Boskovic et al. (2016)
		381.44 €	Based on real data on Catalan municipalities (Sora and González 2014)
		180–270 €	Costs for municipal recyclable waste in the EU cities (Hogg 2002)

(continued)

100   7 Economic Perspective

Table 7.1 (continued)

Waste type/ stream	System type	Cost/waste collected (ton)	Observations
	Door-to-door (plastic bags)	100–180 €	Costs for municipal recyclable waste in the EU cities (Hogg 2002)
	Door-to-door	252.72 €	Based on real data on Catalan municipalities (Sora and González 2014)
		169–300 €	Costs for municipal recyclable waste in the EU cities (Hogg 2002)
Paper and card	n/a	158.03 €	Data from full cost accounting (D'Onza et al. 2016)
	Road containers, door-to-door	173.49 €	Values for Portugal municipalities (Levy 2004)
		193.00 €	Based on measured cost data (Rodrigues et al. 2016a, b)
	Road containers	82–150 €	Costs for municipal recyclable waste in the EU cities (Hogg 2002)
		41.74 €	Values for Denmark municipalities (Larsen et al. 2010)
		49–191 €	Based on measured cost data (Rodrigues et al. 2016a, b)
	Door-to-door	30–146 €	Costs for municipal recyclable waste in the EU cities (Hogg 2002)
		75.30 €	Values for Denmark municipalities (Larsen et al. 2010)
		77.00 €	Based on measured cost data (Rodrigues et al. 2016a, b)
	Door-to-door (plastic bags)	732.00 €	Based on measured cost data (Rodrigues et al. 2016a, b)
Plastic and metal	Road containers, door-to-door	277.00 €	Average of the different types of collection (Rodrigues et al. 2016a, b)
	Road containers	74–269 €	Based on measured cost data (Rodrigues et al. 2016a, b)
	Door-to-door	123.00 €	Based on measured cost data (Rodrigues et al. 2016a, b)
	Door-to-door (plastic bags)	1109 €	Based on measured cost data (Rodrigues et al. 2016a, b)
Plastic	Road containers, door-to-door	504.40 €	Values for Portugal municipalities (Levy 2004)
	Road containers	230–500 €	Costs for municipal recyclable waste in the EU cities (Hogg 2002)
		58.52 €	Values for Denmark municipalities (Larsen et al. 2010)
	Door-to-door	300–750 €	Costs for municipal recyclable waste in the EU cities (Hogg 2002)
		75.30 €	Values for Denmark municipalities (Larsen et al. 2010)

(continued)

7.1 Waste Collection Costs

Table 7.1 (continued)

Waste type/ stream	System type	Cost/waste collected (ton)	Observations
Glass	n/a	157.56 €	Data from full cost accounting (D'Onza et al. 2016)
	Road containers, door-to-door	127.92 €	Values for Portugal municipalities (Levy 2004)
		61.00 €	Average of the different types of collection (Rodrigues et al. 2016a, b)
	Road containers	20–50 €	Costs for municipal recyclable waste in the EU cities (Hogg 2002)
		58.52 €	Values for Denmark municipalities (Larsen et al. 2010)
		26–115 €	Based on measured cost data (Rodrigues et al. 2016a, b)
	Door-to-door	70–194 €	Costs for municipal recyclable waste in the EU cities (Hogg 2002)
		75.30 €	Values for Denmark municipalities (Larsen et al. 2010)
Recyclables without glass and textiles	Door-to-door (rural)	111–133 €	Costs for municipal recyclable waste in the EU cities (Hogg 2002)
	Door-to-door (urban)	171–202 €	Costs for municipal recyclable waste in the EU cities (Hogg 2002)
Organic	n/a	182.75 €	Data from full cost accounting (D'Onza et al. 2016)

n/a: not available

that contribute the most to the total costs of an MSW collection and transport system (Dogan and Süleyman 2003; Arribas et al. 2010).

Miller and Delbridge (1995) indicate a distribution percentage for the waste collection and transport system costs between 47% and 56% for labor and 22% for the fleet, making these two components representing, on average, 80% of the total costs. The distribution of the other costs is the following: 3% for construction and utilities, 12% for administrative services, and 5% for other components. Analyzing the cost distribution of the MSW waste collection and transport system of Lisbon municipality in Portugal, Santos et al. (1994) also reached similar values for the percentage distribution of cost components, which were 66% for labor costs, 24% for fleet, and 10% for other expenses. Labor costs naturally result from the size of the collection and transport team, depending directly on the method of collection, the number of working hours, and the average salary of the employees (Rodrigues 2016).

Waste collection and transport system costs have been organized and presented in three main formats: (a) capital or investment costs (CAPEX) (i.e., acquisition of containers and vehicles) and exploration or operation (O&M) (i.e., human resources, maintenance, fuel) (e.g., Levy et al. 2007); (b) direct costs (i.e., materials, containers, vehicles, direct salaries, and services) and indirect costs (i.e., administrative costs, other general costs, financial charges, and taxes) (e.g., Greco et al. 2015; ERSAR 2014); and (c) initial costs (i.e., initial necessary investments and expenses to implement services), operational costs (i.e., salaries and maintenance of vehicles, energy, and fuel, rent, and leases, contracted services, interest charges), and end-of-life expenses (i.e., end-of-life facilities or workers' retirement pensions) (USEPA 1997).

As mentioned above, the most relevant financial variables that must be considered for the total costs of the MSW collection and transport services are human resources, equipment (vehicles and containers), and fuel consumption. Concerning human resources, the data needed to calculate the associated costs may be limited to the total number of employees, the total number of working hours, and the average value of salaries or can be much more detailed as, for example, considering the type of professional category or the function of the worker, disaggregating the basic salary, overtime, health insurance, night service, social security charges, uniforms, and others components. The number of employees is naturally dependent on the method of waste collection and transport system, the characteristics of the routes, the type of containers and vehicles, and the frequency of collection. These factors are determinant to establish the size of the teams which can have a significant effect on the total costs (O'Leary and Walsh 1995). To compare labor costs, beyond the operational differences of the systems, it is also important to consider the level of remuneration and the labor and trade union laws in each country (i.e., that may have implications for hiring, conditions for daily working hours, working conditions, among others).

About capital goods, the essential elements in MSW collection systems are containers and vehicles, which can be expressed either in the number of existing physical units or their capacity (García-Sánchez 2008). Selecting the right type of container for MSW, adapting its characteristics to the specific needs of the waste collection and transport services, is a critical task, because the amount of waste collected per collection point and the number of collection points per unit of time affect the costs of the service (Hogg 2002). If the separate collection system is door-to-door using bags, the costs will significantly vary if the users themselves provide the bags, systems supply bags free of charge, or the users of the service pay the bags. In the latter two situations, there will be a need to consider the costs of bags acquisition and distribution. Regarding the fleet, although the costs are highly dependent on the type of vehicle, the variables to be considered for the calculation of costs should include the average annual cost of vehicle amortization, fuel consumption, tires, oils and other fluids, maintenance (materials and labor), insurance, accidents, and washing, among others.

Fuel consumption (diesel, biofuel, compressed natural gas (CNG), among others) is a significant part of the fleet costs, depending of course on the type and characteristics of the vehicles, type of collection and transport system (density of collection points per area), distance covered, cycles of acceleration, stop and waste mechanical compaction, terrain orography, traffic conditions, distance traveled between each collection point of the route, legal speed limits, waste characteristics, driver's way of driving, and fuel prices, among others (Larsen et al. 2009; Nguyen and Wilson 2010; Sonesson 2000). With the optimization of the collection and transport routes, it is almost always possible to decrease the number of vehicles or the distances covered, which results in a reduction of fuel consumption, costs, and pollutant emissions (Abdelli et al. 2016; Rizzoli et al. 2007; Sanjeevi and Shahabudeen 2016). The number and capacity of the containers to be installed depends on the quantity of MSW to be collected, their characteristics (e.g., specific weight, fermentation), collection frequency, associated technology (e.g., sensors), and the existence of some local standard regarding the accessibility of equipment. There is an excellent diversity of waste disposal equipment on the market, with very different acquisition costs.

The waste management entities readily supply data regarding the average fuel consumption per 100 km (e.g., actual consumption records of vehicles used, per day/route, year/route). The results have to be interpreted considering the sample of routes analyzed, which are naturally dependent of the routes, the number of stops, and the number of turns of the routes, as defended by Sonesson (2000).

7.1.1 Investment Costs (CAPEX)

For calculation of initial investment costs, it is considered equipment acquisition costs related to MSW collection and transport systems (i.e., undifferentiated or separate collection): vehicles and containers. In the vehicles, it is considered the chassis and structure (commonly referred as "box" or "superstructure"). However, it should be noted that some vehicles allow the structure to be replaced with the same chassis ("multilift" vehicles), or even the replacement of only one component at the end of its life, while the other remains in proper operating conditions.

Typically, an MSW collection and transport system from a specific area or municipality presents a wide diversity of equipment, acquisition dates, and models (purchase or leasing), and it is necessary to carry out an exhaustive data survey to convert into annual capital costs. One way to do so is calculating capital costs from containers and vehicles current investment costs, the methodology used by Larsen et al. (2010). However, equipment acquisition conditions depend on purchased quantities and other particular circumstances offered by suppliers (e.g., "regular" clients or quantity containers acquisition that reduces purchase prices). For that reason, Rodrigues (2016) chose another methodology to calculate investment costs, which consider real costs for vehicles and containers in the acquisition year. Then, costs were updated for the same reference year to obtain a more reliable effective cost for different routes of MSW separate collection. The conversion between current prices to constant prices can be done using national statistics, for example, the consumer price index (CPI) variation rates, which allow the investment cost to be updated between the year of the equipment acquisition and the reference year.

Depreciation is a method to allocate capital expenditure costs over the resource lifetime (USEPA 1997). A "straight-line depreciation method" or "constant share depreciation method" considers that the value of good or service decreases at a constant rate, and depreciation costs are calculated by dividing the capital outlay over the lifetime of the asset acquired. In this method, if the municipality or company made equal annual deposits, at the end of the asset lifetime, this value is exactly fair to the replacement cost of that asset (Assis 2011). For example, a 10 years lifetime vehicle would have an annual depreciation cost of one-tenth of its total capital cost. This method of constant yearly depreciation is indicated by USEPA (1997).

This method contemplates that the capital opportunity costs are null. For that reason, some authors (Rodrigues 2016; Goulart 2003; Lavita 2008) opted for "decelerated depreciation methods," a more appropriate method for goods that lose value more quickly at the beginning than at the end of their lifetime, such as collection vehicles, which is usually accompanied by lower maintenance costs at the beginning and more significant at their end of life. In the case of containers, depreciation is more constant, as well as maintenance costs. Regarding containers, maintenance costs are more constant, although in some models, such as underground models with hydraulic systems, there is also an increase in maintenance costs at the end of life. Rodrigues (2016), Goulart (2003), and Carvalho (2011) calculate the equipment's annual cost according to the following equation:

$$ \mathrm{ac} = \mathrm{ic} \times \frac{i(1+i)^n}{(1+i)^n - 1} \times (1 - \mathrm{fr}) $$

where ac = annual cost (€), ic = initial cost (€), i = social discount rate (%), n = equipment lifetime (yr), and fr = fractional residual value at the end of the equipment lifetime (%).

Regarding social discount tax (i), there is considerable diversity in the amounts that can be considered. Rodrigues (2016) opted to use a rate between 4% and 5%, which is the value indicated by the European Commission (2015). This value is a benchmark parameter for the real opportunity cost for the long-term capital. Larsen et al. (2010) updates the capital costs for waste collection equipment, assuming 10 years for equipment lifetime and a social discount rate of 6%, value recommended by the Danish Environmental Protection Agency. Komilis (2008) uses the value of 3% for the social discount rate applied to MSW collection vehicles. Goulart (2003), Lavita (2008), and Gomes (2008) use a differentiated interest rate, specifically 8% for containers purchase and a value between 8% and 15% for vehicles.

Through the contacts with municipalities and equipment suppliers, Rodrigues (2016) notes the diversity of practices adopted that lead to different equipment lifetimes. These factors are related to equipment design characteristics in its construction allied to its use during the lifetime (e.g., operating limits and vehicle overloading) and different maintenance conditions given to the same equipment (e.g., oil change and other fluids at scheduled time).

Assis (2011) states two factors that must be considered about the equipment lifetime (n): the time during which the equipment is used in proper operating conditions (physical life) and its loss of value resulting from technology obsolescence. Rodrigues (2016) considers that the accounting depreciation used, for example, in municipal services companies or the cases reported by Delloite (2004) and Rhoma (2010), should be avoided. Alternatively, the author proposes that real values of equipment technical depreciation must be used. To do this, the adopted number of years in the formula should be close as possible to reality, based on information provided by equipment suppliers and users, such as municipalities.

In the specific case of Lisbon Metropolitan Area (Portugal), Rodrigues (2016) consulted the municipalities and waste management entities of Lisbon, Cascais, and Sintra to obtain equipment lifetime information. Thereafter, Rodrigues (2016) concludes Lisbon considers 14 years for the chassis and a value between 8 and 10 years for vehicle's superstructure; Cascais assumes 8 years for the chassis and the collection structure; and Sintra uses the equipment's accounting depreciation, which is 5 years for vehicles (intensive use with two shifts per day) and 7 years for containers. In addition, a company that works in the same intervention area indicates that after 4 years a deep repair of the collection structure is required, which entails considerable costs.

Rodrigues (2016) also analyzed different information provided by Portuguese environmental equipment suppliers, representatives of waste collection vehicle's international brands, and containers washing services, among others. The values are between 6 and 12 years, although with some specificities: 8 years for side-loader compactor vehicle amortization, 8–10 years for all compactor vehicles, 10–12 for the multilift vehicles with mobile compactor and crane and for single-compartment open-body vehicles, and 6 years for the rest of the vehicles.

Vehicle life span assumed in the literature mentioned studies is variable, depending on the characteristics of the equipments, although the most common is 5 years: Lavita (2008) for any type of waste selective collection vehicle; Goulart (2003) for closed-body single-compartment vehicle, with intermittent compactor and double crane hook; Rhoma (2010) for vehicles with compaction; and Delloite (2004) for all vehicles. Values between 6 and 7 years are used for closed-body single-compartmented waste collection vehicles, with intermittent compactor and rear-end loading (Gomes 2008; Ricci 2003) and 7 years for the single-compartment open-body vehicles, without compactor, nor loading system (Ricci 2003). In the transport optimization model presented by Komilis (2008), 8 years of amortization for the vehicles are used. USEPA (1997) indicated waste collection vehicle's life span between 5 and 7 years.

Regarding containers literature references, the life span for these equipments varies between 8 and 15 years, according to their characteristics: 8 years for drop-off recycling containers (Delloite 2004) and for surface containers, immobile, without compaction, collection with lift frontal and/or lift side supports (Lavita 2008); 8–10 years to surface containers, wheeled, without compaction, collection with lift frontal and/or lift side supports (Rhoma 2010; Gomes 2008; Goulart 2003); and 15 years for semiunderground, without compaction, compact container, crane one ring (Goulart 2003).

Based on all references consulted, Rodrigues (2016) assumes 8 years of the lifetime for all types of vehicles, except for open-body vehicles, with 10 years of lifetime. For containers, the same author uses 10 years of the lifetime for surface containers and 15 years for semiunderground and underground containers.

Concerning the fractional residual value at the end of equipment lifetime (fr), Rodrigues (2016) used the information collected from Lisbon, Cascais, and Sintra municipalities and MSW management entities, as well as literature references (Goulart 2003; Lavita 2008; Gomes 2008; Carvalho 2011; Rhoma 2010). For vehicles, the author considers that the residual value is 15% of the initial cost. For containers, the value considered is null, since at the end of this equipment lifetime, they have no market value, except in the case of metal containers, which can be sold when their dismantle is compensatory.

7.1.2 Operating and Maintenance Costs (O&M)

For the operating and maintenance costs, only those directly related to the waste collection and transport service are considered, to allow a better comparison between the performance of the waste management systems and avoid the differences that arise from the local contexts in which they are inserted. In this regard, the following costs must be considered: vehicle and container maintenance, fuel consumption, and human resources (excluding technicians and administrative staff).

For maintenance costs analysis, Rodrigues (2016) considered the economic studies published about MSW collection and transport services, as well as the costs data collected in three municipalities of the Lisbon Metropolitan Area (Lisbon, Cascais, and Sintra) and from other service providers of waste management sector.

Regarding the annual maintenance costs related to vehicle repair, the values obtained in the literature vary, although the most consensual is 5% about the acquisition cost (Carvalho 2011; Gomes 2008; Goulart 2003). Nevertheless, Rhoma (2010) indicates a lower annual maintenance cost, between 2.5–3.1% of vehicle acquisition costs. From the contact with service providers, Rodrigues (2016) obtained maintenance values representing 6% of the cost of acquisition, which is a similar value to those indicated in the literature references.

In the process of gathering information carried out by Rodrigues (2016) in the municipalities of Lisbon, Cascais, and Sintra, it was verified in most cases that the costs are disaggregated per type of vehicle, and this control is performed through the vehicle registration. However, the values refer to actual maintenance costs (preventive and corrective), which are different because they are directly related to the regular and standard use and equipment maintenance. Rodrigues (2016) also concludes that the values presented by the municipalities and service providers are higher than those indicated in the literature, which, according to the equipment service providers, is due to failures in preventive maintenance, which does not prevent later serious repairs.

Regarding containers, the fraction of the investment cost spent annually on the maintenance and repair of these equipments assumes different values according to the authors. In the literature review, Goulart (2003) considers the highest value, around 35%. On the other hand, Rhoma (2010) and Lavita (2008) show significantly lower costs for surface containers: Rhoma indicates 4.3% for rear-end loading surface containers with a lifetime of 10 years; Lavita indicates the maintenance cost of 4.6% in relation to the purchase cost for selective collection containers with crane (0.65 €/container/year, in relation to 2500 L containers), also considering the container insurance (19 €/container/year).

Concerning container maintenance costs, data that Rodrigues (2016) tried to collect through the municipalities of Lisbon, Cascais, and Sintra were not adequate for analysis, since the local authorities did not have the analytical accounting. The main reason is the disparate information that does not always allow costs allocation for differentiating container types. Rodrigues (2016) tried to validate data collected in the literature review. From the exercise carried out, a percentage value of cost around 4.3% of the acquisition cost was obtained for container maintenance, which coincides with the value indicated by Rhoma (2010). On the other hand, the service providers also reported values in the same order of magnitude for the rear-end loading surface containers with wheels. Although the most common maintenance costs are obtained for rear-end surface containers, Rodrigues (2016) also obtained data from service providers about the maintenance cost for underground containers, which is around 5.4%, including equipment washing.

Among the existing methods to estimate energy consumption and its costs, the most common is to consider that the fuel consumption, as well as the time required to collection and transport operations, is a function of the distance traveled. These two factors are considered sufficient to calculate both the associated cost and the environmental impact of the transport, as they give a driving conditions estimate of the vehicles (stopping and acceleration times related to traffic or the collection routes and waste compaction operations) (Sonesson 2000).

Fuel consumption usually is an accessible data, with a low associated error, through existing records of the waste management entities responsible for waste collection and transport vehicles. Data are generally disaggregated for each equipment regarding fuel supplies control and their costs. However, as Sonesson (2000) defend, the means of consumption, by type of vehicle, will have to be interpreted considering the size of the sampled routes analyzed, as well as their characteristics.

Another operating cost component is the size of the work team responsible for the waste collection and transport, which can have a significant effect on total costs, depending on labor and equipment costs, collection methods, routes, and union contracts. Costs should also be broken down into categories, such as waste collection workers and truck drivers, as Rodrigues (2016) states, to assess the average values for the team due to its constitution.

Also related to human resources costs, Santos et al. (1994) define the cost components that should be considered to estimate this component: base salary, overtime, healthcare allowance, night service, and other social charges. The uniforms can be considered in a different parcel. In the municipalities consulted by Rodrigues (2016) (Lisbon, Cascais, and Sintra), the costs of health insurance, occupational medicine, work accidents, and life insurance are sometimes accounted in a separate component, corresponding to around 4% of the base salary, the value that must be added to obtain total costs.

According to Rodrigues (2016), other operational costs that must be considered are water and detergents consumed in the washing of containers and vehicles, which may be high. However, Rodrigues (2016) was not able to measure the costs for this parcel.

In conclusion, when the overall costs of SWM have been identified, unit costs (e.g., per ton of waste, per inhabitant, per household, per m^2) can be calculated. It is a powerful tool to obtain information on the relative costs for different alternatives to specific system component or to determine the most appropriate user charge for inhabitants. For example, the evaluation of alternative scenarios for collecting waste and transporting it to waste treatment or disposal sites, with and without transfer stations, can be carried out by directly comparing the unit costs per ton of total waste managed during the operational period of each scenario (GIZ 2015).

The collecting cost data required is not often well known or available in local authority accounting systems. Few jurisdictions use total cost accounting methods, and there is considerable variation in the way municipalities report waste management expenditure in their budget reports. Local governments generally maintain separate accounts for different types of expenses and publish them in financial statements according to line items; expenses are grouped according to the kinds of items or services purchased rather than by activity, not providing information on the waste management service or its components (USEPA 1997; Folz 1999; GIZ 2015; Rodrigues et al. 2016b).

Individually, costs are grouped into expenses with salaries, benefits, services purchased (e.g., consulting and technical assistance, repair and maintenance services, rent), supplies (e.g., energy, food), and property (e.g., building, equipment, and vehicles), but are not discriminated by type of activity (e.g., selective collection, undifferentiated collection, container washing, vehicle washing). When there is no breakdown of costs by type of service or activity, then it is difficult to expect that the management entities will discriminate by collection system or circuit, which would add a second level of detail and would allow a more direct allocation of costs to a specific component of the system. This implies additional collection and processing of data, but any approach necessarily requires a comprehensive collection of information, the definition of which depends on the level or types of costs to be determined (USEPA 1997).

7.2 Financial Concerns of Waste Management Systems and Instruments of Waste Policy

The total cost continuous analysis is essential for waste collection and transport system managers, as it identifies opportunities for service improvement and costs reduction, as well as it evaluates financial needs to ensure economic sustainability.

However, decision-making of collection operational system cannot be made based only on this specific operation cost analysis but must be analyzed in the context of an integrated waste management system.

The existence of separate collection schemes by waste (i.e., undifferentiated further waste streams selective collection), the presence of different types of containers and vehicles, as well as how the collection service is planned, has implications both upstream and downstream of the waste collection system itself. For example, container type and location are determinant factors for end users behavior: (i) efficiency on recycling programs participation and (ii) waste quantity generated by households. To this phenomenon Rathje and Murphy (1992) nominated "Parkinson's Law," and when it applies to waste management means, "waste expands according to available space." According to Rathje and Murphy (1992), when people have small containers they tend to find other destinations for their bulky waste that does not fit in the containers (e.g. garden waste, large packages, furniture), such as keeping it on household garage, sending for repair, selling, offering or transporting it to specific drop-off locations. On the other hand, people with more extensive containers tend to discard bulky waste mixed with other waste without thinking of alternatives and increasing waste generation.

Additionally, simpler and low-priced waste collection systems (i.e., undifferentiated collection devoid of major equipment or technologies) have consequences on treatment system costs, as sorting and recovery technologies are needed. Moreover, the sale of products from waste treatment (e.g., recyclable materials, compost, or energy) may not generate as much revenue because of the lower quality of them.

Because the collection phase is interlinked with the processing and disposal phases in an MSW plan, there is always a trade-off from a system analysis perspective (Chang and Pires 2015). So, financing of waste collection and transport system should be approached from a global perspective of integrated waste management system, taking into account the operational components of the system (i.e., collection, material recovery facility, mechanical and biological treatment facility, energy recovery, and/or landfill), as well as policies, legislation, and regulations at local, regional, and national level.

Waste policies, legislation, and regulations are decisive for MSW management systems, either for the operating system design, for the costs, or for the financing models and sources. Waste hierarchy principle has been recognized for several decades by all European countries. This principle is required under the European Waste Framework Directive (Article 4) and is reflected on policies and specific legislation in several countries, regions, and/or municipalities, with particular goals and targets for reduction, reuse, recycling, and separate collection for some waste streams (e.g., package and packaging waste, bio-waste, WEEE, and batteries and accumulators), and some countries (e.g., Germany) limit or even ban it from waste disposal. MSW selective collection systems are strategic for target accomplishment as well as for current paradigm of the circular economy, once economy needs materials in quantity and quality. The diversity and the highest number of selective collection systems are translated into increasing costs of MSW-integrated management system. These costs cannot be compensated by recyclable material purchase or by energy generated on processes of biological or energy valorization (i.e., anaerobic digestion and incineration), if proper economic instruments are not adopted.

Regulatory, economic and information instruments have been the tools used by several governments to put in practice or encourage established policies for waste to reduce pollution, reduce waste production, improve waste management, and maximize waste recovery. The first waste policy instruments, which began in most developed countries in the 1970s, focused exclusively on the regulation of technical standards to control and prevent pollution caused by incorrect waste management. These regulatory or "command and control" instruments are mostly based on the emission limit values for specific pollutants or guideline value concentration of pollutants in the various environmental compartments (i.e., air, water, soil), both generally established according to their toxicity to public health.

There is also direct imposition of constructive and operational standards for waste infrastructures (i.e., landfills, incinerators), product use (e.g., sludge from the WWTP or compost in the soil), ban marketing of products containing some hazardous substances (e.g., mercury and cadmium in batteries, flame retardants in EEE), fixes targets for selective collection and recycling (e.g., packaging, batteries, biodegradable waste, WEEE, construction and demolition waste, end of life vehicles (ELV), waste oils), and licensing procedures. This "command and control" instruments regulate behavior directly by prescribing specific legislation and standards which must be achieved and enforcing compliance through the use of penalties and fines (Perman et al. 2003).

The application of regulatory instruments to the waste sector has had positive effects on the development and improvement of recovery, treatment, and disposal technologies. However, these measures have not been sufficient to reverse the trend of increasing production of waste, and it is a weak management. The most significant problem with the implementation of these instruments has been that, by chance or exceptionally, the regulated values lead to an efficient economic solution and therefore to an optimal level of externality. In addition, the establishment of standards involves monitoring compliance by a monitoring and control entity. This situation is, in most cases, tough to implement extensively.

The recognition that the traditional "command and control" approach has not been sufficient for the objectives of sustainable waste management and the shared responsibility has led many developed countries, since the 1980s, to integrate economic or market-based instruments into their waste policies. In fact, the European Waste Framework Directive (2008/98/CE) points out the crucial role of economic instruments to reach the goals of an ecologically sound waste management.

In addition to their high flexibility, which contrasts with the rigidity of regulatory instruments, economic instruments have the advantage of being a constant incentive for waste reduction and recovery, promoting the development of cleaner technologies and offering greater cost-effectiveness and better integration with other sector policies. The application of these economic instruments in MSW management has two primary objectives: to cover or improve cost recovery and to change or influence behavior, through their impact on market signals, in order to minimize waste, avoid adverse impacts (e.g., from landfill or incineration), or increase resource recovery and recycling (GIZ 2015).

Today the economic instruments are a critical component of the waste policy mix of many countries and encompass a range of tools. The most widely used have been charged for services (or user charges), product charges, deposit-refund systems, extended producer responsibility (e.g., producer fee schemes for packaging, WEE), landfill and incineration taxes, subsidies, and tax rebates (e.g., VAT) for recyclable and recycled.

The product charges are based on a principle directly related to the principle of producer responsibility, since they seek to internalize in the product the subsequent costs with the collection, treatment, or final disposal of the waste. Generally speaking, the application of such instrument refers to the obligation for producers or importers of a particular product to pay (or agree) an additional fee to ensure that a final safe and non-polluting final destination be equated for the same. This is what is happening today with several integrated systems of specific waste streams (e.g., single-use plastic bags). Also providing a recipe for product management when it becomes waste, these rates are also intended to encourage producers to reduce the amount or hazard of the waste that their products originate through, for example, eco-design.

Deposit-refund systems are to some extent related to the principle of producer responsibility. This is, in fact, a surcharge on the price of the product, but differs in the following issues: the deposit is not based on the weight or volume of the product concerned; the deposit is returned when the product is delivered to the collection operators. This instrument acts as an economic incentive for returning products after the end of their lifetime, thus contributing to the waste reduction and, indirectly, for preserving and recovering resources and energy due to reuse and recycling. In addition to its widespread application to reusable beverage packaging, in some countries, the use of this type of instrument to other non-reusable products, such as packaging, household appliances, automotive batteries, end-of-life vehicles, and tires, has been found. Experience from the application of this instrument indicates that return rates are not very sensitive to the deposit value. Much more important in this context is the number, knowledge, and convenience of collection points for consumers (Pearce and Brisson 1995).

Concerning landfill taxes, they are a capable instrument to correct market failures and help internalize external costs caused by waste disposal, through methane emissions, potential leachate, neighbor communities' amenity costs, and increased transport. Through a price signal, landfill taxes can divert waste streams from landfills to recycling. The effectiveness of the landfill tax environmental incentive depends on the tax rate, very different among countries. Usually landfill taxes are implemented with other command and control instruments, like some substances landfill ban or more ambitions landfill standards, which may have an immediate effect in the recyclable material separate collection. Revenues from landfill taxes can be used to fund activities improving waste management and recycling activities, as at household level, landfill diversion is mostly an indirect result. In other words, a national level tax will impact local authorities, which are responsible for promoting awareness and information campaigns for citizens. Thereby, that is a good way to get to know the taxes impact on the amount charged of households' waste management and consequently reduce waste generation.

The extended producer responsibility is a policy approach under which producers are given a significant responsibility – financial and physical – for the treatment or disposal of postconsumer products. According to OECD (2018), assigning such responsibility could provide incentives to prevent waste, promote eco-design of products, and support the achievement of public recycling and material management goals.

Product charges, deposit-refund systems, landfill taxes, and extended producer responsibility have a significant impact on the collected MSW quantities and composition, as well as on waste collection and transport costs. For example, in deposit-refund schemes for specific products, the administrative and logistical costs of collection and preparation for reuse and recycling are usually covered by industry. The same applies to MSW stream collection and treatment costs covered by extended producer responsibility, such as packaging waste, batteries, accumulators, and electrical and electronic equipment. In most European countries, waste stream management is totally or partially supported by private management entities. The tax paid by producers to finance the end-of-life products management system is an economic instrument which enables municipalities to be relieved of the financial burden of collecting and managing such waste: the charge would be transferred to waste managers, ensuring sufficient financial support for its waste management activities (Chang and Pires 2015).

The user charges may be the most critical economic instrument to cover directly total or partial costs of waste collection and transportation, especially the O&M costs, since specific government subsidies or funding programs can more easily cover investment costs. User charges applied to end users/consumers must achieve two critical goals. On the one hand, the economic and financial cost's gradual recovery incurred by service providers, in a scenario of productive efficiency, to ensure service provided quality, and the operators' economic and financial sustainability. On the other hand, allow economic access of services for all population and act as an incentive to adopt more environmentally friendly behavior.

Waste management services are often financed by revenue from local authorities, especially in low- and middle-income countries, such as the share of property tax revenue or other local taxes that also serve to finance waste services (GIZ 2015). In other countries, the payment of waste services is charged on the water or electricity bill, as a function of this utility consumption. The user-charge billing can perform using indirect billing linked to an existing utility bill or tax instrument, such as the local property tax or water/energy bills.

The property tax, water or electricity charges, and other charge collection rates, which are a simple way to earn revenue, do not usually cover the total costs of the waste management system and are insufficient to really tackle the waste hierarchy, as it does not provide an incentive for reduction or source separation (Dahlén and Lagerkvist 2010). In this situation, the marginal cost of putting an extra waste bag in the container is zero, and there is no economic incentive to reduce the amount of waste.

There is a relationship between water/electricity consumption and the number of inhabitants per household, and also between waste generation and the number of inhabitants per household. In this way, theoretically, the tax will be proportional to the amount of waste generated per family. This system, however, is not the fairest and presents several problems, namely, families that have water-/energy-saving systems do not necessarily generate less waste; families with garden or backyards consume more water but may generate less waste as, for example, some organic waste can be used for animal feed or composting; households that reduce and recycle their waste also have lower MSW quantities to collect but end up paying the same as the others.

The need of a fairer tax system creation that works as an incentive for citizens to find alternative ways to reduce continuously the amount of waste they generate resulted on a large number of local communities introducing tax proportional to the amount of waste produced by each family, also known as pay-as-you-throw (PAYT) systems, unit-pricing models, or direct charging. PAYT schemes should include a fixed fee element and lower or zero fee for recyclable waste streams (i.e., green waste, kitchen waste, and dry recyclables), encouraging home composting and source separation (Watkins et al. 2012). According to Connecticut Department Environmental Protection (2004), the main advantages associated to PAYT schemes are in general:

- Fees may be calculated in such a way as to cover all or part of the collection and treatment/disposal costs.
- They can be an incentive for waste prevention and reduction, since households and businesses can divert part of the waste generated to alternative systems. The reduction can represent between 15% and 50% of the undifferentiated MSW.

PAYT schemes can also present certain disadvantages (Bilitewski et al. 2004; Connecticut Department Environmental Protection 2004), crucial to know when implementing these systems, namely:

- Potential waste illegal disposal/dumping – it is not a problem when precautions are taken.
- Increased administrative costs with changes of waste collection and transport system, waste generation quantification, monitoring, and population awareness.
- Instability risk in the revenues, if the tariff has been poorly designed.
- The need to build public consensuses, not only on the tariff type but also on its value. This situation can be aggravated in cases where levels of public perception of total costs are low concerning waste management, so intolerance for cost increase is predictable.
- Initial political resistance to PAYT programs is not uncommon. Understanding the concerns and the decision-making process is critical in moving the project forward successfully.

As referred by Kling et al. (2016), PAYT systems are the first instrument to be prioritized based on utility criteria, closely followed by landfill tax in case of economic instruments assessment for countries with low MSW management performance. However, Kling et al. (2016) also pointed that PAYT seems to be one of the most expensive instrument, an issue that should be taken into account.

As pointed by Bilitewski et al. (2004), there is no "best" approach for user charges, and existing legal and institutional arrangements influence decision-making. As GIZ (2015) referred, when designing a user-charge system, it is crucial to assure that the tariff structure is socially acceptable, equitable, and fair and that the billing system is adequate. Support PAYT systems by citizens will be higher if costs are transparent and there are ample opportunities for recycling and composting. PAYT schemes can be implemented through the following modalities (Reichenbach 2008):

- Per-user identifier: volume-based accounting (choice of container size); weight-based accounting
- Per bin identifier (individually or collectively assigned bins): volume-based accounting (identification system); weight-based accounting
- Pre-paid systems: pre-paid sack, tag, sticker, or token

In the volume-based systems, the consumer can choose and pay according the number or size of cans/containers used or, in the case of collective containers, using a prepaid card to open the container lid and that allows to put a specific volume of waste bag, through an installed reading "chip" that counts the frequency/volume use. In the weight-based accounting, each container has a chip or barcode that identifies its owner, and the collection vehicle is equipped with computerized systems that weight and record it for automated customer billing. In the bag prepayment systems, whose price will be fixed or variable according to its capacity, it can use labels, barcode, chips, or prepaid bags or tags/stickers which can be acquired to serve the same purpose. The bags or tags/stickers are sold at retail stores or municipal offices. These systems require the adoption of suitable provider structures for the pre-paid equipment (e.g., waste bags or stickers for bags), as well as the mechanism through which these service providers pass that costs to public authorities (GIZ 2015).

Table 7.2 describes some of the advantages and disadvantages indicated by Connecticut Department of Environmental Protection (s.d.) for bag or tags/sticker that are sold at retail stores or municipal offices, cans subscription (payment system according to containers capacity or frequency of collection), and cans used with a "pay-as-you-go" pricing system (under which residents are billed based on the number and size of cans they set out for collection).

PAYT systems vary across the EU. A study conducted by Watkins et al. (2012) for the EU in 2012 has found that 17 MS employ PAYT systems for municipal waste, but only three MS (AT, FI, IE) have PAYT schemes in place in all municipalities. The most usual forms of PAYT schemes are volume-based schemes (16 MS), frequency-based schemes (15 MS), weight-based schemes (9 MS), and sack-based schemes (6 MS). The authors note that several MS use a mixture of different types of schemes. Nowadays almost all European countries, at least some

Table 7.2 Advantages and disadvantages of different PAYT systems

Advantages/disadvantages		PAYT system type			
		Bag systems	Tags or sticker systems	Cans (subscription)	Can systems (pay-as-you-go)
Advantages	Waste reduction incentive	Stronger than can system	Stronger than can systems		
	Collection efficiency	Faster and more efficient than can systems Easy adaptation and monitoring	Faster and more efficient than can systems Easy adaptation and monitoring	Can work with (semi)automated collection systems Cans are reusable and prevent animals	Can work with (semi)automated collection systems Cans are reusable and prevent animals
	Convenience and costs to residents	Convenient and easy	Convenient and easy Lower costs than bags		Flexible to residents
	Costs for the system	Lower	Lower		
	Revenues			Stable and easy to forecast	
Disadvantages	Waste reduction incentive			Does not provide a significant incentive	
	Collection inefficiency	Animals can tear bags and bags can break during lifting Often incompatible with (semi)automated equipment	Size-limit compliance	Greater collection time Complex tracking and billing system needed Alternative needed for bulky waste	Greater collection time Complex tracking and billing system needed Alternative needed for bulky waste
	Inconvenience and costs to residents	Purchasing and storing bags	Purchasing tags/stickers Tags/stickers can fall off, be stolen or be counterfeited		
	Costs for the system			Implementation costs are required	Greater start-up costs
	Revenue	Greater uncertainty	Greater uncertainty		Greater uncertainty
	Waste management entities		Administration program required to purchase tags/stickers	Lag time between collecting waste and receiving payment	Lag time between collecting waste and receiving payment

Source: Connecticut Department of Environmental Protection (2004)

municipalities, have some unit-based pricing system to charge for municipal waste produced and collected. In sum, all the economic instruments addressed in this chapter need to be put in a country-specific context and other factors influencing policy-making need to be considered before practical decisions on their combined or isolated implementation are made (Kling et al. 2016).

7.3 Public and Private Sector Financing

The collection and transport of MSW are a responsibility of the local authorities and are organized and managed locally by municipalities, either through public or private waste management services. Collection and transport costs of MSW may represent 80–90% (in low-income countries), 50–80% (in middle-income countries), and 5–25% (in high-income countries) of the municipal solid waste management budge, due to the high costs of investment, operational costs, and human resources (Ghose 2006; Hoornweg and Bhada-Tata 2012; McLeod and Cherrett 2011). Besides being an expensive procedure, it is also the municipal sector that generates more employment, and therefore it has an equally important social dimension.

The issue of financing collection and transport systems is critical for many municipalities, exceptionally small- and medium-sized ones or those in low- and middle-income countries. According to GIZ (2015) in low- and middle-income countries, the O&M costs often constitute 60–85% of total waste management costs, and if it is relatively easy to find sponsors or funding sources for infrastructure and equipment (e.g., government grants, subsidies), it's very difficult or even impossible to find donors willing to participate in meeting to discuss O&M costs. For this reason, the cost recovery policy should at least cover the O&M costs, but in the longer term, full cost recovery is desirable to ensure a sustainable financing system. As already mentioned in the previous section, the costs of the waste management service must be supported by its users. However, in the case of a service with clear public utility, the price to be charged must be calculated under conditions of efficient performance, based on professional management principles and tools, and not penalize the user for any system inefficiencies (Rogge and De Jaeger 2012). The financing model should take into account two of the fundamental principles: the polluter pays principle and full cost recovery (Reichenbach 2008).

However, it is often observed that this service, which is generally ensured by municipalities, quickly distances it from the concept of efficiency, being needed the introduction of market forces to reach cost savings (Girth et al. 2012). Collection provided by local authorities also involves the typical constraints that can be faced in the public sector, such as overworked teams, obsolete equipment, burdensome hiring procedures, inflexible working hours, limitations on management changes, inadequate supervision, and active worker unions (Massouda et al. 2003). These difficulties have in practice been translated in many situations by the weak support of the solutions adopted for waste collection systems, often based on experience and intuitive methods, which result in inefficient and costly management practices Arribas et al. (2010).

7.3 Public and Private Sector Financing

Another difficulty is that the procurement process is more elaborate and more time-consuming in the public sector than in the private sector, since it involves first approving the budget and releasing the funds, followed by a lengthy public tender procedure. This is one of the reasons that can lead to the use of the equipment beyond its useful life, in particular, the vehicles, which leads to an increase in maintenance costs. In the private sector, the equipment is usually purchased by funds from commercial banks, being the acquisition faster. In the study carried out in Gaborone (Botswana) by Bolaane and Isaac (2015), the authors reported that the average age of municipal vehicles was 8 years, while the average age of private sector vehicles was 4 years.

Although some municipalities continue to prefer to manage collection waste themselves, through municipal departments, or through municipally owned waste management companies, due to these problems and the increasing cost of collection waste, many have opted for indirect management models, such as outsourcing or even privatization of the waste collection and treatment service.

The models of global management of services of public interest can be categorized in different ways. For example, based on the classification made by Van Dijk and Schouten (2004) for the existing European models of water supply and the sanitation management, Adamsen et al. (2016) proposed the following four main models for the waste sector:

(a) Direct public management – the responsible authority assumes full responsibility and executes the service itself, usually through one of its departments, which includes public-public cooperation and the use of "in-house entities."
(b) Delegated public management – the responsible entity appoints a managing entity to execute the public tasks (public-private partnerships).
(c) Direct private management – the public authority puts the responsibility in the private party, which assumes full responsibility for the provision of services.
(d) Delegated private management – the public authority appoints a private company for the management of tasks, through a time-bound contract in the form of public contract or concession contract, following procurement procedures (outsourcing).

According to Adamsen et al. (2016) during the 1980s and 1990s, due to political, legal, economic, and fiscal factors, almost all European member states underwent a process of public-private partnerships or outsourcing of waste management services (through public procurement or concession procedures), as a promising alternative to reduce costs and improve MSW management performance.

In the collection and transport of the MSW sector, in-sourcing and outsourcing management models are the most usual. The main drivers that can explain the choice between municipalization and privatization may differ from country to country, reflecting the differences in the political, economic, and social contexts. Among the main advantages pointed to the benefits of outsourcing, several authors point out the following (Adamsen et al. 2016; Bel and Miralles 2003; Bouhamed and Chaabouni 2008; Greaver 1999; Jacobsen et al. 2013; Kakabadse and Kakabadse 2005; Kinyua 2015; Post 2004):

- Greater potential for cost savings and cost-efficiency advantages, due to economies of scale
- More purchasing power which may enable the level of capital investment required to be reduced, resulting in a lower contract price
- Higher quality of services and better performance indicators of services
- Lower bureaucratic and legislative barriers, for example, not being subject to the mandatory use of public procurement to provide procedures: public procurement rules
- Access to specialized know-how and greater facility to invest in state-of-the-art waste collection and treatment technology
- Flexibility in recruiting human resources, better possibilities for career advancement, and less pressure from unions
- Greater flexibility to hire human resources, more career prospects for workers, and less pressure from unions

However, not always these advantages are confirmed in practice, and the outsourcing solution has not always achieved a municipality's goal of high-quality services and reduced cost. By nature, the outsourcing process is a high-risk and requires proper management to ensure it is successful. However, some municipalities do not have sufficient know-how, human resources or information technology that allows them to plan their public offerings and select the best private company (Gamberini et al. 2009; Padovani and Young 2008; Rodrigues 2016). Also, most private concession contracts rely on the weight of waste as the primary billing base, encouraging companies to maximize waste collected and discouraging reduction and selective collection (Massouda et al. 2003; Chowdhury 2009; Rodrigues 2016).

The management and prevention of outsourcing risks require that municipalities focus on the following (Adamsen et al. 2016): (a) a set of operational and economic performance indicators to measure the results; (b) a well-established process for communication and cooperation that fills the inevitable gaps in any high-risk contract; and (c) a management control process that includes quality measures, outcome measures, and process measures concerning. For these reasons, the changing from private providers to direct provision by the municipality or provision by in-house arrangements or public-public cooperation (i.e., the re-municipalization of municipal waste services) has been increasing in some countries, as in Germany (Adamsen et al. 2016).

Despite the recurrent debate around which the best management model for waste collection – public or private – the literature review is no clarity about if private entities always provide a better service managing. Specific national and local contexts (e.g., political, regulation, fiscal) will dictate which is the best, and public, private, and hybrid ways to organize municipal waste services can be equally effective and efficient (SusValueWaste 2017). The different actors play complementing roles that collectively lead to innovative and sustainable waste management.

References

Abdelli I, Abdelmalek F, Djelloul A, Mesghouni K, Addou A (2016) GIS-based approach for optimised collection of household waste in Mostaganem city (Western Algeria). Waste Manag Res 34:417–426

Adamsen C, Blagoeva T, Ilisescu AR, Le Den X, Nielsen SS (2016) Legal assistance on the application of public procurement rules in the waste sector – final report. Publications Office of the European Union, Luxembourg

Allesch A, Brunner PH (2014) Assessment methods for solid waste management: a literature review. Waste Manag Res 32:461–473

Arribas C, Blazquez C, Lamas A (2010) Urban solid waste collection system using mathematical modelling and tools of geographic information systems. Waste Manag Res 28:355–363

Assis R (2011) Depreciation methods (in Portuguese: Métodos de amortização). http://www.rassis.com/artigos/Economia/Metodos%20de%20Amortizacao.pdf. Accessed 10 Sep 2015

Bel G, Miralles A (2003) Factors influencing the privatisation of urban solid waste collection in Spain. Urban Stud 40:1323–1334

Bilitewski B, Härdtle G, Maek K, Weissbach A, Boeddicker H (1994) Waste management. Springer, Berlin

Bilitewski B, Werner P, Reichenbach J (2004) Handbook on the implementation of Pay-As-You-Throw as a tool for urban waste management. Eigenverlag des Forum für Abfallwirtschaft und Altlasten e. V., Dresden

Bohm R, Folz D, Kinnaman T, Podolsky M (2010) The costs of municipal curbside recycling and waste collection. Resour Conserv Recycl 54:864–871

Bolaane B, Isaac E (2015) Privatization of solid waste collection services: lessons from Gaborone. Waste Manag 40:14–21

Bosch N, Pedraja F, Suárez-Pandiello J (2001) The efficiency of refuse collection services in Spanish municipalities: do non-controllable variables matter? Institut d'Economia de Barcelona, Centre de Recerca en Federalisme fiscal I Economia regional, Barcelona

Boskovic G, Jovicic N, Jovanovic S, Simovic V (2016) Calculating the costs of waste collection: a methodological proposal. Waste Manag Res 34:775–783

Bouhamed A, Chaabouni J (2008) Partenariat public-privé en Tunisie: Les conditions de succès et d'échec. Int J Technol Manag Sustain Dev 7:71–89

Callan S, Thomas J (2001) Economies of scale and scope: a cost analysis of municipal solid waste services. Land Econ 77:548–560

Carvalho JM (2011) Annex 5: municipal solid waste management costs (in Portuguese: Anexo 5: Custos de sistema de gestão de resíduos urbanos). www.ua.pt/ReadObject.aspx?obj=18467. Accessed 23 Sep 2015

Chang NB, Pires A (2015) Sustainable solid waste management: a systems engineering approach. IEEE Wiley, Hoboken

Chowdhury M (2009) Searching quality data for municipal solid waste planning. Waste Manag 29:2240–2247

Connecticut Department of Environmental Protection (2004) Smart (Pay-as-You-Throw) implementation handbook. http://www.ct.gov/deep/lib/deep/reduce_reuse_recycle/payt/ct_dep_smart_implementation_handbook_parts_1_thru_4.pdf. Accessed 4 May 2018

D'Onza G, Greco G, Allegrini A (2016) Full cost accounting in the analysis of separated waste collection efficiency: a methodological proposal. J Environ Manag 167:59–65

Dahlén L, Lagerkvist A (2010) Pay as you throw: strengths and weaknesses of weight-based billing in household waste collection systems in Sweden. Waste Manag 30:23–31

Dahlén L, Åberg H, Lagerkvist A, Berg PE (2009) Inconsistent pathways of household waste. Waste Manag 29:1798–1806

Delloite (2004) Model of calculation of counterpart amount (in Portuguese: Modelo de cálculo do valor de contrapartida - versão final). Instituto dos Resíduos, Lisboa

Dijkgraaf E, Gradus RHJM (2004) Cost savings in unit-based pricing of household waste: the case of The Netherlands. Resour Energy Econ 26:353–371

Dogan K, Süleyman S (2003) Report: cost and financing of municipal solid waste collection services in Istanbul. Waste Manag Res 21:480–485

Dubin J, Navarro P (1988) How markets for impure public goods organize the case of households refuse collection. J Law Econ Organ 4:217–241

Entidade Reguladora dos Serviços de Águas e Resíduos (ERSAR) (2014) Tariff regulation for municipla solid .waste management service (in Portuguese: Regulamento tarifário do serviço de gestão de resíduos urbanos). http://www.ersar.pt/. Accessed 3 Jun 2015

European Commission (2015) Guide to cost-benefit analysis of investment projects – economic appraisal tool for Cohesion Policy 2014–2020. Publications Office of the European Union, Luxembourg

Faccio MP (2011) Waste collection multi objective model with real time traceability data. Waste Manag 31(12):2391–2405

Folz DH (1999) Recycling policy and performance: trends in participation, diversion, and costs. Public Work Manag Policy 4:131–142

Gamberini R, Galloni L, Rimini B, Beltrami FO (2009) Evaluation and comparison of waste collection services. In: Abstracts of the twelfth international waste management and landfill symposium, Sardinia, 9–11 Oct 2009

Gamberini R, Del Buono D, Lolli F, Rimini B (2013) Municipal solid waste management: identification and analysis of engineering indexes representing demand and costs generated in virtuous Italian communities. Waste Manag 33:2532–2540

García-Sánchez MI (2008) The performance of Spanish solid waste collection. Waste Manag Res 26:327–336

Ghose MD (2006) A GIS based transport model for solid waste disposal – a case study on Asansol municipality. Waste Manag 26:1278–1293

Girth AM, Hefetz A, Johnston JM, Warner ME (2012) Outsourcing public service delivery: management responses in noncompetitive markets. Public Admin Rev 72:887–900

GIZ GmbH (2015) Economic instruments in solid waste management: applying economic instruments for sustainable solid waste management in low and middle-income countries. https://www.giz.de/en/downloads/giz2015-en-waste-management-economic-instruments.pdf. Accessed 4 May 2018

Gomes AP (2008) Separate collection of the biodegradable fraction of MSW: an economic assessment. Waste Manag 28:1711–1719

González-Torre, P. L., Adenso-Díaza, B., Ruiz-Torres, R. (2003). Some comparative factors regarding recycling collection systems in regions of the USA and Europe. Journal of Environmental Management, 69 (2), 129–138

Goulart A (2003) Comparison of deep collection system with traditional system. Dissertation, Tampere University of Technology (TUT)

Greaver F (1999) Strategic outsourcing – a structured approach of outsourcing decisions and iniatives. AMACOM, New York

Greco G, Allegrini M, Del Lungo C, Savellini PG, Gabellini L (2015) Drivers of solid waste collection costs Empirical evidence from Italy. J Clean Prod 106:364–371

Hage OS (2008) An econometric analysis of regional differences in household waste collection: the case of plastic packaging waste in Sweden. Waste Manag 28:1720–1731

Hogg D (2002) Costs for municipal waste management in the EU: final report to directorate general environment, European Commission. Eunomia Research & Consulting, Ltd, on behalf of Ecotec-Research & Consulting

Hoornweg D, Bhada-Tata P (2012) What a waste: a global review of solid waste management. Urban development series knowledge papers, nr. 15. Urban Development & Local Government Unit, World Bank, Washington, DC

Huang Y, Pan T, Kao J (2011) Performance assessment for municipal solid waste collection in Taiwan. J Environ Manag 92:1277–1283

Jacobsen R, Buysse J, Gellynck X (2013) Cost comparison between private and public collection of residual household waste: multiple case studies in the Flemish region of Belgium. Waste Manag 33:3–11

References

Johansson O (2006) The effect of dynamic scheduling and routing in a solid waste management system. Waste Manag 26:875–885

Kakabadse A, Kakabadse N (2005) Outsourcing: current and future trends. Thunderbird Int Bus Rev 47:183–204

Karadimas N, Papatzelou K, Loumos V (2007) Optimal solid waste collection routes identified by the ant colony system algorithm. Waste Manag Res 25:139–147

Kinyua BK (2015) Determinants of outsourcing services as a cost reduction measure in devolve government: a case study of Narobi city country, Kenya. Eur J Bus Soc Sci 4:12–23

Kling M, Seyring N, Tzanova P (2016) Assessment of economic instruments for countries with low municipal waste management performance: an approach based on the analytic hierarchy process. Waste Manag Res 34:912–922

Komilis DP (2008) Conceptual modeling to optimize the haul and transfer. Waste Manag 28:2355–2365

Larsen A, Vrgoc M, Christensen T, Lieberknecht P (2009) Diesel consumption in waste collection and transport and its environmental significance. Waste Manag Res 27:652–659

Larsen A, Merrild H, Moller J, Christensen T (2010) Waste collection systems for recyclables: an environmental and economic assessment for the municipality of Aarhus (Denmark). Waste Manag 30:744–754

Lavita, M (2008) Multi-material door-to-door collection routes (in Portuguese: Circuitos de recolha selectiva multi-material porta-a-porta). Dissertation, IST/UTL – Instituto Superior Técnico, Lisbon

Levy J (2004) National paradigm and 3R policy (reverse logistics) (in Portuguese: Panorama 999 nacional e a política dos 3R (logística inversa)). In: Abstracts from the Recursos, resíduos e 1000 reciclagem, IST, Lisbon, 25 Oct 2004

Levy J, Oliveira R, Brito J (2007) Reverse logistics (in Portuguese: A logística inversa). http://seminarios.ist.utl.pt/03-04/des/IST-seminario/sprod/3sessao. Accessed Dec 26 2013

López JV, Soriano F, Aguilar M, León B, Ramos-Catalina P, Carretero C (2007) Metodología de Contenerización para residuos de envases ligeros: caso de Aranjuez. E.T.S.I. de Montes. Universidad Politécnica de Madrid, Dpto. Ingeniería Forestal. Madrid: Universidad Politécnica de Madrid

Massouda M, El-Fadelb M, Malak AA (2003) Assessment of public vs private MSW management: a case study. J Environ Manag 69:15–24

McLeod F, Cherrett T (2011) Waste collection. In: Letcher TM, Vallero DA (eds) Waste: a handbook for management. Elsevier, Burlington, pp 61–76

Miller L, Delbridge P (1995) Waste management in Hampshire feedback on the strategy, December 1995. Miller Associates and PDA International

Nguyen TTT, Wilson GB (2010) Fuel consumption estimation for kerbside municipal solid waste (MSW) collection activities. Waste Manag Res 28:289–297

O'Leary P, Walsh P (1995) Decision maker's guide to solid waste management, 2nd edn. United States Environmental Protection Agency, Office of Solid Waste, RCRA Information Center, Washington, DC

Ohlsson H (2003) Ownership and production costs: choosing between public production and contracting-out in the case of Swedish refuse collection. Fisc Stud 24:451–476

Organisation for Economic Co-operation and Development (OECD) (2018) Extended producer responsibility. http://www.oecd.org/env/tools-evaluation/extendedproducerresponsibilityhtm. Accessed 4 May 2018

Padovani E, Young DW (2008) Toward a framework for managing high-risk government outsourcing: field research in three Italian municipalities. J Public Procure 8:215–247

Pearce D, Brisson I (1995) The economics of waste management. In: Hester RE, Harrison RM (eds) Waste treatment and disposal. The Royal Society of Chemistry, London

Perman R, Common M, McGilvray J, Ma Y (2003) Natural resource and environmental economics, 3rd edn. Pearson Education, Harlow

Post J (2004) Evolving partnerships in the collection of urban solid waste in the developing world. In: Isa B, Post J, Furedy C (eds) Solid waste management and recycling: actors, partnerships and policies in Hyderabad, India. Kluwer, Dordrecht, pp 21–36

Rada E, Ragazzi M, Fedrizzi P (2013) Web-GIS oriented systems viability for municipal solid waste selective collection optimization in developed and transient economies. Waste Manag 33:785–792

Rathje WL, Murphy C (1992) Rubbish! The archaeology of garbage. Harper Collins Publishers, New York

Reichenbach J (2008) Status and prospects of pay-as-you-throw in Europe – a review of pilot research and implementation studies. Waste Manag 28:2809–2814

Rhoma FZ (2010) Environmental & economical optimization for municipal solid waste collection problems, a modeling and algorithmic approach case study. In: WSEAS international conference on mathematical methods, computational techniques and intelligent systems, Sousse, Tunisia, May 3–6, 2010, pp 205–211

Ricci M (2003) Economic assessment of separate collection cost: tools to optimize it and the advantage of operative integration. In: Abstracts from the future of source separation of organic waste in Europe. Barcelona, Spain, 15–16 Dec 2003

Rizzoli A, Montemanni R, Lucibello E, Gambardella L (2007) Ant colony optimization for realworld vehicle routing problems – from theory to applications. Swarm Intell 1:135–151

Rodrigues S (2016) Classification and benchmarking of urban waste collection systems (in portuguese: Classificação e benchmarking de sistemas de recolha de resíduos urbanos). Dissertation, NOVA University of Lisbon

Rodrigues S, Martinho G, Pires A (2016a) Waste collection systems part A: a taxonomy. J Clean Prod 113:374–387

Rodrigues S, Martinho G, Pires A (2016b) Waste collection systems part B: benchmarking. J Clean Prod 139:230–241

Rogge N, De Jaeger S (2012) Evaluating the efficiency of municipalities in collecting and processing. Waste Manag 32:1968–1978

Sanjeevi V, Shahabudeen P (2016) Optimal routing for efficient municipal solid waste transportation by using ArcGIS application in Chennai, India. Waste Manag Res 34:11–21

Santos RF, Santana F, Antunes P, Martinho MG, Jordão L, Sirgado P, Neves AG (1994) Municipal solid waste management system of Lisbon – Cost analysis (in Portuguese: Sistema de resíduos sólidos do município de Lisboa – Análise da estrutura de custos do sistema de resíduos sólidos). Department of Environmental Science and Engineering, NOVA University of Lisbon

Shekdar A, Mistry P (2001) Evaluation of multifarious solid waste management systems – a goal programming approach. Waste Manag Res 19:391–402

Sonesson U (2000) Modelling of waste collection – a general approach to calculate fuel consumption and time. Waste Manag Res 18:115–123

Sora M, González J (2014) Economic balance of door-to-door and road containers waste collection for local authorities and proposals for its optimization. Commissioned by the Association of Catalan Municipalities for Door-to-Door Selective collection to Fundació ENT

Stevens B (1978) Scale, market structure, and the cost of refuse collection services: the case of Ireland. Econ Soc Rev 31:129–150

SusValueWaste (2017) Private or .public waste management? http://www.susvaluewaste.no/2017/10/12/private-or-public-waste-management/. Accessed 4 May 2018

Tchobanoglous G, Theisen H, Vigil S (1993) Integrated solid waste management engineering: principles and management issues. McGraw-Hill, New York

United States Environmental Protection Agency (USEPA) (1997) Full cost accounting for municipal solid waste management: a handbook. United States Environmental Protection, Washington, DC

van Dijk MP, Schouten M (2004) The dynamics of the European water supply and sanitation market. In: Abstracts from AWRA international specialty conference 2004, University of Dundee, 29 Aug – 2 Sept 2004

Vidal RG (2001) Integrated analysis for pre-sorting and waste collection schemes implemented in Spanish cities. Waste Manag Res 19:380–390

Watkins E, Hogg D, Mitsios A, Mudgal S, Neubauer S, Reisinger H, Troeltzsch J, Van Acoleyen M (2012) Use of economic instruments and waste management performances – final report. http://ec.europa.eu/environment/waste/pdf/final_report_10042012.pdf. Accessed 4 May 2018

Chapter 8
Environmental Context

Abstract Today's environmental concerns are related to the population and its consumption of resources, which have led to significant environmental and global changes, such as climate change and resource overexploitation. The solid waste management, in an integrated way, has been capable of influencing and contributing to the solution of such challenges. The purpose of this chapter is to discuss the environmental context which is the role of waste management, focusing on waste collection, in such context. It leads to further magnifying the importance of the contributions that each waste management operation – collection, recycling, treatment, recovery, and disposal – alone and together, can bring to the environmental challenges.

Keywords Waste collection · Circular economy · Zero waste · Sustainable development · Eco-cities · Climate change · Resource scarcity

8.1 Environmental Context of Twenty-First Century

In the last decades, the growth of population, rapid urbanization, industrialization of the emerging economies, and increasing shortage of resources, energy, and consumption are leading to a challenging world. In this respect, strategies that are being studied to help face those challenges are the development of regional economies, promotion of rural development, production processes with more efficient use of resources, minimization of waste generation and stablization of cycling of ecosystems, and creation of cities that serve as role models of environmental protection (Chang 2016). In this respect, solid waste management is starting to have an essential role in the attempt to deal with such global challenges. The reason is related to the amount of waste generated and to its composition. Current European municipal solid waste (MSW) generation is around 242 million tons per year; China generates 300 million tons per year; and the USA generates around 229 million tons per year (Waste Atlas 2017; Eurostat 2017). The last estimate of global MSW generation is 1.9 billion tons with almost 30% remaining uncollected; from the collected, only 30% is recycled or recovered, including energetically recovered (Waste Atlas 2017).

© Springer International Publishing AG, part of Springer Nature 2019
A. Pires et al., *Sustainable Solid Waste Collection and Management*,
https://doi.org/10.1007/978-3-319-93200-2_8

Concerning the environmental stress from waste (the amount of waste generated in a country divided by country's area), Singapore presents around 9.9 thousand tons MSW/km^2, Macao SAR around 8.9 thousand tons MSW/ km^2, and Hong Kong SAR 3.1 thousand tons MSW/ km^2 (Waste Atlas 2017). All these indicators highlight how waste is damaging the environment, needing more efforts to reduce them and, when not possible, to manage them most sustainably.

In the next few sections, the environmental context will be detailed. The way how the context influences waste and how waste management is influencing the context will also be approached.

8.1.1 Globalization and Economic Growth

According to Gupta (2015):

> Economic growth refers to an increase in the level of goods and services produced by an economy.

The ways to measure such level of economic growth (that should lead to economic development) are commonly by gross domestic product (GDP) and gross national income (GNI). Definitions from World Bank for this economic concepts are (World Bank 2018):

> GDP at purchaser's prices is the sum of gross value added by all resident producers in the economy plus any product taxes and minus any subsidies not included in the value of the products. GNI is the sum of value added by all resident producers plus any product taxes (less subsidies) not included in the valuation of output plus net receipts of primary income (compensation of employees and property income) from abroad.

Economic growth and waste generation (and environmental degradation) are strongly correlated; once economic growth results from the production of goods and services, renewable and non-renewable raw materials are being needed. Because no process is 100% efficient (according to the second law of thermodynamics), waste is generated, causing environmental degradation and pollution. There is the need to separate economic growth from the use of biophysical resources, through eco-economic decoupling or dematerialization to reduce the waste generated and environmental impacts from the economy (Van der Voet et al. 2004; van Caneghem et al. 2010; Verfaillie and Bidwell 2000). Decoupling economic growth from negative environmental impact is one of the leading objectives of the OECD Environmental Strategy for the First Decade of the Twenty-First Century (OECD 2002).

Besides economic growth, globalization is also responsible for misleading waste management. The term globalization is referent to a process of international integration of markets for goods, products, services, means of production, financial systems, technology, and industries, with increase of capital mobility, of technological innovation propagation, also with the influences of other aspects related to cultural and social environment and interdependency of national markets (United Nations

8.1 Environmental Context of Twenty-First Century

et al. 2002; Surugiu and Surugiu 2015). According to ISWA (2012), globalization influences and changes waste management practices, but also waste management practices impose new global markets for waste products (or secondary materials from waste).

Due to globalization, waste managers' practices have originated misleading management of waste which may be costly to be treated in developed countries, but shipping them to areas without such environmental concerns makes it less expensive. Countries have signed Basel Convention to reduce such transfer of waste (namely, hazardous waste). More recently, e-waste (also named waste of electric and electronic equipment and e-scrap) is being sent from developed countries as used equipment, i.e., they are being traded to developing countries, not being under the scrutiny of Basel Convention, because they are supposedly being sold as secondhand devices, not waste. However, there is no proof that they are still working. In 2012, 1.3 million tons of WEEE were exported from European Union countries in undocumented exports, being likely classified as illegal exports, where they do not adhere to the guidelines for differentiating used equipment from waste, being estimated that 30% of this volume is e-waste (Huisman et al. 2015). In developing countries, the demand for inexpensive second-hand device raw materials is the most significant driver for the global trade of e-waste (Baldé et al. 2015).

Concerning global markets for waste products, in 2015, around 180 million metric tons of waste were recycled as secondary commodities, valued at more than 86 USD billion according to United Nations (2017). The most tradable scrap in 2016 was ferrous metal (53%), followed by paper (26%), nonferrous metals (10%), plastics (8%), glass (2%), and others including proecios metals (1%) (United Nations 2017). The USA is the most significant exporter of such recycled items, and China is the most relevant consumer of such materials (ISRI 2015). The trade of such commodities, and the high quality (and potentially high price), depends on the existence and performance of source separation and collection of those materials. Besides the global market of ferrous, non-ferrous metals, paper, and cardboard, plastics and textiles, there are also regional waste markets for glass waste and solid recovered fuels, and local waste markets for compost and construction and demolition waste aggregates (UNEP and ISWA 2015). Global and regional trades of waste products can bring economic benefits once that they are a low-cost resource comparatively to virgin materials, and environmental benefits by reducing the depletion of natural resources, reducing landfill space, and increasing energy savings (ISRI 2015). However, if countries that are remanufacturing and recycling those materials in countries without requirements for health and environment protection, recycling can cause injuries to workplace and all around the recycling activity. The trade of waste products is made from developed countries to developing countries, where they recycle waste at a lower cost than in developed countries. Developed countries, mostly Asian, have high recycling costs to meet environmental compliance in recycling sector, sending waste to other countries with lower environmental control and enforcement, lower operating costs for recycling industry, and lower quality standards (UNEP and ISWA 2015).

The price volatility of waste products at a global scale is a problem to the recycling industry and to the market itself. Some factors can explain such volatility, namely, the fact that recycled plastic price is affected by oil price (Angus et al. 2012) and economic crisis (UNEP and ISWA 2015). To solve the problem of price volatility, the European Commission intended to transform the vision of waste products, in such way that they could be secondary resources or products and no more defined as waste. The end-of-waste criteria were published in Waste Framework Directive (European Parliament and Council 2008) with the aim to define the requirements to establish such criteria. Until now, the criteria elaborated and regulated were for glass cutlet and copper scrap, being in development the regulations for waste plastics, compost/digestate from biodegradable waste, copper and copper alloy, and waste paper (JRC 2015).

8.1.2 Megacities, Eco-cities, and Industrial Symbiosis at Cities

Urban population is proliferating, expecting to grow by another 4000 million by 2050, being the growth mainly in developing countries (United Nations 2015). Such growth will result in megacities, which will demand substantial infrastructure services. Also, new urban areas will emerge, where it is expected that infrastructure development near those new centers will remain neglected (Vaiude et al. 2017). Scale urbanization induced a series of environmental problems, including landscape fragmentation due to massive land use, regional climate change, biodiversity loss, and ecosystem degradation, not only due to massive land use but also the overriding of MSW generation and of wastewater (Bai et al. 2012; Gaubatz 1999; Liu et al. 2005, 2014; Jha et al. 2008), which will damage land and coastal ecosystems near sea since most megacities are coastal cities.

According to the definition of United Nations (2015), cities can be divided in cities (until 5 million), large cities (5–10 million habitants), and megacities, with 10 million or more inhabitants. The number of large cities is growing to count, expecting to reach 63 by 2030, representing 9% of the global urban population; megacities are home to about 1/8 of world's urban dwellers, expecting to be 41 cities in 2030, representing near 14% at that year (United Nations 2015).

Several countries (by their governments) have tried to find solutions to the increasing growth of their cities, by the development of eco-cities or sustainable cities. The term "eco-city" was reportedly coined during the winter of 1979–1980 by the members of a voluntary organization Arcology Circle (Dongtan 2008). The term features prominently in a 1987 book (Register 1987) and is used interchangeably with the term "sustainable city" (Eryildiz and Xhexhi 2012). The concept "sustainable city" was used by Girardet (1999):

> Sustainable city is organized to enable all its citizens to meet their own needs and to enhance their well-being without damaging the natural world or endangering the living conditions of other people, now or in the future.

8.1 Environmental Context of Twenty-First Century

Eco-city projects are in a small number all over the world. Completed "eco-city" projects such as Vauban Freiberg (Germany) and Hammarby Sjöstad (Sweden) and uncompleted projects, for example, Masdar City (UAE) and Tianjin Eco-City (China), are designed to offer a good quality of life (Zaman and Lehman 2011). The eco-cities are needed to ensure a sustainable expansion of cities and to reach global sustainability (Zaman and Lehman 2011). In this respect, Premalatha et al. (2013) have compiled ten attributes assigned to eco-cities (Roseland 1997, 2001):

1. Should have land-use priorities such that it creates compact, diverse, green, and safe mixed-use communities around public transportation facilities.
2. Should have transportation priorities such that it will discourage driving and emphasize "access by proximity".
3. Should restore damaged urban environments.
4. Should create affordable, safe, convenient, and economically mixed housing.
5. Should nurture social justice and create improved opportunities for the underprivileged.
6. Should support local agriculture, urban greening, and community gardening.
7. Should promote recycling and resource conservation while reducing pollution and hazardous waste.
8. Should support ecologically sound economic activities while discouraging hazardous and polluting ones.
9. Should promote simple lifestyles and discourage excessive consumption of material goods.
10. Should increase public awareness of the local environment and bioregion through educational and outreach activities.

The attributes of an eco-city include the waste reduction, existing in some cities which have intended to reach a zero-waste city. Two eco-cities which have tried to reach zero waste are Dongtan in China and Masdar City near Abu Dhabi. The strategy defined in Dongtan was that all waste should be collected and processed, that is, MSW to be sorted, up to 80% of waste to be recycled, and organic waste and human waste to be digested and composted to be used in farmlands (Cheng and Hu 2010). In this city, the eco-industry (waste management, wind, and solar technology) will be a significant component of the economy (Premalatha et al. 2013). In the Masdar City, the predicted measures taken to reach a zero-waste city were to have a vacuum system for waste under the city, with transportation of waste into a central facility where waste would be sorted as much as possible, being the nonrecyclable waste gasified and the recyclable waste incorporated into building materials (Bullis 2009).

The purpose of such cities was ambitious and unfortunately was not successful. The zero waste is not reachable because of the second law of thermodynamics. Over the "life, rebirth, and final death cycle" of any product, waste generation is an absolute certainty. However, in most situations, recovery, recycle, or reuse involves one or other kinds of processing which are inevitably accompanied by consumption of energy and wastage of some material, in conformity with the second law of thermodynamics (Premalatha et al. 2013).

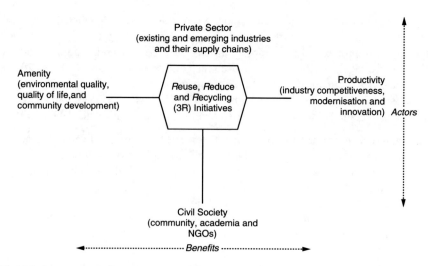

Fig. 8.1 Impact areas of eco-towns. (Source: van Berkel et al. (2009))

Besides zero waste, the zero-carbon goal is unattainable, unless a dramatic reduction of energy and material consumption occurs, together with the sequestration of anthropogenic carbon emitted (Premalatha et al. 2013):

> Even the low-carbon low waste lifestyle appeared increasingly difficult to achieve because it was more expensive than conventional lifestyle and more restrictive as well.

These two cities have failed because they were not capable of implementing a holistic vision to achieve the zero-waste goal (Premalatha et al. 2013); they failed because they have not applied the concept of "urban symbiosis." Urban symbiosis is an extension for industrial symbiosis, where waste or by-products from the city (or urban areas) metabolism are used as secondary raw materials for industrial operations (van Berkel et al. 2009) (Fig. 8.1). The concept is based on the synergistic opportunity of proximity to transfer the secondary materials, which has been the basis of MSW management in Japan (management of waste near the source), where eco-town projects from 1997 have embedded the concept of urban symbiosis (Okuda and Thompson, 2007; Desrochers and Cities 2001). The eco-towns have been capable of increasing the number of recycling plants, to the diversification and sophistication of recycling technologies, with the capacity to treat two million metric tons of waste per annum (in 2009) (van Berkel et al. 2009).

8.1.3 Climate Change

United Nations Framework Convention on Climate Change (UNFCCC, Article 1) defines that (United Nations 1992):

> Climate change means a change of climate which is attributed directly or indirectly to human activity that alters the composition of the global atmosphere and which is in addition to natural climate variability observed over comparable time periods.

Waste management is, by definition, resulting from human activity and is contributing to climate change in many different ways. The release of greenhouse gases (GHG) from landfills due to degradation of biodegradable waste (mostly methane), the emissions resulting from energy recovery and thermal treatment of waste, and the emissions from the waste collection due to energy use are contributing to the release of GEE to the atmosphere. However, the recycling of waste, the production of compost from biodegradable waste, the sequestration of carbon due to compost application into the soil and to the avoided consumption of virgin paper due to paper recycling (Ackerman 2000) are reducing the presence of GHG in the atmosphere.

The climate change can affect the way how today waste managers plans and operates waste collection systems. According to Zimmerman and Faris (2010) and USAID (2013), the climate change can affect waste collection in five aspects including temperature change, precipitation, sea level rise, storm surge, and extreme wind. Regarding temperature, the collection will need to be more frequent due to the increased odor and pest activity, e.g., higher exposure to flies, which are a vector to disseminate diseases; also overheating of collection vehicles will require additional cooling capacity. In the case of precipitation change, the flooding of collection routes and accesses to landfill and waste treatment facilities can be inaccessible, and also the collection of waterlogged waste can be another factor of stress for vehicles and workers. Concerning sea level rise, the effects can be at the collection routes being more narrowed, the increasing of waste in a concentrated area as people go to higher elevations, without forgetting the permanent inundation of collection, processing, and disposal infrastructure which will make the collection difficult. For storm surge, situations of temporary flooding of and diminished access to transport infrastructure will not only make the waste collection difficult but also can lead to the closure of facilities due to infrastructure damage. The extreme wind can disperse waste both from temporary storage and collection vehicle as well as can make the access to collection routes and waste facilities difficult due to damage and debris caused by the wind.

Waste collection is a stage of waste management which is critical to protecting human health and local resources and reduces exposure to contaminated waste and disease-carrying rodents and insects (USAID 2013). Several adaptive measures should be taken to minimize the effects of climate change in waste collection.

8.2 Sustainability and Circular Economy Considerations

Due to the global changes and challenges identified in the previous sections, the society is requesting for a more sustainable living and developing systems. The concept of the circular economy – while not entirely new – has gained relevance on the agendas of policymakers, academia, and companies, being presented as the new economy that could face climate change problems and scarcity of resources at the same time (Brenan et al. 2015; Gregson et al. 2015; Geissdoerfer et al. 2017).

The circular economy is capable of offering a bottom-up industrial perspective, where recycling and markets for secondary resources are the pillars (Welfens et al. 2017). According to the European Commission (2015), circular economy is an economic model where the product, material, and resource value keeps in the economy for as long as possible and the generation of waste minimized, in such way that contributes to the development of a low sustainable carbon, resource-efficient, and competitive economy. Several benefits from the circular economy occur in European Union: protect the business against resource scarcity and create local jobs, focusing on the production design of products, production processes, consumption, waste management, and the market for secondary raw materials and water reuse (European Commission 2015). Quite before European Commission have established such strategy, Germany was the pioneer in integrating the concept of the circular economy on their legislation in 1996, followed by Japan in 2002 and China in 2009 (Su et al. 2013; METI 2004; Lieder and Rashid 2016). International initiatives exist, like the Sino-European Circular Economy and Resource Efficiency (SINCERE) project, which intends to develop new economic modeling tools to understand the resource use patterns of China and the EU (Welfens et al. 2017).

Although circular economy is gaining importance in the society, the conceptual relationship between circular economy and sustainability is not clear, with similarities and differences pointed out by Geissdoerfer et al. (2017) (Table 8.1). In another way, circular economy contribution still needs to be investigated shortly, precisely to the waste management sector. A circular economy is focused on waste flows and closing loops flows, through the economic system, where the destinations are the players that origintaed the product at the first place. Also replace the sentence before: "In another way, circular economy contribution still needs to be investigated shortly, precisely to the waste management sector.

Table 8.1 Similarities and differences between sustainability and circular economy

Main similarities	Main differences
Commitments intra- and intergenerational level	Sustainability goals are open-ended, multitude of goals/circular economy goals are closed loop, ideally eliminating all resource input into and leakage out of the system
Integrating noneconomic aspects into development	
System change/design and innovation at the core	
Multi−/interdisciplinary research field	Sustainability prioritizes the triple bottom line, and circular economy prioritizes the economic system
Cooperation of different stakeholders necessary	
Regulation and incentive as core implementation tools	Responsibilities are shared but not clearly defined in sustainability; when in circular economy, the responsibilities are of private business and regulators/policymakers
Technological solutions are important but often pose implementation problems	

Source: adapted from Geissdoerfer et al. (2017)

Looking at waste collection, the circular economy package elaborated by European Commission (2015) proposes the following vital elements (Wielenga 2016): a requirement for separate collection of biowaste; a ban on landfilling of separately collected waste fractions; and a sharp increase of the targets for recycling, making visible the role of separate waste collection to reach the such targets, as well as to reduce GHG emissions, although requiring the participation of citizens. The need to reinforce separate collection is to ensure high rates of recycling and high quality of materials, increasing the number of paths to make waste reintroduced again in the same economy that originated the product (and consequently, the waste itself), including domestic sector (European Commission 2015). If it is true that separate waste collection is needed to obtain the waste/resources to be reintroduced again on the economy, such is valid for non-durable products; however, the circular economy package omits the role of tack-back schemes to promote the refurbishment and durability of products, mainly for durable products (before becoming waste). This critic is also pointed out by Singh and Ordoñez (2016), which have analyzed practical examples of circular economy in Sweden, where the recirculation of resources to make different types of products only occurs and no recirculation of products, which is the goal of circular economy, where manufacturers must take back their products to secure their material sources.

Zaman and Lehmann (2011) proposed that, to make zero waste cities, the methods, tools, or strategies have to be affordable in the socioeconomic context, regulatory or manageable in the sociopolitical context, applicable in terms of policy and technological context, effective or efficient at economy and technology, and all be related to environmental sustainability. The proposed zero-waste city of the future by Zaman and Lahmann (2011) is like:

- Retrofitting existing communities, infrastructure, and building fabric at the same time as we develop new ones.
- Food production will be brought back into the city with urban farming, building efficiency will be improved, and public transport will have priority over private vehicles.
- Sustainable designs inspired by nature, where waste is a resource and organic waste is a fertilizer, new building materials constructed from recycled waste, and the potential for renewable energy fully unleashed, harnessing wind, geothermal, solar, and biomass resources to feed renewable energy into a smart grid.

8.3 Adaptive Management Strategies for Waste Collection Systems

The challenges identified so far will require that decision-makers and waste managers to look at the waste collection as the pivotal role to ensure an adequate and sustainable solid waste management. Shaping advanced waste collection system requires having in consideration the sustainable criteria as technical, economic, environmental, and social dimensions. Adaptive management in this context focuses on the concept that future influences/disorders to a collection system are predictable.

To identify the adaptive management strategies, models and tools at all stages of a solid waste collection system should be applied, being those models and tools introduced in the next chapters. Geographic information systems, route analysis and optimization, life cycle assessment, carbon footprint, indicators, behavior studies and awareness campaigns, decision trees, and multi-criteria decision-making are all effective options to analyze adaptive management strategies. Dedicated collection system study requires financing option analysis. Economic instruments such as pay-as-you-throw and deposit-refund schemes are policy tools that promote the better collection of recyclables, for example. Policy and regulatory measures, such as targets for source separation of recyclable fractions (packaging, waste of electric and electronic equipment, just to name a few), as well as regulation to promote the market for source-separated waste fractions (which is the case of the end-of-waste criteria for waste products) can bring the separate waste collection system into reality but also can be constraint by limitation of land use to receive specific waste treatment and disposal solutions. Minor environmental impacts resulting from the waste collection can result from the use of fewer polluter vehicles and routes optimization. Considering the behavior of users to make a successful waste collection system will ensure that the amount and quality of the collected waste are capable of supporting the collection system.

> **Box 8.1: Glass Recycling: The Case of Separate Collection that Potentiates Circular Economy and Reduces Global Warming (ACR+ and FEVE 2012)**
>
> Glass stands out as one of the best cases of closed loop production model at Europe due to the recycling rate reached – 67%. First, glass is a natural material being 100% recyclable and infinitely recyclable, with a well-established source separation collection system. The environmental benefits are related to the recycling process of glass, which avoids the extraction of new material and consumes less energy in producing new products, which releases less CO_2. To improve results, the quality of source-separated glass needs to be increased, to ensure bottle to bottle recycling, which is better to a downcycling.

Cities are now looking for adaptive measures on the challenges presented so far. The newest adaptive measure is related to the application of circular economy to cities, named "circular cities." Circular city concept is so recent that is still being in the discussion. Prendeville et al. (2017) tried to define circular cities as "is a city that practices circular economy principles to close resource loops, in partnership with the city's stakeholders (citizen, community, business and knowledge stakeholders), to realize its vision of a future-proof city" although they recognize that the concept is still under development.

Implementing the concept of the circular economy into a city may be hard, existing both bottom-up and top-down initiatives. Bottom-up initiatives are social movements, including entrepreneurial activities, and run by civil society, non-governmental organizations, communities, and business, characterized by being hard to be identified and includes company collaborations, supply chain efforts, product design, information, and communication technology (Krauz 2016; Pomponi and Moncaster 2017; Ghisellini et al. 2016; Lieder and Rashid 2016; Prendeville et al. 2017). Top-down initiatives applied to a city are driven by the municipality/local government such as strategy and policy decisions including public-private partnership projects concerned with developing and facilitating initiatives, including regulatory frameworks, awareness/information campaigns, collaboration platforms, business support schemes, procurement and infrastructure projects, and fiscal frameworks (Krauz 2016; 2017; Ghisellini et al. 2016; Lieder and Rashid 2016; EMF 2015).

Several cities are implementing circular economy (Table 8.2). One which has implemented a zero-waste approach inside of the circular economy was Amsterdam,

Table 8.2 Circular cities

Cities features	Circular economy initiatives
Amsterdam (Dutch capital, population = 800.000, average income/household = €31,400)	Published CE strategy, several real estate projects including CE plans, a large number of community-owned CE initiatives. Multiple knowledge development projects concerning CE (including the development of an independent institute for urban sustainability research)
Rotterdam (population = 600.000, average income/household = €31,600)	Published comprehensive sustainability strategy including CE- and bio-based economy plans. A large number of community-owned CE initiatives, mostly bio-based. Active involvement with Port of Rotterdam for CE. A serious commitment to developing CE further by commissioning celebrity economist to create CE future vision
Glasgow (population = 600.000, average disposable income/household = €39,400)	The Scottish government has shown significant commitment to developing CE through Zero Waste Scotland. City council published sustainability strategy. Chamber of Commerce commissioned circle economy (circular consultants) to perform extensive research on material flows and potential of CE
Haarlemmermeer (population = 144.000, average income/household = €39,400)	Published a "scenario study" exploring future challenges and broad sustainability agenda with accompanying project plans. Strong involvement with Schiphol Group (airport in the region) creating an interesting dynamic. Member of the EMF's CE100 group
The Hague (population = 520.000, average income/household = €32,600)	Chosen as a city that has not yet made progress or taken concrete steps toward a CE. Recently (2015) published their first sustainability agenda and are currently in the process of taking an inventory of smaller CE projects, initiatives, and enthusiasts within the organization
Barcelona (population = 1600.000, average income/household = €22,101)	Barcelona has taken the lead in developing a smart city, through a top-down and comprehensive master plan. It has also been a pioneer in exploring the concept of a Fab City, for local urban production systems (e.g., food, energy)

Source: Prendeville et al. (2017)

in Buiksloterham district. van der Leer (2016) defined seven interventions to reach zero-waste goal:

- Resource market: the place where all the household waste is collected, transporting the waste (or resources) to other facilities or recycling industries.
- Productive street: is a street where small-scale production facilities can be located, with mixed use.
- Biorefinery: is a place where digestion of organic waste, black water, and yard waste into biogas, heat, and fertilizer occurs.
- Helophyte filters: are vertical flowing reed fields treating gray water.
- Home composting: to be implemented in two ways – by a wormery or a compost bin.
- Smart collection points: two types of collection, one for daily waste flows (organic waste, glass, paper, cardboard, and other waste + metals, plastics + drinking cartons and sanitary waste) and another central collection for monthly waste (textiles and bulky waste), being a total of eight fractions.
- Separate sewage system: to be implemented in new housings, occurring separate collection and transport of wastewater to support the helophyte filters and biorefinery (Table 8.2).

8.4 Final Remarks

The way how society has encountered to deal with global challenges is not being capable of dealing with all the challenges at the same time. There is the need for cities to be sustainable, to work as eco-cities, to reduce GHG emissions, and to generate less waste, circularly, not allowing waste to get outside the frontiers of the city. However, those challenges require financial resources, acknowledgment, and desire of stakeholders to solve them.

There is the need to look at the lessons learned so far, namely, for the cases which have failed to avoid that the same mistakes occur again and again. Transforming high-consuming cities into "zero-waste cities" is quite challenging and a long-term process. At first, the lifestyle of inhabitants and consumer behaviors needs to be understood to act to change the way how resource consumption and waste occurs. At last, when concepts like circular cities, eco-cities, sustainable cities, and zero-waste and zero-carbon cities are implemented poorly, they can lose credibility, only remaining as buzzwords or greenwashing. The scientific community has the responsibility to help the rest of the society to clarify concepts, to elaborate guidelines, and to help to perform the concepts in the field.

References

Ackerman F (2000) Waste management and climate change. Local Environ 5:223–229

ACR+, FEVE (2012) Good practices in collection and closed-loop glass recycling in Europe. http://wwwacrplusorg/images/glass_recycling/Good_Practices_in_collection_and_closed-loop_glass_recycling_in_Europe_REPORT_-_ACR_FINAL_DOCpdf. Accessed 10 Dec 2018

Angus A, Casado MR, Fitzsimons D (2012) Exploring the usefulness of a simple linear regression model for understanding price movements of selected recycled materials in the UK. Resour Conserv Recycl 60:10–19

Bai XM, Chen J, Shi PJ (2012) Landscape urbanization ad economic growth in China: positive feedbacks and sustainability dilemma. Environ Sci Technol 46:132–139

Baldé CP, Wang F, Kuehr R, Huisman J (2015) The global e-waste monitor – 2014. United Nations University, IAS-SCYCLE, Bonn

Brennan G, Tennant M, Blomsma F (2015) Business and production solutions: closing the loop. In: Kopnina H, Shoreman-Ouiet E (eds) Sustainability: key issues. EarthScan, Routledge, pp 219–239

Bullis K (2009) A zero-emissions city in the desert – Oil-rich Abu Dhabi is building a green metropolis – should the rest of the world care? In: Technol Rev Available via MIT Technology Review http://www.dartmouth.edu/~cushman/courses/engs44-old/Zero-Emissions-City-2009.pdf. Accessed 10 Dec 2018

Chang G (2016) Environmentally friendly cities. In: Li J, Yang T (eds) China's eco-city construction, research series on the Chinese dream and China's development path. Springer, Berlin, pp 135–154

Cheng H, Hu Y (2010) Planning for sustainability in China's urban development: status and challenges for Dongtan eco-city project. J Environ Monit 12:119–126

Desrochers P (2001) Cities industrial symbiosis: some historical perspectives and policy implications. J Ind Ecol 5:29–44

Dongtan CH (2008) China's flagship eco-city. Archit Des 78:64–69

EMF (2015) Delivering the circular economy: a toolkit for policymakers. In EMF. Available via https://www.ellenmacarthurfoundation.org/assets/downloads/publications/EllenMacArthurFoundation_PolicymakerToolkit.pdf. Accessed 10 Oct 2017

Eryildiz S, Xhexhi K (2012) Eco cities under construction. Gazi Univ J Sci 25:257–261

European Commission (2015) Communication from the Commission to the European Parliament, the Council, the European Economic and Social Committee and the Committee of the Regions: Closing the loop – An EU action plan for the Circular Economy (COM/2015/0614 final). https://eur-lex.europa.eu/legal-content/EN/TXT/?uri=CELEX:52015DC0614. Accessed 10 Oct 2017

European Parliament, Council (2008) Directive 2008/98/EC of the European Parliament and of the Council of 19 November 2008 on waste and repealing certain Directives. Off J Eur Union L312:3–30

Eurostat (2017) Municipal waste statistics. http://appssoeurostatceuropaeu/nui/showdo?dataset=env_wasmun&lang=en. Accessed 18 Apr 2018

Gaubatz P (1999) China's urban transformation: patterns and processes of morphological change in Beijing, shanghai and Guanzhou. Urban Stud 36:1495–1521

Geissdoerfer M, Savaget P, Bocken NMP, Hultink EJ (2017) The circular economy – a new sustainability paradigm? J Clean Prod 143:757–768

Ghisellini P, Cialani C, Ulgiati S (2016) A review on circular economy: the expected transition to a balanced interplay of environmental and economic systems. J Clean Prod 114:11–32

Girardet H (1999) Creating sustainable cities, Schumacher briefings 2. Green Books, Dartington

Gregson N, Crang M, Fuller S, Holmes H (2015) Interrogating the circular economy: the moral economy of resource recovery in the EU. Econ Soc 44:218–243

Gupta S (2015) Decoupling: a step toward sustainable development with reference to OECD countries. Int J Sustain Dev World Ecol 22:510–519

Huisman J, Botezatu I, Herreras L, Liddane M, Hintsa J, Luda di Cortemiglia V, Leroy P, Vermeersch E, Mohanty S, van den Brink S, Ghenciu B, Dimitrova D, Nash E, Shryane T, Wieting M, Kehoe J, Baldé CP, Magalini F, Zanasi A, Ruini F, Bonzio A (2015) Countering WEEE illegal trade summary report. The CWIT Consortium, Lyon

ISRI (2015) The scrap recycling industry: commodities. http://www.isri.org/recycling-commodities/international-scrap-trade-database. Accessed 20 Oct 2017

ISWA (2012) Globalization and waste management – Phase 1: concepts and facts. ISWA

Jha AK, Sharma C, Singh N, Ramesh R, Purvaja R, Gupta PK (2008) Greenhouse gas emissions from municipal solid waste management in Indian mega-cities: a case study of Chennai landfill sites. Chemosphere 71:750–758

Joint Research Centre (JRC) (2015) Waste and recycling. http://susprocjrceceuropaeu/activities/waste/indexhtml. Accessed 18 Apr 2018

Krauz A (2016) Transition management in Montreuil: towards perspectives of hybridization between 'top-down' and 'bottom-up' transitions. In: Loorbach D, Wittmayer J, Shiroyama H, Fujino J, Mizuguchi S (eds) Governance of urban sustainability transitions. Springer, Tokyo, pp 137–154

Lieder M, Rashid A (2016) Towards circular economy implementation: a comprehensive review in context of manufacturing industry. J Clean Prod 115:36–51

Liu JY, Zhan JY, Deng XZ (2005) Spatiotemporal patterns and driving forces of urban land expansion in China during the economic reform era. Ambio 34:450–455

Liu ZF, He CY, Zhou YY, Wu JG (2014) How much of the world's land has been urbanized, really? A hierarchical framework for avoiding confusion. Landsc Ecol 29:763–771

METI (2004) Handbook on resource recycling legislation and 3R initiatives. Japanese Ministry of Economy, Trade and Industry, Tokyo

OECD (2002) Indicators to measure decoupling of environmental pressure from economic growth SG/SD(2002)1/FINAL. http://wwwoecdorg/officialdocuments/publicdisplaydocumentpdf/?cote=sg/sd(2002)1/final&doclanguage=en. Accessed 3 Jan 2014

Okuda I, Thomson VE (2007) Regionalization of municipal solid waste management in Japan: balancing the proximity principle with economic efficiency. Environ Manag 40:12–19

Pomponi F, Moncaster A (2017) Circular economy for the built environment: a research framework. J Clean Prod 143:710–718

Premalatha M, Tauseef SM, Abbasi T, Abbasi SA (2013) The promise and the performance of the world's first two zero carbon eco-cities. Renew Sust Energ Rev 25:660–669

Prendeville S, Cherim E, Bocken N (2017) Circular cities: mapping six cities in transition. Environ Innov Soc Transit 26:171–194

Register R (1987) Building cities for a healthy future. Berkeley North Atlantic Books, Berkeley

Roseland M (1997) Dimensions of the eco-city. Cities 14:197–202

Roseland M (2001) How green is the City? Sustainability assessment and the management of urban environments. Columbia University Press, New York

Singh J, Ordoñez I (2016) Resource recovery from post-consumer waste: important lessons for the upcoming circular economy. J Clean Prod 134:342–353

Su B, Heshmati A, Geng Y, Yu X (2013) A review of the circular economy in China: moving from rhetoric to implementation. J Clean Prod 42:215–227

Surugiu MR, Surugiu C (2015) International trade, globalization and economic interdependence between European countries: implications for business and marketing framework. Procedia Econ Financ 32:131–138

UNEP, ISWA (2015) Global waste management Outlook 2015. UNEP, Japan

United Nations (1992) United nations framework convention on climate change. https://unfccc.int/files/essential_background/background_publications_htmlpdf/application/pdf/conveng.pdf. Accessed 10 Apr 2018

United Nations (2015) World urbanization prospects – the 2014 revision. United Nations, New York

United Nations (2017) UN Comtrade Database. https://comtrade.un.org/. Accessed 7 Aug 2018

References

United Nations, European Commission, International Monetary Fund, Organisation for Economic Co-operation and Development, United Nations Conference on Trade and Development, World Trade Organization (2002) Manual on statistics of international trade in services. United Nations, New York

USAID (2013) Addressing climate change impact on infrastructure: preparing for change. http://wwwadaptationlearningnet/sites/default/files/resource-files/Addressing-Climate-Change-Impacts-on-Infrastructure-reportpdf. Accessed 12 Mar 2018

Vaiude N, Deshpande G, Deosthali AM (2017) Eco-City or environmentally sustainable villages. In: Seta F, Sen J, Biswas A, Khare A (eds) From poverty, inequality to smart city. Transactions in civil and environmental engineering. Springer, Singapore, pp 161–169

Van Berkel R, Fijita T, Hashimoto S, Geng Y (2009) Industrial and urban symbiosis in Japan: analysis of the eco-town program 1997–2006. J Environ Manag 90:1544–1556

Van Caneghem J, Block C, Van Hooste H, Vandecasteele C (2010) Eco-efficiency trends of the Flemish industry: decoupling of environmental impact from economic growth. J Clean Prod 18:1349–1357

Van der Leer J (2016) Zero waste Buiksloterham - an integrated approach to circular cities. Dissertation, Delft University of Technology

Van der Voet E, Van Oers L, Nikolic I (2004) Dematerialization: not just a matter of weight. J Ind Ecol 8:121–137

Verfaillie A, Bidwell R (2000) Measuring eco-efficiency: a guide to reporting company performance. World Business Council for Sustainable Development. Available via https://www.gdrc.org/sustbiz/measuring.pdf. Accessed 1 Aug 2014

Waste Atlas (2017) Municipal solid waste generation. http://wwwatlasd-wastecom/. Accessed 1 Aug 2014

Welfens P, Bleischwitz R, Geng Y (2017) Resource efficiency, circular economy and sustainability dynamics in China and OECD countries. Int Econ Econ Policy 14:377–382

Wielenga K (2016) Separate waste collection in the context of a circular economy in Europe. Report for FFact, Delft

World Bank (2018) Metadata glossary. https://data.worldbank.org/. Accessed 30 Apr 2018

Zaman AU, Lehmann S (2011) Urban growth and waste management optimization towards 'zero waste city'. City, Cult Soc 2:177–187

Zimmerman R, Faris C (2010) Chapter 4: infrastructure impacts and adaptation challenges. Ann New York Acad Sci 1196:63–85

Part II
Models and Tools for Waste Collection

Chapter 9
Design and Planning of Waste Collection System

Abstract The purpose of this chapter is to present the principal factors and objectives to consider when planning a collection system. The prediction and estimation of the amount of waste and the type of waste collection service that is intended to be provided, together with the help of geographic information systems (GIS) to locate containers and design routes, are tools to be used during the adequate design and planning of a waste collection system. Here a specific focus is on waste prediction models, due to its importance on planning, operating, and optimizing waste management system, as well as in the difficulty in predicting, directly, waste generation and its dependence on numerous factors, directly and indirectly, related with the consumption patterns, disposal habits, and urbanization.

Keywords Forecasting models · Time series · Waste generation estimations · Trucks · Containers · GIS · Routing

9.1 Waste Generation Estimation

Forecasting of generation of municipal solid waste (MSW) is the first and fundamental step in planning waste collection systems (Rimaityte et al. 2012). Forecast data is the baseline for the development of solid waste management systems and their continuous improvement and optimization toward sustainability (Beigl et al. 2008).

The majority of the national waste legislations require reliable information on waste quantity and composition, which is difficult to accomplish since waste generation cannot be measured directly (as gas consumption, for instance), to which adds the existence of several parallel channels in waste disposal systems (e.g., public curbside collection, private collectors, take back by retailers, among others) and the scarcity of the system to measure the waste arising on a single household (only areas where pay-as-you-throw systems have been installed). Thus, waste generation cannot be measured on a detailed basis, which would allow further evaluation of disposal habits, changes, and trends. This is one of the cases where modeling is of

---

© Springer International Publishing AG, part of Springer Nature 2019 141
A. Pires et al., *Sustainable Solid Waste Collection and Management*,
https://doi.org/10.1007/978-3-319-93200-2_9

particular significance. According to Armstrong (2001), the selection of the method to perform a waste generation forecast depends on some criteria:

- Amount and quality of available data.
- Type of the data.
- Interactions among waste generation parameters and socioeconomic indicators.
- Predictable changes in solid waste management systems.

Beigl et al. (2008) concluded after reviewing a considerable amount of published works that models applied were very different even when tackling the same kind of waste generation issue. These authors suggest three criteria to help to identify the adequate approaches: the regional scale (from household up to country perspective), the type of waste stream, and the availability of data. However, the first aspect that should be taken into account before selecting a method is the type of waste stream. In general, correlation and regression analyses (to test mainly the affluence-related impacts by analysis consumption-related variables and consumption and disposal related variables) and group comparisons (to test the effect of waste management activities on recycling quotas) have been the most used modeling methods, both to test the relationship between the level of influence and the generation of total MSW (or a material-related fraction) and to identify significant effects of waste management activities on recycling quotas. The application of time-series analyses and input-output analyses is advantageous for special information needs (e.g., assessment of seasonal effects for short-term forecasts) or for appropriate data availability.

The timescale is also an important aspect one cannot forget (Hyndman and Athanasopoulos 2014). Forecasting waste generation for 2 years is entirely different from forecasting the weekly generated amounts. Short-term forecasts (day, week) are needed concerning collection, personnel and truck utilization, transport to landfill, and final disposal (Navarro-Esbri et al. 2002). Long-term forecast supports strategic decision-making and should take into account macro aspects such as population behavior and environmental factors. In the middle, there is medium-term forecast supporting strategic decision-making (determine future resources needed, hire personnel, buying new equipment, among others).

In long-term forecast or when there is no adequate data available, the adequate methods to perform good forecast are *qualitative* approaches. These are not "simple guesses"; they are well-structured and validated methods. Nonetheless, they do not come free of bias and subjectivity since they depend profoundly on human judgment. Examples of such methods are the Delphi method, forecasting by analogy, and scenario-based forecasting. In this section the focus will be the quantitative methods; therefore for further detail refer to Hyndman and Athanasopoulos (2014).

Quantitative methods are the adequate methods to use when two criteria are met: the existence of useful quality data and when it is reasonable to assume that past trends/patterns will continue in the future. Most quantitative methods use time-series data (measurements taken at equally spaced movements, such as hours, days, months) or cross-sectional data (several measurements taken at a single period). The short-term prediction of future MSW generation rates can facilitate planning

9.1 Waste Generation Estimation

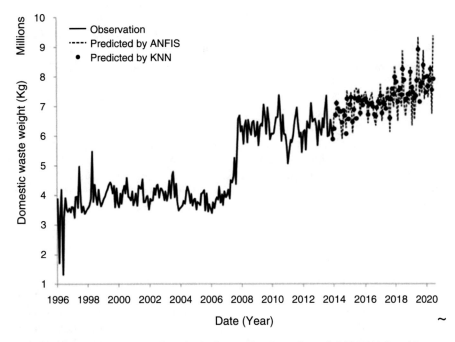

Fig. 9.1 Monthly solid waste generation in Logan City, Australia: real (1996–2014) and forecast (2015–2020). (Source: Abbasi and Hanandeh (2016))

concerning the collection, personnel and truck utilization, transport to landfill, and final disposal (Navarro-Esbri et al. 2002).

Time-series forecasting models are a useful tool to forecast MSW generation when historical data is available (Winston and Goldberg 2004). These methods are often called as extrapolation methods since they use previously observed data to infer about future data. More precisely, they assume that past patterns and trends in data will continue in the future. They do not take into account what are the causes of such data; they merely assume past patterns and trends will go on in the future. To understand the causes, one should choose correlation and regression methods since they make use of several different types of variables (e.g., population growth) to explain the behavior of one single variable (named as the dependent variable).

9.1.1 Time Series

Figure 9.1 shows the monthly MSW generated in Logan City, Australia, between 1996 and 2014. It also shows the monthly forecast for a 5-year period (2015–2020) predicted by two different methods (Abbasi and Hanandeh 2016). Notice how the forecasts follow the seasonal pattern seen in the historical data and replicate it in the next years.

The most common components present in a time series are trend, seasonality, and cyclicity. A trend exists when there is a long-term increase or decrease in the data. It does not have to be linear and may change directions (moving from increasing to decreasing trend). Seasonality appears when there are factors of fixed period that influence the series behavior (e.g., day of the week, or the month, or the quarter of the year). Seasonality has always a known period. Lastly, a cyclic pattern exists when data exhibit rises and falls that are not of fixed period. The duration of these fluctuations is usually of at least 2 years. In general, the average length of cycles is longer than the length of a seasonal pattern, and the magnitude of cycles tends to be more variable than the magnitude of seasonal patterns.

These components should be modeled separately to better capture the influence of each of them. The most common decomposition models are the additive and the multiplicative models. The following equations are possible representations of each model, respectively.

$$x_t = S_t + T_t + \varepsilon_t \quad \text{and} \quad x_t = S_t \cdot T_t \cdot \varepsilon_t$$

where x_t is the data at period t, S_t is the seasonal term at period t, T_t is the trend-cycle term at period t, and ε_t is the error term at period t. When the seasonal fluctuations magnitude or the variation around the trend-cycle is independent from the level of the time series, the additive model is the most suitable one, while the multiplicative model is more suitable when those phenomena (variation in the seasonal pattern and variation around the trend cycle) appear to be proportional to the level of the time series.

Autoregressive integrated moving average models, best known as ARIMA models, provide another approach to time-series forecasting. Exponential smoothing and ARIMA models are the two most widely used approaches to time-series forecasting and provide complementary approaches to the problem. While the former describes the trend and seasonality observed in the data, the latter seeks to describe the autocorrelations among data values.

In Appendix A.1 to A.6, the most common approaches to model time-series components are presented with detail: naïve model, moving average models, exponential smoothing, Holt's model, and Holt-Winters model, respectively. A brief introduction to ARIMA models is also presented.

9.1.2 Forecast Accuracy

The accuracy of forecasts, i.e., how good is a forecast model, should be determined by studying how the model performs on new data that were not used to develop the model parameters. A common practice is to split the available data into two subsets: training and test data. The training data is used to estimate all parameters of the model (or models since one should study some different options), and the test data

9.1 Waste Generation Estimation

should be about 20% of the total sample and is used to assess its "quality" (Hyndman and Athanasopoulos 2014). Since test data has not been used in parameters estimation, it should provide an accurate indication of how well the model is likely to forecast on new data. The test set should ideally be at least as large as the maximum forecast horizon required.

For period t, the forecast error e_t is given by

$$e_t = x_t - \bar{x}_t,$$

where $\bar{x}_t$ is the forecasted value for period t. There are several methods to measure forecast accuracy. When these measures are based on e_t, they are scale dependent. The most commonly used scale-dependent measures are based on absolute errors or on squared errors: mean absolute error, geometric mean absolute error, and mean square error (see Appendix B.1 to B.3 for the respective mathematical expressions). The use of absolute or squared values stops negative and positive errors from balancing each other. A different kind of accuracy measures involves the percentage error p_t defined by

$$p_t = \frac{e_t}{x_t} \cdot 100.$$

These errors have the advantage of not being scale dependent, so they are adequate to compare forecast accuracy between different time series. The most common measure is the mean absolute percentage error, MAPE (in Appendix B.4). Since the percentage error p_t is divided by the observed value x_t, it may produce infinite or undefined values when the observed values are zero or close to zero. When the observed values are very small, MAPE yields extremely large percentage errors.

Lastly, the relative errors are also scaled independently and may be used to compare different timescales. These accuracy measures involve dividing each error by the error obtained using some benchmark method of forecasting. Let r_t be the relative error defined by

$$r_t = \frac{e_t}{e'_t}$$

where e'_t is the forecast error obtained from the benchmark method. The most used method as a benchmark is the naïve method (see Appendix A.1). Two measurements can be defined with r_t: geometric mean relative absolute error (in Appendix B.5). The former is the median of the absolute value of r_t. The latter is computed as the geometric mean absolute error.

9.1.3 Linear and Multiple Linear Regression Models

Regression models are used when the variable to be forecast (dependent variable) reveals a relationship with one or more other response variables (independent variables). The purpose of such models is to describe the linear relationship between the dependent and independent variables and use it to forecast values of the dependent variable that have not been observed. Under this model, any change in response will affect the output of the system expectedly.

The most used way to evaluate how well a linear regression model fits the data is using the coefficient of determination, R^2. Using forecast variable $\hat{y}$ and the observed value y, this coefficient measures the proportion of the variation in the forecast variable that is accounted for (or explained) by the regression model. The given expression assures the $R^2 \in [0, 1]$. If the forecasts are close to the observed values, then R^2 is close to 1. On the contrary, if forecasts and observed values differ considerably, then R^2 is close to 0. Although commonly used, there are no rules for what is a good value for R^2 when applied to forecasting (Hyndman and Athanasopoulos 2014).

Assuming that the regression errors are normally distributed, an approximate $\alpha\%$ forecast interval (a prediction interval) associated with this forecast can be computed, providing a more in-depth knowledge regarding possible variations in the forecast value.

9.1.4 Advanced Forecast Models

Artificial intelligent approaches have recently been used to forecast waste generation: artificial neural networks (ANNs), support vector machine (SVM), k-nearest neighbors algorithms (kNN), and other machine learning techniques. ANNs are algorithms inspired by the way the brain system works and have the "ability" to improve their performance; often one says artificial neural networks are algorithms with the capacity to learn. This capacity is one of the most beneficial and significant features of ANNs when addressing forecasting issues (Abbasi and Hanandeh 2016). ANNs can construct a complex nonlinear system through a set of input/output examples, which makes them suitable candidates for forecasting waste generation. However, ANNs' performances may suffer given their movement to overfitting training. SVM algorithms aim at separating data so that they are as far apart as possible. These algorithms find a hyperplane by selecting the most appropriate points in a training subset (the supporting vectors), that split data into categories. They are less prone to overfitting than other machine learning techniques (Han et al. 2011). As a forecasting method, k-nearest neighbors algorithm seeks to identify the past sequence, in the time series, that is most similar to the one to be forecast (Arroyo and Maté 2009).

The SVM was used to forecast weekly MSW generation in Tehran City, Iran by Abbasi et al. (2013). The authors concluded that SVM could predict MSW generation in a short-term scale with reasonable accuracy. When applying k-nearest neighbors algorithm to univariate time series, one aims at finding if consistent data-generating processes produce observations of repeated patterns of behavior. If that is the case and a previous pattern can be identified as analogous to the current behavior of the time series, the subsequent behavior of the previous pattern can provide valuable information to predict the behavior in the immediate future. Abbasi and Hanandeh (2016) compared ANNs, SVM, and kNN and a fourth artificial intelligence approach (Adaptive Neuro-fuzzy Inference System) to access their accuracy of predicting waste generation rates. Their results showed that artificial intelligence models provide promising tools that may allow decision-makers to successfully forecast future trends in MSW generation for planning and design of MSW management purposes. More precisely, it was observed that the Adaptive Neuro-fuzzy Inference System accurately estimated waste generation at peaks, while k-nearest neighbors algorithm was successful in estimating monthly average values. The machine learning application to MSW prediction was reported by Johnson et al. (2017). These authors linked historical MSW collection data supplied by the New York City Department of Sanitation (DSNY) with other datasets related to New York City (NYC) to forecast MSW generation throughout the city. Spatiotemporal tonnage data from the DSNY was combined with external datasets (as the Longitudinal Employer-Household Dynamics data, the American Community Survey, and historical weather data) to develop a Gradient Boosting Regression model. This machine learning-based model has been firstly introduced by Friedman (2001), and according to Johnson et al. (2017), it is an excellent choice for this kind of studies because of "their interpretability, ability to handle non-linear and complex relationships between data and have demonstrated higher prediction accuracy compared to traditional time-series models such as ARIMA." Historical data from 2005 to 2011 was used to train and validate the model performance. With this model, they were able to accurately forecast weekly MSW generation tonnages for each of the 232 geographic sections in NYC across three waste streams of refuse, paper, and metal/glass/plastic. "Importantly, the model identifies regularity of urban waste generation and is also able to capture very short timescale fluctuations associated to holidays, special events, seasonal variations, and weather related events. This research shows New York City's waste generation trends and the importance of comprehensive data collection (especially weather patterns) in order to accurately predict waste generation." (Johnson et al. 2017).

9.1.5 Case Study: Using Time-Series Models to Estimate MSW in Kaunas, Lithuania

Rimaitytė et al. (2012) modeled MSW generated in some neighborhoods in Kaunas, Lithuania, to access adequate forecasting methods for this urban area in a fast developing economy. Kaunas has 353,000 inhabitants and is the second largest city in Lithuania. It occupies an area of 157km^2. In 2008, about 460 kg per capita and year of MSW were generated which amounts to 165,500 metric tons. Data concerning waste generated in households (as glass, metal, plastic, paper, and biowaste) was collected, and the forecast was performed for the period between 2008 and 2017.

Methods of descriptive statistics were applied with the aim of describing the variation observed in MSW generation rates. Authors assumed the waste was collected at precise points in time, and the failure to collect waste during a single week did not influence the amount of waste collected during the next week. Given the character of the historical data available for the analysis and the amount and its quality, time-series models were the appropriate techniques to be applied. Data presented trend since a substantial increase in MSW generation occurred in a selected time horizon, and a cyclic variation of MSW generation was visible throughout different months.

Several types of time-series analysis were applied. Seasonal exponential smoothing (SES) and ARIMA models were selected as best in predicting the data. The former showed an increasing trend, caused mainly by the MSW growth from 2004 to 2006, while the latter underestimated this rising trend in the same period. Since these two models perform differently in the modeling of the data randomness (white noise), a weighted linear combination of models was tested. The estimate $\widehat{Y}_t$ was defined as the combination of both models:

$$\widehat{Y}_t = \alpha \widehat{Y}_t^{\text{ARIMA}} + (1 - \alpha) \widehat{Y}_t^{\text{SES}}$$

where $\widehat{Y}_t$ is the forecast for time t and $\widehat{Y}_t^{\text{ARIMA}}$ and $\widehat{Y}_t^{\text{SES}}$ are the estimates of the ARIMA and SES models also at the time t, respectively; the weight coefficient α was chosen to best fit the observed data. The model was validated with the data from 2008 to 2009.

9.2 Waste Collection System Planning and Selection

The best operating solid waste systems engage all the stakeholders in planning, implementing, and monitoring changes to the system (Wilson and Scheinberg 2010). However, only with acknowledgment of the options involved in the different collection systems and optimized costs can be expected to see appropriate solid waste management techniques applied in all countries of the world (Ross 2010).

9.2.1 Factors to Consider When Planning a Collection System

According to Bilitewski et al. (1994), the choice of a waste collection system depends on waste composition, existing waste collection system, existing waste treatment, recovery and disposal, the availability of citizens to pay for the waste collection and to participate on the collection system, and awareness of citizens. In a more detailed view, Lechner (2004) defined the distance to collection centers, simplicity, easiness of utilization by citizens, the prevention of malodors and parasites, and the design of devices as prerequisites to consider when planning a waste collection system. Choosing a waste collection system has to consider the territorial distribution and urban constraints, population density, pedestrian areas, street width, and the existence of residential areas, mostly (Satué 2000). All these features are specific to each city, leading to different waste collection and management systems (Scharff and Vogel 1994). Legal aspects are also determinant for the operation of the waste collection system, influencing the planning and design (Tchobanoglous et al. 1993). More generically, the strategy for a waste collection system must include the following goals (Pferdehirt et al. 1993; Tchobanoglous et al. 1993):

- The system must supply adequate levels of services, established to meet political regulations and sanitary and environmental requirements, in such way that could be safe for human health and the environment, trustable, and efficient.
- The system must try to reach the lowest cost possible.
- The system must develop local partnerships between public and private sectors.
- The system must be flexible to allow to answer to the changes occurring in the city and on waste generation.
- The system must support policies promoting the reduction and prevention of waste.
- Recycling targets to accomplish.
- Financial support to acquire innovative waste collection technology, to ensure best available techniques implemented on the field.
- Prescribe patterns established for service, performance, and citizens' satisfaction.

Most of the time, the goals of the waste collection system conflict, for example, the need to collect recyclables of high quality and collection efficiency. Recycling needs clean materials without contaminants, forcing for a separate collection, which increases collection costs, in opposition to the collection of mixed waste. Incentives to help financing separate waste collection of recyclables, together with increasing waste tariffs, can lead to better management of those different goals (Tchobanoglous et al. 1993):

(a) Waste collection has to be ensured by public or private entities, being private entities helpful financially to solve issues related to the improvements in the waste management system.

(b) Management concerns in the mechanization of collection relate primarily to the adaptability of the new containers to the needs of the user and the preparation of workers for the new system; these concerns must first be analyzed in tests, selecting a given pilot circuit and evaluating variables such as the times of loading, capacities, costs, and adhesion of the users.

(c) In the waste sector, collective labor contracts and trade union organizations are common: Mechanization and modernization of collection systems lead to the use of less labor, and communication with trade unions is therefore important to include them in the assessment and testing of collection systems, training of employees, and adjustment of salaries in order to recognize the increase in the skills and efficiency of the workers who received training and the reduction of the workforce by redeployment, rather than dismissals.

(d) Purchasing the new waste collection system to be implemented requires the selection of a financing method that is adapted to the financial availability of users (considering that the waste management system must be supported by users); many projects to modernize collection systems have been made unfeasible by the lack of funding, which is usually obtained through the municipal budget or loans, with leasing being an alternative; the purchase of new equipment in phases and the use of old equipment until it is amortized or through the simultaneous payment of new and old equipment must also be considered.

The selection of a waste collection system must be made in an integrative way for all waste streams to be collected, looking for synergies, being relevant for entities with low dimension (Satué 2000). However, for the chosen system, the goals are always the safe and efficient removal of waste, ensuring the health patterns and public/private needs, not forgetting the waste collection frequency and containers location, because they will influence how equipment will be used, the efficiency of the collection work, and customers' satisfaction (Bilitewkshi et al. 2010; Tchobanoglous et al. 1993).

Concerning devices, the set container-vehicle choice is one of the decisions of technicians of the waste department must deal with, being the correct selection one of the resources to increase separation at source (Tanskanen and Melanen 1999). The factors that must be considered in the selection of the set container-vehicle are addressed in the next sections.

9.2.2 Factors to Consider When Selecting WCS Devices

Vehicles

Concerning the technology and devices used, an important goal on the selection of vehicles is the amount of waste that can be collected respecting maximum legal weights (O'Leary and Walsh 1995). Mechanized vehicles are chosen because they reduce collection time, but the opinion of vehicle drivers and waste workers must be consulted, since they are familiarized with the device (O'Leary and Walsh 1995).

Generically, factors affecting the evaluation of a waste vehicle are the collection method and elevation method (elevator lateral/frontal/background), vehicle capacity, the velocity of transportation until discharge point, and costs of operation and costs capital. Pferdehirt (1994) and Urban Upgrading (2001) have identified the following factors to consider when selecting waste collection vehicles:

- Loading collection (side, back, or front) and loading mechanisms, considering the manual work of the team.
- Truck body capacity: to select the optimum capacity for a given community, the trade-off between labor and equipment costs should be determined.
- Chassis selection: although the chassis is similar to all collection vehicles, they must be large enough to support the truck body with waste.
- Purchase, operation, and maintenance costs.
- Spare parts obtaining delays.
- Loading height: the lower the loading height, the easier can waste be loaded into the vehicle, saving time, strain, and injuries in the work team.
- Loading and unloading mechanisms: they should be considered to minimize labor costs over capital costs and should be easy for workers.
- Truck turning radius: to allow the truck to turn over.
- Watertightness: to ensure that liquids from waste do not run off from the vehicle.
- Safety and comfort: vehicles should be designed with ergonomics principles.
- Speed: the vehicles should have an appropriate performance at a different range of speed.
- Adaptability to other uses: other uses may be given to collection vehicles, like maintenance of gardens, for example.

Containers

One of the factors influencing the MSW composition reported by Dahlén and Lagerkvist (2010) is the container type and bag used, together with the functionality and attractiveness of collection points. Vijay et al. (2008) conducted studies related with the location of collection points and their storage capacity, and Kaliampakos and Benardos (2013) highlighted the use of underground space for the development of infrastructure capable of bringing the solution to the limitations of existing MSW systems. From a technical point of view, the most relevant criteria to select containers are (O'Leary and Walsh 1995; WRAP 2013):

- Site performance: to ensure that containers are functional for the volume of materials that will be collected and ease of emptying.
- Space available on site: to make containers accessible, responding to servicing requirements,
- Containers that could reduce site footprint.
- Ease of relocating containers in other sites, i.e., modular containers, compatible with vehicles to be transported to other places and to be discharged by a large number of mechanisms that exist in collection vehicles.
- Accessibility for all site users, including disabled users.
- Select containers that could be used for all materials.
- Consider the aesthetics of the site and for the street when selecting containers.

- Safety and security of containers regarding vandalism and damage, by selecting containers with appropriate-sized apertures to reduce contamination and littering and use of locks if needed to be safe for users and collection crews.
- Economically affordable includes capital costs but also maintenance and repair costs. To reduce capital costs, options are available like buying in bulk and acquiring containers with a high level of adaptation to avoid acquiring new containers. Maintenance and repair costs relate to container cleaning; repair or replacement arising from damage from arson or vandalism, or just wear, where containers which are resistant and durable are preferred.The choice of containers is directly related to the fleet required, already highlighted by ISWA (2007), Rhyner et al. (2017), and WRAP (2013). The decision to install a waste collection system requires to have the suitable collection vehicles to collect the proposed containers, all being planned in conformity (containers and vehicles together) (Pieber 2004; WRAP 2013). The financing limitations may limit the choice of containers, looking firstly for the existing vehicles, and only then select the adequate container, considering the amounts of waste generated and available space at a street and selected places to locate containers (Rhyner et al. 2017). If a municipality has different contractors for different waste materials, probably those contractors need to be involved in any decisions about containers (WRAP 2013).

9.3 The Role of GIS in Waste Collection Planning

GIS is one of the most high-level spatial technologies. Having a computer-based information system, it can assemble, store, manage, integrate, analyze, and display spatial data known as geographically referenced data. Commonly, data is organized into thematic layers and forms digital maps. This is one of the most interesting features of GIS systems. Analyzing data by visualizing helps in identifying trends, patterns, and interactions that might not be perceived if data is displayed in table or written form Vijay et al. (2008).

The evolution of the quality and accessibility of geographic and population database (e.g., distances, coverage, or the number of inhabitants to serve) allowed collection sizing and planning more accurately. Waste collection routing software use GIS in scheduling and optimizing MSW collection routes, based on a container's location in a coordinate system and combining with street maps, having the shortest route as criterion, but the inputs required by this software still rely heavily on statistics and empirical data. Significant assumptions are usually made because of the lack of input data.

There are key database that are fundamental to feed the optimization MSW collection routes software: the net capacity of the container and vehicle, specific weight for recyclable material in the different types of container sizes and shapes, and emptying time for the different types of coupling system with the vehicle or pickup method for unloading the container.

The most widespread use of GIS on MSW lies in the areas of site selection for landfill, trash bin, and transfer stations (Vijay et al. 2008; Tralhão et al. 2010) and routing and scheduling optimization based on historical or predicted data (Tavares et al. 2009). Other applications of GIS in MSW can be found when addressing local management planning (Hrebicek and Soukopova 2010) on waste generation estimation through the use of socioeconomic and local demographic data (Karadimas and Loumos 2008) and integrated MSW establishment (Tao 2010).

9.3.1 Routes Definition

Route definition is one of the most challenging issues faced by a waste collection manager. Not only one has to assure that all waste is collected correctly, but one has also to set a system that is cost-efficient. Resources, both human and machinery, are limited and should be used in the most efficient way. However, not only cost should be the goal when designing collection routes. Environmental and social aspects should also be taken into consideration.

To define adequate collection routes, data, goals, and constraints have to be adequately defined. In the following paragraphs, three questions will be addressed when designing waste collection routes: What goals should one seek to optimize? What are the constraints that describe the system? And, what data needs to be taken into account?

What Goals Should One Seek to Optimize?
The goals are organized according to the sustainability pillar they relate to (economic, environmental, and social).

- Economic perspective.

  - The minimization of distance, costs, and time are the most frequent goals when planning collection routes. However, the minimum distance is not always a guarantee for minimum costs. For instance, collection during the night will be more efficient since less traffic exists, but it represents higher labor costs (Viotti et al. 2003). Therefore, a close analysis should be made to investigate the trade-offs between collection costs during the day and during the night.
  - The minimization of the number of vehicles is one objective that should be taken into account when no fleet of collection vehicles is available. By focusing on this objective, investments can be reduced. It is not sure, however, that costs are minimized this way (Beliën et al. 2014). Notice that the lack of vehicle capacity may lead to overtime work by the employees. A trade-off arises between the number of vehicles and overtime costs.
  - The maximization of route compactness: route has the best compactness if it has the smallest number of crossovers among the routes (Kim et al. 2006). Not only compact routes tend to be cheaper, but they are also more transparent for the collection route planners and drivers. One measure to assess route compactness could be the number of crossovers between different routes (Sniezek and Bodin 2006).

- The best location for collection bins and disposal containers: this objective is decomposed into two conflicting objectives: (1) minimum desired distance from a dwelling to its nearest bin location and (2) maximum desired walking distance to a bin location.

- Environmental perspective

  Although recycling contributes positively to the environment, collecting waste activity is mainly a transportation activity. Therefore, the use of vehicles generates greenhouse gas emissions, resource consumption, land use, acidification, toxic effects on ecosystems and humans, noise, and other negative impacts on the environment.

  - Minimization of CO_2 emissions is the most common goal used when minimizing environmental impacts.
  - Minimization of the energy requirements: when a vehicle travels from stop point a to stop point b, the greenhouse emissions depend on the fuel consumption, which in turn, is directly connected with the energy requirements; these energy requirements are established through fuel consumption that is a function of several aspects such as vehicle load (curb weight plus load), speed, road slope, engine features, vehicle frontal surface area, coefficients of rolling resistance and drag, and air density. So this objective is a more comprehensive way to model environmental impacts than CO_2 emissions (Ramos et al. 2014a).

Bektas and Laporte (2011) performed some computational experiments to access the trade-offs between distance, load, emissions, and costs in vehicle routing. Authors concluded that distance minimization does not necessarily lead to the minimization of fuel cost or driving cost. This work also suggests that in contrast, minimizing cost leads to solutions where more energy is consumed (hence yielding an increased amount of fuel consumption and emissions) to bring down driving costs. In a cost-minimized solution, the savings in driving costs can be up to 20%, which translates in a 5–8% reduction in the total cost (depending on the engine efficiency, and on time window restrictions) when compared to solutions provided by other objective functions.

- Social perspective:

  This is still the understudied perspective, and therefore only a few metrics have been proposed to address this issue.

  - Balance work among workers or equity (Ramos et al. 2014b).
  - Safety: these concerns are related to potential accidents in the workplace due to loading, unloading, or handling activities (Faulin et al. 2012).

Frequently, environmental and social issues are taken into consideration as constraints in the system. They are not viewed as goals but as limitations within the collection systems (e.g., labor hours, maximum route length, minimum collection fractions imposed by law).

9.3 The Role of GIS in Waste Collection Planning

What Are the Constraints That Describe the System?

- Capacity constraint: the capacity of a vehicle is limited; quite often, the amounts to be collected in a route do not exceed the vehicle capacity; then this capacity constraint is no longer and issue.
- Labor constraints: the duration of the shift of each crew has to be taken into account; these constraints can influence the problem in another way: for instance, it is possible that there are enough vehicles left but that all drivers are already assigned to a vehicle; in this case, the labor constraints rule out the override capacity constraints; mandatory lunch breaks is another common constraint.
- Demand constraints: it is not acceptable that refuse bags are left on the street because the waste management company fails to service all its customers; therefore, one has to assure at least one collection vehicle passes by every customer or street that has waste to be collected.
- Environmental constraints such as noise control, traffic congestion, and hazardous materials; noise control is designed primarily for the busy city centers where the amount of noise for the surrounding neighborhoods can be alarming; traffic congestion constraints reduce the burden of traffic in rush hours; hence, preference may be given to serving the busy roads of the network in periods of low traffic density.
- Political constraints: certain recycling percentages are imposed by law; therefore, when designing, for instance, the minimum cost routes, one must assure that these recycling percentages are met.
- Time windows imposed by law, for picking up certain types of waste; the waste bins and waste disposal site have given time windows in which they must be visited.
- Forbidden turns: addresses one-way streets or streets that are too small to be entered by the vehicle; they can also be used to consider traffic lights that if avoided may save time.

What Data Should Be Stored?
According to Nuortio et al. (2006), at least the following data should be in a database or in Excel worksheet. This data is stored once and does not need to be regularly updated.

- Bins: for each bin its type, location, associated stop point of the vehicle, waste type, mass and volume of the waste, service time, and the required time interval between visits.
- Stop points: a stopping point is a location where the vehicle is parked for loading or unloading; frequently more than one waste bin is collected in the same stop point; for each stop point, the precise location should be registered.
- Distance and time matrices: distance and drive time for paths connecting the stop points.
- Vehicles: for each vehicle, the volume and weight capacities for different waste types.
- Facilities: the depot and waste disposal sites location, facility type (depot or waste disposal site), service time, and operating time windows.

- Human resources: allowed maximum daily and weekly working hours and working days. Ideally, to efficiently plan collection routes, data concerning the amounts collected in each bin should also be stored, together with the date, time, and the visiting vehicle. This type has to be stored daily or whenever a collection route is performed. Although a very demanding task, with the new technologies, this data gathering is a much easier task. For instance, RFID[1] identify and tracked different waste streams collected in different bins, and image sensing can differentiate different waste types. Linking together RFID and real-time location sensors (as GPS[2], Wi-Fi, and other sensor nodes), the waste flow can be accurately and efficiently controlled both geographically and chronologically (Lu et al. 2015).

9.3.2 Case Study: Minimizing Operational Costs and Pollutant Emission in Collection Routes Using GIS

Zsigraiova et al. (2013) propose a "methodology for the reduction of the operation costs and pollutant emissions involved in the waste collection and transportation. It combines vehicle route optimization with that of waste collection scheduling. The latter uses historical data of the filling rate of each container individually to establish the daily circuits of collection points to be visited, which is more realistic than the usual assumption of a single average fill-up rate common to all the system containers. Moreover, this allows for the ahead planning of the collection scheduling, which permits a better system management. The optimization process of the routes to be travelled makes recourse to GIS and uses interchangeably two optimization criteria: total spent time and travelled distance. Furthermore, rather than using average values, the relevant parameters influencing fuel consumption and pollutant emissions, such as vehicle speed in different roads and loading weight, are taken into consideration. The established methodology is applied to the glass-waste collection and transportation system of Amarsul S.A., in Barreiro, Portugal. Moreover, to isolate the influence of the dynamic load on fuel consumption and pollutant emissions a sensitivity analysis of the vehicle loading process is performed. For that, two hypothetical scenarios are tested: one with the collected volume increasing exponentially along the collection path; the other assuming that the collected volume decreases exponentially along the same path. The results evidence unquestionable beneficial impacts of the optimization on both the operation costs (labor and vehicles maintenance and fuel consumption) and pollutant emissions, regardless the optimization criterion used. Nonetheless, such impact is particularly relevant when optimizing for time yielding substantial improvements to the existing system: potential reductions of 62% for the total spent time, 43% for the fuel consumption and 40% for

---

[1]RFID: Radio-frequency identification
[2]GPS: Global positioning system

the emitted pollutants. This results in total cost savings of 57%, labor being the greatest contributor, representing over €11,000 per year for the two vehicles collecting glass-waste. Moreover, it is shown herein that the dynamic loading process of the collection vehicle impacts on both the fuel consumption and on pollutant emissions."

9.4 Conclusion

The selection of a well-performing waste and recyclable material collection system is the basis for a circular economy, where high-quality materials are separately collected from nonrecyclable materials. When designing and planning a waste collection system, aspects like the type of container, the type of waste stream to be collected, and the frequency need to be combined to decide the waste collection system, respecting the characteristics and preferences of each territory. Selecting the best collection system for each area of municipality requires considering the climate, type of urbanization (buildings, the density of population), demographics, and infrastructures, just to name a few. All stakeholders should be engaged in the waste collection design and planning process.

According to IMPACTPapeRec (2016), a European project, to improve the source separation of paper and cardboard, design and planning of a waste collection system should start by:

- Define the baseline: through collecting reliable data on existing waste and recyclable situation, to provide a realistic and quantitative basis for the development of the plan and to prioritize requirements and needs. Here the waste estimation tools are useful to understand the actual situation.
- Identify the roles and responsibilities of key stakeholders: to understand who is responsible for the collection and treatment of waste according to the daily generation and the existence of private contracts on waste collection.
- Identify the strong and weak points of the current SWM system, regarding the lack of equipment or planning capacity, factors influencing waste generation increase or decrease, external problems of uncontrolled urbanization, population, explosions are a few to be identified.
- Prepare the appropriate SWM action plans: to be the core of the planning procedures, which will include regulation, market-based, information, and voluntary instruments.
- Provide guidelines on how to pass from planning to implementation. To implement the waste collection system for the several waste streams (from residual waste to recyclable waste materials), it is required to:

  - Define properly the specific characteristics of the waste streams to be collected.
  - Define the system size: make an estimation of the amount of waste and resources generated.

- Select the adequate and most promising collection system for the territory.
- Conduct an information campaign addressed to citizens.
- Define a monitoring performance plan of waste collection system and implement it.

The basis to define those aspects is the waste generated forecast since it will dictate the relationship of the dimension of container versus the frequency (the bigger the container, the lower the frequency) and consequently, the collection system. Choosing vehicles and containers requires a profound knowledge on territorial features, and GIS can help to deeper such knowledge.

Appendix A: Forecasting Methods

Let $x_1, x_2, \ldots, x_t, \ldots$ the observed values of the times series, where x_t is the value observed in period t.

A.1: Naïve Forecast Model

This is the simplest forecasting model. It assumes the value for the next period will equal the one last observed. Let $f_{t, 1}$ be the forecast for period $t + 1$ after observing x_t, then

$$f_{t,1} = x_t.$$

A.2: Moving-Average Models

Among the simpler and commonly used methods are the moving-average methods. These forecast period t as the average of the last N observed values (with N a given parameter). Let $f_{t, 1}$ be the forecast for period $t + 1$ after observing x_t.

$$f_{t,1} = \frac{1}{N} \sum_{k=t-N}^{t} x_k$$

The choice of N depends on the deviation of the forecast regarding the observed value (the forecast error). For period t, the forecast error e_t is given by

Appendix A: Forecasting Methods 159

$$e_t = x_t - f_{t-1,1}.$$

There are several ways to model the forecast accuracy (see Appendix B). To find the adequate value for N, one has to choose one of such measures and determine the value that minimizes the accuracy measure.

These methods are adequate for time-series data that fluctuate around a base value b:

$$x_t = b + e_t.$$

A.3: Exponential Smoothing Model

This model is also adequate for time series that may be written as

$$x_t = b + e_t.$$

Again consider $f_{t,1}$ is the forecast for period $t + 1$ after observing x_t. The simple exponential smoothing method "says" the next forecast ($f_{t+1,1}$) is a weighted average between the observed value x_t and the forecast at period t:

$$f_{t+1,1} = \alpha\, x_t + (1 - \alpha) f_{t,1}$$

where $\alpha \in [0, 1]$ is the smooth constant.

Let $f_{t,k}$ be the forecast for period $t + k$ at the end of period t, then

$$f_{t,k} = f_{t,1}.$$

A.4: Holt's Model

The Holt's method divides the time-series data into two components: the level, L_t, and the trend, T_t. These two components can be calculated by the expressions below:

$$L_t = \alpha x_t + (1 - \alpha)(L_{t-1} + T_{t-1})$$
$$T_t = \beta(L_t - L_{t-1}) + (1 - \beta)T_{t-1}$$

where L_t denotes an estimate of the level of the series at period t, T_t denotes an estimate of the trend of the series at period t, and α and β are the smoothing parameters for the level and the trend, respectively, $0 \le \alpha, \beta \le 1$.

Let $f_{t,k}$ be the forecast for period $t + k$ at the end of period t, then

$$f_{t,k} = L_t + k \cdot T_t.$$

A.5: Holt-Winters Method

The Holt-Winters (seasonal) model divides the time-series data into three components: the level, L_t; the trend, T_t; and the seasonal component S_t. These three components are given by:

$$L_t = \alpha\frac{x_t}{S_{t-c}} + (1 - \alpha)(L_{t-1} + T_{t-1})$$
$$T_t = \beta(L_t - L_{t-1}) + (1 - \beta)T_{t-1}$$
$$S_t = \gamma\frac{x_t}{L_t} + (1 - \gamma)S_{t-c}$$

where L_t denotes an estimate of the level of the series at period t, T_t denotes an estimate of the trend of the series at period t, and S_t denotes the seasonal component at period t; c is the frequency/pattern of the seasonality (i.e., for quarterly pattern $c = 4$; for a yearly pattern $c = 12$). Lastly, α, β, and γ are the smoothing parameters for the level, the trend, and the seasonality, respectively, $0 \leq \alpha, \beta, \gamma \leq 1$.

Let $f_{t,\,k}$ be the forecast for period $t + k$ at the end of period t, then

$$f_{t,k} = (L_t + k \cdot T_t) \cdot S_{t+k-c}.$$

A.6: ARIMA Models

Many other forecasting methods are available in the literature. Among the best known are autoregressive integrated moving average (ARIMA) models also named as Box-Jenkins models Hoffman et al. (2013). The general form for this family of models is

$$\left(1 - \phi_1\beta - \phi_2 B^2 - \ldots - \phi_p B^p\right)(1 - B)^d x_t$$
$$= \theta_0 + \left(1 - \theta_1 B - \theta_2 B^2 - \ldots - \theta_q B^q\right) \cdot \varepsilon_t$$

where ϕ_k is the autoregressive parameter, θ_k is the moving average parameter, B is a backshift operator defined so that $B^r x_t = x_{t-R}$, $\Delta^d = (1 - B)^d$ is the backward difference operator, and ε_t is an uncorrelated sequence of random errors with mean zero and variance σ^2.

This generic model can be extended to incorporate seasonal behavior (Box et al. 2015). One chooses a model by specifying the integers p, d, and q, resulting in an

Appendix B: Measures of Accuracy

ARIMA(p, d, q) model. These integer parameters are determined by examining the sample autocorrelation and partial autocorrelation function. For additional detail see Hyndman and Athanasopoulos (2006).

Appendix B: Measures of Accuracy

In this section, some accuracy measures will be presented. More details can be found in Hyndman (2006).

Let the forecast error e_t be

$$e_t = x_t - \bar{x}_t,$$

where x_t and $\bar{x}_t$ are, respectively, the observed and the forecasted values for period t and the percentage error p_t defined by

$$p_t = \frac{e_t}{x_t} \cdot 100.$$

Let r_t be the relative error defined by

$$r_t = \frac{e_t}{e'_t}$$

where e'_t is the forecast error obtained from the benchmark method.

B.1: Mean Absolute Error[3] (MAE)

$$\text{MAE} = \frac{1}{m} \sum_{t=1}^{m} |e_t|$$

B.2: Geometric Mean Absolute Error (GMAE)

$$\text{GMAE} = \sqrt[m]{|e_1 \cdot e_2 \cdot \ldots \cdot e_m|}$$

---

[3] Also known as absolute mean deviation (MAD)

B.3: Mean Square Error (MSE)

$$\text{MSE} = \frac{1}{m} \sum_{t=1}^{m} e_t^2$$

B.4: Mean Absolute Percentage Error (MAPE)

$$\text{MAPE} = \frac{1}{m} \sum_{t=1}^{m} |p_t|.$$

B.5: Geometric Mean Relative Absolute Error (GMRAE)

$$\text{GMRAE} = \sqrt[m]{|\, r_1 \cdot r_2 \cdot \ldots \cdot r \,|}$$

Appendix C: Linear Regression Models

C.1: Simple Linear Regression Model

Let (x_i, y_i), $i = 1, \ldots, n$, be a set of n observations. The simple linear regression model is given by

$$y_i = \beta_0 + \beta_1 x_i + \varepsilon_i$$

where ε_i is the residual value and β_0 and β_1 are the least squares estimators computed as.

$$\beta_1 = \frac{\sum_{i=1}^{n} (y_i - \bar{y})(x_i - \bar{x})}{\sum_{i=1}^{n} (x_i - \bar{x})^2} \quad \text{and} \quad \beta_0 = \bar{y} - \beta_1 \bar{x}.$$

with $\bar{x}$ and $\bar{y}$ the averages of x and y, respectively. Residuals ε_i should have mean zero and be uncorrelated with each other and with the independent variable.

Appendix C: Linear Regression Models

Let the forecast variable be $\hat{y}$ and the observed value y, the coefficient of determination R^2 is given by:

$$R^2 = \frac{\sum_{i=1}^{n} \left(\hat{y}_i - \bar{y}\right)^2}{\sum_{i=1}^{n} \left(y_i - \bar{y}\right)^2}.$$

Standard deviation of the residuals, S_ε:

$$S_\varepsilon = \sqrt{\frac{1}{n-2} \sum_{i=1}^{n} \varepsilon_i}$$

$100 \cdot (1 - \alpha)\%$ prediction interval:

$$\hat{y} \pm z_{1-\alpha/2} s_\varepsilon \sqrt{1 + \frac{1}{n} + \frac{\left(x - \bar{x}\right)^2}{(n-1)s_x^2}}$$

where, $z_{1-\alpha/2}$ is $(1 - \alpha/2)$ the critical value of the standard normal distribution, s_ε is the standard deviation of the residuals, x is the value used to calculate $\hat{y}$, $\bar{x}$ is the mean value of all x observed values, and s_x is standard deviation of all x observed values.

C.2: Multiple Linear Regression Model

The multiple linear regression models are an extension of the simple linear regression. It assumes that the dependent variable is explained by more than one factor (the independent variables). Given $(x_{1i}, x_{2i}, \ldots, x_{ki}, y_i)$, $i = 1, \ldots, n$, be a set of n observations. The multiple linear regression model is given by

$$y_i = \beta_0 + \beta_1 x_{1i} + \beta_2 x_{2i} + \ldots + \beta_k x_{ki} + \varepsilon_i.$$

The coefficients β_0, β_1, $\ldots$, β_k measure the effect of each independent variable after taking into account the effect of all other independent variables in the model.

Again, residuals ε_i should have mean zero and be uncorrelated with each other and with each independent variable.

The selection of the independent variables to use in the model is not a straightforward process. Measures of predictive accuracy should be used (e.g., adjusted R^2, cross-validation, Akaike's information criterion, Schwarz Bayesian information criterion ...). For all technical details about multiple Linear Regression model refer to Jobson (1991).

References

Abbasi M, El Hanandeh A (2016) Forecasting municipal solid waste generation using artificial intelligence modelling approaches. Waste Manag 56:13–22

Abbasi M, Abduli MA, Omidvar B, Baghvand A (2013) Forecasting municipal solid waste generation by hybrid support vector machine and partial least square model. Int J Environ Res 7:27–38

Armstrong JS (Ed.). (2001). *Principles of forecasting: a handbook for researchers and practitioners* (Vol. 30). Springer Science & Business Media

Arroyo J, Maté C (2009) Forecasting histogram time series with k-nearest neighbours methods. Int J Forecast 25(1):192–207

Beigl P, Lebersorger S, Salhofer S (2008) Modelling municipal solid waste generation: a review. Waste Manag 28:200–214

Bektaş T, Laporte G (2011) The pollution-routing problem. Transp Res B Methodol 45 (8):1232–1250

Beliën J, De Boeck L, Van Ackere J (2014) Municipal solid waste collection and management problems: a literature review. Transp Sci 48:78–102

Bilitewski B, Härdtle G, Marek K (1994) Waste management. Springer, Berlin

Bilitewski B, Wagner J, Reichenbach J (2010) Best practice municipal waste management (INTECUS Dresden GmbH - Abfallwirtschaft und umweltintegratives Management). Federal Environmental Agency, Intecus

Box GE, Jenkins GM, Reinsel GC, Ljung GM (2015) Time series analysis: forecasting and control, 4th edn. Wiley, Hoboken, N.J

Dahlén L, Lagerkvist A (2010) Evaluation of recycling programmes in household waste collection systems. Waste Manag Res 28:577–586

Faulin J, Juan A, Lera-López F (2012) Optimizing routes with safety and environmental criteria in transportation management in Spain: a case study. In: Wang J (ed) Management innovations for intelligent supply chains. IGI Global Books, Pennsylvania, pp 144–165

Friedman JH (2001) Greedy function approximation: a gradient boosting machine. Ann Stat 29:1189–1232

Han J, Pei J, Kamber M (2011) Data mining: concepts and techniques. Elsevier, Waltham, USA

Hoffman KL, Padberg M, Rinaldi G (2013) T. In Encyclopedia of Operations Research and Management Science (pp. 1573–1578)

Hrebicek J, Soukopova J (2010) Modelling integrated waste management system of the Czech Republic. In: Proceedings of the latest trends on systems: 14th WSEAS international conference on systems (part of the 14th WSEAS CSCC multiconference), pp 510–515

Hyndman R (2006) Another look at forecast-accuracy metrics for intermittent demand. Foresight Int J Appl Forecast:43–46

Hyndman RJ, Athanasopoulos G (2014) Forecasting: principles and practice. OTexts

IMPACTPapeRec (2016) Best practice handbook – Selection trees Available at: http://impactpaperec.eu/en/best-practices/selection-tree/. Accessed March 2018

ISWA (2007) ISWA TECHNICAL POLICY NO. 5 - Storage, Collection, Transportation and Transfer of Solid waste. 2ª versão available at : http://www.iswa.org/en/76/publications.html accessed January 2015

Jobson JD (1991) Multiple Linear Regression. In: Applied Multivariate Data Analysis. Springer Texts in Statistics. Springer, New York, NY

Kaliampakos D, Benardos A (2013) Underground solutions for urban waste management: status and perspectives. ISWA Report ISWA

Karadimas NV, Loumos VG (2008) GIS-based modelling for the estimation of municipal solid waste generation and collection. Waste Manag Res 26(4):337–346

Kim B-I, Kim S, Sahoo S (2006) Waste collection vehicle routing problem with time windows. Comput Oper Res 33:3624–3642

Lechner P (2004) Kommunale Abfallentsorgung. Wien: Facultas Verlags- und Buchhandels AG

References

Lu JW, Chang NB, Liao L, Liao MY (2015) Smart and green urban solid waste collection systems: advances, challenges, and perspectives. IEEE Syst J:1–14

Navarro-Esbrí J, Diamadopoulos E, Ginestar D (2002) Time series analysis and forecasting techniques for municipal solid waste management. Resour Conserv Recycl 35:201–214

Nuortio T, Kytöjoki J, Niska H, Bräysy O (2006) Improved route planning and scheduling of waste collection and transport. Expert Syst Appl 30:223–232

O'Leary P, Walsh P (1995) Decision Maker's guide to solid waste management. United States Environmental Protection Agency, Office of Solid Waste, RCRA Information Center, Washington, DC

Pferdehirt W (1994) University of Wisconsin–Madison Solid and hazardous waste education center

Pferdehirt W, O'Leary P, Walsh P (1993) Developing an integrated collection strategy. Waste recycling collection course, lesson one. Waste Age 1:25–38

Pieber M (2004) Waste collection from urban households in Europe and Australia. Waste Manag World:111–124

Ramos TRP, Gomes MI, Barbosa-Póvoa AP (2014a) Economic and environmental concerns in planning recyclable waste collection systems. Transp Res E Logist Transp Rev 62:34–54

Ramos TRP, Gomes MI, Barbosa-Póvoa AP (2014b) Planning a sustainable reverse logistics system: balancing costs with environmental and social concerns. Omega 48:60–74

Rhyner CR, Schwartz LJ, Wenger RB, Kohrell MG (1995) Waste management and resource 568 recovery. CRC Press/Lewis Publishers, Boca Raton

Rimaityté I, Ruzgas T, Denafas G et al (2012) Application and evaluation of forecasting methods for municipal solid waste generation in an eastern-European city. Waste Manag Res 30:89–98

Ross DE (2010) Editorial - affordability is key to proper solid waste management. Waste Manag Res 28:287–288

Satué S (2000) Gestión Eficiente: fase de recogida selectiva

Scharff C, Vogel G (1994) A comparison of collections systems in European cities. Waste Manag Res 12(5):387–404

Sniezek J, Bodin L (2006) Using mixed integer programming for solving the capacitated arc routing problem with vehicle/site dependencies with an application to the routing of residential sanitation collection vehicles. Ann Oper Res 144:33–58

Tanskanen J, Melanen M (1999) Modelling separation strategies of municipal solid waste in Finland. Waste Manag Res 17:80–92

Tao J (2010) Reverse logistics information system of e-waste based on internet. In: Proceedings of the 2010 international conference on challenges in environmental science and computer engineering (CESCE), pp 447–450

Tavares G, Zsigraiova Z, Semiao V, Carvalho MDG (2009) Optimisation of MSW collection routes for minimum fuel consumption using 3D GIS modelling. Waste Manag 29(3):1176–1185

Tchobanoglous G, Theisen H, Vigil S (1993) Integrated solid waste management. Engineering principles and management issues. McGraw-Hill International Editions, New York

Tralhão L, Coutinho-Rodrigues J, Alçada-Almeida L (2010) A multiobjective modeling approach to locate multi-compartment containers for urban-sorted waste. Waste Manag 30 (12):2418–2429

Urban Upgrading (2001) Waste collection. http://web.mit.edu/urbanupgrading/upgrading/issues-tools/issues/waste-collection.html. Assessed March 2018

Vijay R, Gautam A, Kalamdhad A et al (2008) GIS-based locational analysis of collection bins in municipal solid waste management systems. J Environ Eng Sci 7:39–43

Viotti P, Polettini A, Pomi R, Carlo I (2003) Genetic algorithms as a promising tool for optimisation of the MSW collection routes. Waste Manag Res 21(4):292–298

Wilson DC, Scheinberg A (2010) What is good practice in solid waste management? Waste Manag Res 28:1055–1056

Winston WL, Goldberg JB (2004) Operations research: applications and algorithms. Thomson Brooks/Cole, Belmont

WRAP (2013) Bring site recycling. Project code: BHC002–207. Available at: http://www.wrap. org.uk/sites/files/wrap/Bring%20Site%20Draft%20Report%20v5%20JB%20amends_0.pdf. Accessed March 2018

Zsigraiova Z, Semiao V, Beijoco F (2013) Operation costs and pollutant emissions reduction by definition of new collection scheduling and optimization of MSW collection routes using GIS, the case study of Barreiro, Portugal. Waste Manag 33(4):703–806

Chapter 10
Operation and Monitoring

Abstract The activities related with the waste collection and how they impact on the waste management systems need to be conducted in a sustainable and efficient way. The waste collection needs to be affordable for users, although respecting all regulations applicable to the waste collection, with low environmental impact and respecting workers. Efficient collection of waste requires that data is collected and processed to analyze the actual situation and where improvements can be made to reach the efficiency needed. The purpose of this chapter is to present the existent instruments to analyze the operation and monitor a waste collection system: route analysis and optimization tools and performance indicators.

An indicator is an elementary datum or a simple combination of data capable of measuring an observed phenomenon, providing information that is typical of, and critical to the quality of target issues (Peterson and Granados, Environ Sci Pollut Res 9:204–214, 2002; Resour Conserv Recycl 52:1322–1328, 2008). A suitable indicator must fulfill criteria as relevant, credible, functional, quantifiable, and comparable within different time and space scales (EEA, EEA core set of indicators – guide. Copenhagen, 2005; Key environmental indicators). Indicator sets have been developed to evaluate aspects such as the state and evolution of general and specific environments, policy objectives, the environmental behavior of individual technologies and products, critical economic individual sectors, multi-sectorial analysis, global manufacturing, and management systems. Recently they have also been applied to organizational methods, products, services, and systems in an eco-innovation context (OECD, Eco-innovation in industry enabling green growth; Eco-innovation: when sustainability and competitiveness shake hands. Palgrave Macmillan, Hampshire, 2009).

Keywords Benchmarking · Collection rate · Economic indicators · LCA · Logistics · Participation rate · Performance · Social indicators · Waste generation

10.1 Descriptive Indicators

According to Bertanza et al. (2018), descriptive indicators intend to identify the characteristics of the collection system in terms of the waste amount collected and its composition and of the collection service, in terms of work load for personnel,

© Springer International Publishing AG, part of Springer Nature 2019 167
A. Pires et al., *Sustainable Solid Waste Collection and Management*,
https://doi.org/10.1007/978-3-319-93200-2_10

Table 10.1 Descriptive indicators for waste collection

Descriptive indicator	Description
Waste characterization (%)	Percentage composition of the collected waste in terms of, respectively, separately collected fractions (as a whole), bulky waste, street waste, and unsorted waste
Waste collected per capita (kg/(inhabitants y))	Annual amount of collected waste (or waste component) per capita
Labor per waste (h_{man}/(in y))	Annual amount of man-hour per capita spent for the collection of a waste fraction
Vehicle labor per waste ($h_{vehicle}$/(in y))	Annual amount of vehicle-hour per capita spent for the collection of a waste fraction
Volume available per capita (m^3/(inhabitants y))	Annual available volume per capita for the collection of a waste fraction (i.e. volume of containers, bins, bags)

Source: Bertanza et al. (2018), Martinho et al. (2017)

vehicles, and containers volume capacity. These descriptive indicators are in line with parameters used in design and planning of a waste collection system. In Table 10.1 are presented those indicators.

10.2 Performance Indicators

10.2.1 Technical-Operative and Logistics Indicators

Waste collection systems are diverse and must be studied in detail to determine which systems perform best. Municipal solid waste (MSW) collection and planning requires robust operational, economic, and environmental solutions, so MSW collection operators and decision-makers need effective methodologies and tools to support management options under uncertain and complex operational issues, such as population, costs, equipment, and human resources (Teixeira et al. 2014a).

To date, waste collection has been evaluated through the use of indicators or indexes. The analysis of published research shows that the application of performance indicators in the area of MSW management has evolved over the last five decades (Sanjeevi and Shahabudeen 2016); however, a significant number of authors use indicators for the presentation and evaluation of data without an adequate benchmark (Zaman 2014), the majority of which are basic and one-dimensional quantitative statements that focus on waste generation rates and prevalence of options of treatment and elimination (Fragkou et al. 2010).

One of the prerequisites for better management is the ability to identity and measure the performance of various operating elements. Indicators provide a means of assessing the performance of economic, social, and environmental aspects, with the advantage of being able to summarize, focus, and condense information about complex systems, and highlight trends or phenomenon, which are not immediately detectable through basic data collection (Arendse and Godfrey 2001).

10.2 Performance Indicators

Performance indicators are measurement tools used by organizations to evaluate the success or failure of a given activity and lead to identification of the improvements needed in the system (Sanjeevi and Shahabudeen 2015). By presenting several data in one number that commonly is easier to interpret than complex statistics, indicators can facilitate communication between stakeholders. Often the municipalities have to report performance indicators to a regulatory authority, and also the communication between experts and non-experts is easier when using performance indicators. But the principal role of performance indicators in waste collection management is to provide data when planning, designing, and monitoring waste collection systems.

Different types of performance indicators exist in literature. Berg (1993) proposed indicators to evaluate source-sorting systems focused on amounts collected, quality, recycling rate, and participation rate. Courcelle et al. (1998) used diversion rate and residue ratio in material recovery facilities to assess the performance of sorting programs. Dahlén and Lagerkvist (2010) applied specific waste generation rate, source-sorting ratio, ratios of materials in the residual waste, and ratio of mis-sorted materials indicators to evaluate recycling programs. Petersen and Berg (2004) analyzed the characteristics of the waste in the different containers of the recycling centers, namely, by estimating specific container weights. Zaman (2014) identifies as priority indicators for collection the types of waste collected separately, the distance traveled to collect one ton of waste, the frequency of collection, the type of collection – formal or informal – and the amount of waste collected per unit of collection costs. Federico et al. (2009) propose as collection efficiency indicators the recyclable waste collection rate, the number of containers per 1000 inhabitants, the ratio of the quantity collected to the number of employees used, the distribution of multi-material collection containers, and the amount of recyclables obtained per service in relation to the total capacity of the system. Gallardo et al. (2010) applied annual collection rate and quality in container rate. Passarini et al. (2011) applied the separate waste collection rate and Del Borghi et al. (2009) the frequency of waste collection and waste transport distances.

A benchmarking study carried out in Scotland (Accounts Commission 2000) evaluated the evolution of the collection service using as indicators the amount of waste collected, the gross cost of collection, the number of employees, the main method of collection, the collection frequency, the gross cost per ton collected, the tons collected per vehicles per day (including reserve vehicles), and tons collected per operator per year. The same study highlights the importance of assessing the absenteeism rate, the bonuses paid to workers, and the percentage of fleet reserve vehicles.

To evaluate the productivity of collection routes, Moreira (2008) proposes a large diversity of operational and productivity indicators, using common circuit characterization indicators, such as the installed capacity, the average quantity of waste collected per route, the number of freights per route, and the fuel consumption per route. Moreira (2008) also characterizes the different phases that can be identified in a circuit, differentiating the total time and distances per circuit from the effective time and distances per circuit, adding the transport time, the average collection time spent per point, and time and distance from and to the garage. In the productivity indicators, Moreira (2008) proposes the amount of waste removed per kilometer, per collection point, and per effective working hour, the concentration coefficient of the circuit, the average speed, the number of points and turns per circuit, the fuel consumption per kilometer, and the ratio of actual working hours to normal working hours.

To evaluate the efficiency of the collection services in 75 municipalities in the Catalonian region of Spain, Bosch et al. (2001) defined as operational performance indicators the ratio of the following variables with the amount of waste collected: number of containers, total number of vehicles, and total number of direct workers (expressed on the number of full working days). In addition to these variables, Bosch et al. (2001) mentioned the importance of adopting the number of kilometers performed by the collection vehicles, their capacity, and technical characteristics as well as the number of working hours.

To present some examples of PI, Table 10.2 and 10.3 present, respectively, two types of PI: technical-operative indicators (Bertanza et al. 2018) and logistics indicators (Martinho et al. 2017). Such classification synthetize briefly the PI already applied to waste collection.

Table 10.2 Examples of technical-operative indicators for waste collection systems

Technical-operative indicators	Description
Waste collected per volume (kg/m^3)	Amount of material collected per unit volume of containers
Waste collected per labor (kg/h$_{man}$)	Amount of solid waste (or waste fraction) collected per man-hour
Waste collected per vehicle labor (kg/h$_{vehicle}$)	Amount of solid waste (or waste fraction) collected per vehicle-hour
Separation waste collection rate (%)	Percentage of sorted waste with respect to the total amount of waste generated.
Waste generation per capita per day (kg·person^{-1}·day^{-1})	Per capital per day waste generated by each collection stream (kg/population in the neighborhood)
Recycling rate (%)	Specific waste stream collected sent for recycling/total specific waste stream generated
Contaminants rate in separate waste collection (%)	Refuse and rejects from separate waste stream collected/total waste stream collected
Participation rate (%)	Families participating at least once per month in recycling/total number of families
Maximum total weight (kg)	Sum of waste weight when the filling container rate is 100% and container tare weight. Represents the total weight to be lifted
Waste volume weight inside the container (kg/m^3)	Ratio between waste weight when the filling container rate is 100% and container net capacity
Emptying time (min)	Time interval between vehicle's stops to collect the container and starts to move to the next point
Maximum capacity collected per time unit (m^3/min)	Ratio between container net capacity and emptying time, for a container filling rate of 100%
Maximum weight collected per time unit (kg/min)	Ratio between waste weight and emptying time, for a container filling rate of 100%
Maximum weight collected per distance (kg/km)	Ratio between waste weight and effective waste collection distance
Crew size (no unit)	Total number of workers needed

Source: Bertanza et al. (2018), Martinho et al. (2017), Rodrigues et al. (2016), Rodrigues (2016)

10.2 Performance Indicators

Table 10.3 Logistics indicators

Distance ratio (%)	Distance traveled by vehicle to collect waste (effective distance)/total distance traveled by vehicle from leaving the garage to its return
Total collection distance ($km \cdot t^{-1}$)	Total distance traveled/amount of waste collected
Effective collection distance ($km \cdot t^{-1}$)	Distance traveled to collect waste/amount of waste collected
Total collection time ($h \cdot t^{-1}$)	Collection time from garage to vehicle's return/amount of waste collected
Effective collection time ($h \cdot t^{-1}$)	Collection time of vehicle/amount of waste collected
Effective worktime (%)	Total collection time/scheduled time of workers
Workers dedicated to collection ($worker \cdot t^{-1}$)	Number of workers/amount of waste collected
Crew productivity ($t \cdot worker^{-1} \cdot h^{-1}$)	Amount of waste collected/(workers $\times$ total collection time)
Effective fuel consumption ($l \cdot t^{-1}$)	Amount of fuel consumed by the collection vehicle per unit of waste collected

Source: Martinho et al. (2017), Teixeira et al. (2014a)

In summary, most of the operational indicators found in the literature are based on the quantities collected, which can be referred to in terms of time (shift, day, or year) and/or distances traveled. In addition to these indicators based on quantities collected, there are also indicators related to the evaluation of the geographic coverage of the service. In fact, for selective deposition in collective deposition points, the most popular indicator for measuring the coverage of the service is the density of the selective collection point, defined on the basis of the ratio of the number of residents living in a particular area and the number of selective collection points available in that area for the selective collection (Waite 1995). García-Sánchez (2008) also used the density of collection points as an indicator, expressed as the number of collection points per square kilometer.

10.2.2 Case Study: Calculating Waste Volume Weight Inside the Container and Emptying Time in Greater Lisbon Area, Portugal

Rodrigues et al. (2016) developed 12 indicators to characterize the technological aspect of 22 waste collection systems divided into three groups: container design, container capacity, and operation. Indicators were tested for a case study of packaging source-separated waste collection systems operating in the Greater Lisbon area, Portugal, where source separation is made through three streams: yellow for plastic and metal, green for glass, and blue for paper and cardboard (including nonpackaging).

Waste volume weight inside the container and emptying time were two relevant studied technical-operative indicators, measured in the field, in three municipalities in the Greater Lisbon area—Lisbon, Cascais, and Sintra. The choice of these municipalities was justified by city structure heterogeneity, resulting in a diversity of container and vehicle components, to include the ten possible key container component cases and the ten key vehicle components described in Chap. 3.2., representing all possible taxonomic WCS components and collection method (Tables 10.4 and 10.5).

Table 10.4 Classification of the containers, vehicles, and WCS from case study

Reference	*Container component* (container capacity, m^3)
C1	Surface, without wheels, without compaction, without vehicle coupling (0.03)
C2	Surface, immobile, without compaction, crane one ring (2.06)
C3	Surface, immobile, without compaction, lift side supports (C3.1 = 1.00;C3.2 = 2.14;C3.3 = 2.73;C3.4 = 2.55)
C4	Surface, wheeled containers, without compaction, lift frontal supports (C4.1 = 0.12; C4.2 = 0.24)
C5	Surface, wheeled containers, without compaction, lift frontal or lift side supports (1.10)
C6	Semiunderground, without compaction, compact container, crane one ring (5.00 for blue and yellow waste stream; 3.00 for green waste stream)
C7	Underground, without compaction, compact container, crane one ring (3.00)
C8	Underground, without compaction, compact container, crane mushroom (5.00 for blue and yellow waste stream; 3.00 for green waste stream)
C9	Underground, without compaction, electrohydraulic open and elevating platforms, lift side supports (4.00)
C10	Underground, without compaction, gas cylinders, only opening platform waste recipient, crane one ring (5.00 for blue and yellow waste stream; 3.00 for green waste stream)
C11	Underground, without compaction, hydraulic only opening platform waste recipient, crane one ring (5.00)
Reference	*Vehicle component*
V1	Single-compartment open body, without mechanization, crane double hook, not specific loading location
V2	Single-compartment closed body, intermittent compactor, simple hook, rear-end loading
V3	Single-compartment closed body, intermittent compactor, crane double hook, not specific loading location
V4	Single-compartment closed body, mechanized packer grid, fork lift assisted, rear-end loading
V5	Single-compartment closed body, intermittent compactor, fork lift assisted, side loading [1]
V6	Single-compartment closed body, intermittent compactor, fork or bars lift assisted, rear-end loading
V7	Single-compartment closed body, intermittent compactor, automated arm, side loading

Source: Adapted, Rodrigues et al. (2016)

10.2 Performance Indicators

Table 10.5 Characterization of the WCS from case study

Collection method	Waste collection systems
Manual	C1V5
Assisted	C4V6, C5V4, C5V6
Semiautomated	C2V1, C2V2, C2V3, C6V1, C6V2, C6V3, C7V1, C8V1, C8V3, C10V1, C10V2, C11V1
Fully automated	C3V7, C9V7

Source: Rodrigues et al. (2016)

Volume weight or mass volume, defined as the ratio between waste weight when the filling container rate is 100% and container net capacity, are determinant data when designing and sizing WCS. Usually, available volume weight data are derived from physical waste composition characterization studies held in the treatment facilities, not by measuring the waste inside the containers, so the data are not container specific. Similarly, the most common volume weight data obtained from literature are presented by measuring specific materials, not from waste collection flow. Measuring volume weight in the field is difficult, however, especially for larger containers considering the manpower and equipment required. Also, planning WCS requires data related to the time spent in pickup and unloading containers, the *emptying time*, which is also WCS specific. The emptying time of containers for each pickup and unloading method, defined as the time interval between vehicle's stops to collect the container and starts to move to the next point, are input data for waste collection routing software. When related to the amount of waste collected in the containers (volume and weight), emptying time can be used to evaluate the equipment effectiveness.

A field campaign was conducted to collect volume, mass, and time data for each type of waste stream—yellow, green, and blue—by container and WCS. For volume of containers, Petersen and Berg (2004) method was used, which consists of measuring the height of the contents in each container before being emptied; the waste volume in the container is then calculated based on the design and total volume of the container. For container weight, different scales were used depending on the type of container: (i) a hook dynamometer for higher capacity containers between 1 and 5 m^3, $\pm$ 1 kg of precision; (ii) a platform scale up to 60 kg for weighing smaller capacity containers between 0.09 and 0.360 m^3, $\pm$ 0.01 kg of precision; and (iii) a manual precision hook balance, suitable for weighing up to 50 kg with an accuracy of 0.01 kg, to accommodate the range of curbside collection bag weights.

For emptying time, the campaign was distributed equally during the week and including both day and night shifts. A chronometer was used to record the time interval from when the vehicle stopped to collect the container and started to move to the next point, $\pm$ 1 second of precision. If more than one container was collected, this information was also registered. To summarize and analyze data collected from field campaign and to measure statistical variability (heterogeneity), descriptive statistics were used, including mean, standard deviation (SD), standard error of the mean (SEM), and relative standard deviation (RSD).

Table 10.6 Waste volume weight inside the container for blue (B), yellow (Y), and green (G) waste stream

Ref.	Mean (kg/m^3)			SD (kg/m^3)			RSD (%)			SEM (kg/m^3)		
	B	Y	G	B	Y	G	B	Y	G	B	Y	G
C1	78.78	38.96	–	43.45	12.01	–	55	31	–	8.87	2.36	–
C2	36	25	254	16	4	8	45	15	3	4	1	4
C3.1	42	27	295	17	12	52	39	44	18	5	3	18
C3.2	50	28	278	36	4	46	73	14	17	8	1	21
C3.3	32	25	310	7	7	65	23	27	21	2	2	29
C3.4	31	18	293	8	3	19	26	16	7	2	1	9
C4.1	74.87	42.56	–	50.07	17.35	–	67	41	–	8.99	5.78	–
C4.2	87.59	32.78	–	49.84	5.05	–	57	15	–	12.09	1.91	–
C5	35	33	206	24	7	23	68	20	11	4	2	10
C6	36	19	245	8	3	49	23	15	20	2	1	16
C7	46	23	203	52	9	47	111	37	23	16	3	19
C8	26	22	278	8	8	42	31	36	15	2	2	13
C9	28	36	180	12	17	22	42	47	12	5	8	13
C10	32	25	214	13	3	45	41	12	21	3	1	18
C11	49	38	308	14	13	35	29	34	11	4	3	11

Source: Adapted from Rodrigues et al. (2016)

Results of waste volume weight inside the container are presented in Table 10.6. The average volume weights are 255 kg/m^3 for glass, 29 kg/m^3 for plastic/metal, and 46 kg/m^3 for paper/cardboard. The containers with higher volume weight values for paper/cardboard were C4.2 (88 kg/m^3), C1 (79 kg/m^3), and C4.1 (75 kg/m^3), and for plastic/metal were C4.1 (43 kg/m^3), C1 (39 kg/m^3), and C11 (38 kg/m^3). With the exception of C11, these containers were not those with the largest net capacity; however, they could store higher amount of waste per volume. Containers C1, C4.1, and C4.2 could be easily handled and manually compressed by the user, and the open lid design (no specific deposition openings) allowed the material to be relatively homogeneous and evenly distributed. Just as important as having free access to the inside of the container was the fixed collection frequency of these exclusive curbside containers. Because the available deposition capacity was limited (unlike the others containers), efficient management (manual compression) of the deposition capacity by the user may be required to accommodate more waste. High C11 capacity values were also related to deposition methods: these containers had a rotating deposition drum with a fixed capacity, which can also promote waste compaction. The possible compression effect from the waste itself in higher capacity underground containers was not observed.

C11 also had a high volume weight for glass waste (308 kg/m3), as did C3.3 (310 kg/m3) and C3.1 (295 kg/m3). Although it was expected the lower height on underground containers to result in more glass hull (from broken glass packages) and therefore higher glass volume weights, that effect was not observed. The access to the container content and the small, limited deposition capacity seemed to affect the

10.2 Performance Indicators

Table 10.7 Emptying time for blue (B), yellow (Y), and green (G) waste stream

Ref.	Mean (min)			SD (min)			RSD (%)			SEM (min)		
	B	Y	G	B	Y	G	B	Y	G	B	Y	G
C1V5	0.17	0.16	–	0.14	0.13	–	84	80	–	0.02	0.01	–
C2V1	2.37	1.95	2.06	0.66	0.29	0.48	28	15	24	0.08	0.04	0.05
C2V2	1.77	1.63	–	0.37	0.30	–	21	19	–	0.03	0.02	–
C2V3	2.71	2.59	–	0.85	0.85	–	31	33	–	0.10	0.09	–
C3.1V7	0.85	0.86	0.84	0.11	0.12	0.09	12	14	11	0.01	0.01	0.01
C3.2V7	0.82	0.84	0.79	0.20	0.24	0.13	24	29	16	0.02	0.01	0.01
C3.3V7	0.75	0.89	0.71	0.08	0.64	0.05	10	72	7	0.01	0.05	0.01
C3.4V7	0.75	0.82	0.76	0.07	0.27	0.09	9	32	12	0.01	0.03	0.01
C4.1V6	0.31	0.31	–	0.15	0.18	–	49	57	–	0.00	0.00	–
C4.2V6	0.42	0.38	–	0.50	0.24	–	121	63	–	0.02	0.01	–
C5V4	–	–	1.38	–	–	0.44	–	–	32	–	–	0.04
C5V6	0.81	0.81	–	0.27	0.24	–	33	30	–	0.02	0.01	–
C6V1	4.49	4.27	4.07	0.97	0.68	0.27	22	16	7	0.28	0.18	0.07
C6V2	5.45	–	–	0.58	–	–	11	–	–	0.08	–	–
C6V3	4.08	5.37	5.06	0.52	0.51	0.60	13	10	12	0.23	0.07	0.08
C7V1	4.93	6.98	5.93	–	2.76	1.34	0	40	23	0.00	1.59	0.60
C8V1	3.79	3.37	3.50	0.91	0.54	0.61	24	16	17	0.26	0.12	0.12
C8V3	3.52	3.45	3.52	0.77	0.52	0.73	22	15	21	0.13	0.13	0.12
C9V7	1.86	2.13	2.02	0.44	0.25	0.45	24	12	22	0.25	0.11	0.26
C10V1	–	–	4.97	–	–	0.83	–	–	17	–	–	0.22
C10V2	4.49	4.83	–	0.95	0.71	–	21	15	–	0.17	0.15	–
C11V1	5.16	5.56	6.94	1.33	1.23	1.66	26	22	24	0.29	0.30	0.43

Source: Adapted from Rodrigues et al. (2016)

volume weight. The containers with lower volume weight for paper/cardboard were C9 (28 kg/m3), C3.4 (31 kg/m3), and C3.3 (32 kg/m3). C9 also had a reduced volume weight for glass (180 kg/m3), as did C7 (203 kg/m3) and C5 (205 kg/m3). Containers with the lowest volume weights for plastic/metal were C3.4 and C6 (18 and 19 kg/m3), which were surface, semiunderground, and underground containers with medium capacities and reduced access for the user to compact waste.

Emptying time results are presented in Table 10.7. The results from linear regression of emptying time and number of containers collected were verified for WCS with collection of multiple containers (applicable to C1V5, C4V6, C5V4, and C5V6 WCS). For these WCS, around 71% of the emptying time could be explained by the number of containers, reflecting that the higher the number of containers collected, the more time spent collecting it, if the outliers of C1V5 WCS values are removed. For C1V5, a manual WCS, the R^2 decreased from 65% to 34% if no removal of outliers occurred. The results seem to indicate that other variables may be influencing the emptying time in this manual and assisted type of WCS (i.e., where the influence of manual labor during the collection was expected).

Data presented in Tables 10.6 and 10.7 provide useful information for the optimization models to be used in the WCS planning phase: most optimization models devoted to optimizing collection include aspects such as weight of waste inside the bin, volume of waste inside the bin, and fixed time to unload a bin (Faccio et al. 2011), which are not available for planners who design the WCS and can only be obtained after implementation.

10.3 Economic Indicators

There are several types of economic indicators, which intends to evaluate the principal cost drivers (Bertanza et al. 2018). The diversity of indicators relates with the focus on the waste collection system. In one hand, it is important to look for indicators related with each waste collection system element, like the container and the vehicle. In other hand, the costs and revenues of the waste collection system functioning are also relevant to analyze and monitor the system. Hage and Söderholm (2008) used cost indicators to compare household waste plastic and packaging waste collection in Sweden. Larsen et al. (2010) used collection costs, recycling rates, sorting efficiencies, and waste amounts. Gamberini et al. (2013) presented demand profiles and costs indexes of MSW management in several Italian communities. A study based on 81 Catalan municipalities with the goal of providing local authorities real and accurate information on waste management costs, comparing the door-to-door and collective collection, the indicator that was considered the most accurate was the overall per capita management cost (Sora and González 2014). Teixeira et al. (2014b) define as economic indicators the total cost of collection (annual investment, capital requirements, and operational expenses of collection – placement of containers and maintenance, collection and transportation to the site) quotient with the quantity of collected waste (per circuit), the total cost of collection per inhabitant, and the total cost of collection per dwelling. Any of these indicators can then be divided into undifferentiated and selective collection or even broken down by waste stream, depending on the objectives and available information.

There is no waste collection system that is more efficient in economic terms whatever the context, since the economic efficiency depends on many factors. For example, the costs of recyclable collection vary depending on the size of the collection team, vehicle capacity, participation rate, and distance between stops, among other factors. Factors such as location, amount of waste, its composition, social context, the type of technology used on its collection, the distances traveled, and the human resources used are among the conditioning factors of management system costs (Karadimas et al. 2007). Greco et al. (2014) also refer to the multiple factors that affect the costs of waste collection, indicating the characteristics of the municipality, such as its size, population density, characteristics of the area to be served (e.g., distances, altitude, road network), the quantity and quality of the waste to be collected, and the mechanisms and technologies used to collect and transport them.

Data collected from indicators is helpful to optimize waste collection systems. Optimization in waste collection systems is mainly achieved by minimizing total costs, with most systems being static instead of dynamic (Badran and El-Haggar 2006). When accurate cost data is available, the cost optimization model is more reliable compared to models based on time data (Komilis 2008), although the latter are easier to develop since they are based on readily available data (distances between points and average speeds). According to Viotti (2003), optimization algorithms for collection circuits aim to minimize activity and related costs, such as the total distance traveled by the collection vehicles or the total cost of collection. Hashimoto et al. (2006) generalize the standard problem of defining vehicle routes, allowing restrictions in the window of time and travel time, in which both restrictions are treated as cost functions. In the developed cost optimization model, Badran and El-Haggar (2006) used fixed and variable costs to model the total cost of collection as a function of the quantity of waste, and incorporated the transport cost coefficients separately, one expressed by time and weight transported ($/ton.h), to express the cost of labor, and other expressed per distance and freight carried ($/km.t), to express fuel cost and maintenance.

One of the variables sued on optimization models is the amount of fuel consumed during collection. Several models provide diesel consumption during waste collection on the basis of detailed information on the number of points (stops), number of containers per point, distance between points, and others (Sonesson 2000). Nguyen and Wilson (2010) studied fuel consumption rates of two different vehicle types during different collection activities (or circuit phases) to assess the potential effectiveness of some possible methods to reduce fuel consumption by using the amount of fuel consumed per collected ton and kg CO_2 equivalent per collected ton. Larsen et al. (2009) provided data on diesel consumption per ton of collected waste for a series of waste fractions and collection systems and evaluated emissions from diesel combustion through a life cycle analysis approach. In this study, the number of trips per day, the total diesel consumption per day, the total distance traveled per day, the weight of each waste load, and the net weight of the used vehicle were used indicators. In the same study, the amount of diesel used for the transport phase was estimated by transport simulation software based on the assumption of linear correlation between vehicle gross weight and diesel consumption. Examples of economic indicators applied to solid waste collection systems are presented on Table 10.8.

Table 10.8 Examples of economic indicators applied to solid waste collection systems

Economic indicators	Description
Containers cost per ton (€/t)	Containers cost per amount of collected waste (or fraction of waste)
Personnel cost per ton (€/t)	Personnel cost per amount of collected waste (or fraction of waste)
Vehicle cost per ton (€/t)	Vehicles cost per amount of collected waste (or fraction of waste)
Total cost per ton (€/t)	Total costs (sum of containers, personnel, vehicles) per amount of waste collected or waste fraction
Cost per route (€/route)	Cost of equipment and collection (vehicles, containers and workers) spent on one collection route
Cost per ton (€/t)	Cost of the collection per ton of waste collected
Final cost per ton (€·t − 1)	Cost of collection and treatment, including revenues from the sale of recyclables per ton of waste collected
Break-even point per route (t·route−1)	The amount of waste needed to be collected to make a route economically viable. It is calculated by dividing the cost and revenues from processing the packaging waste by the cost per route
Final cost per inhabitant (€·inhab−1·year−1)	The cost of collection and treatment, including revenues from the sale of recyclables, per inhabitant

Source: Bertanza et al. (2018), Martinho et al. (2017)

10.4 Environmental Indicators

Most of environmental indicators used to assess waste collection systems results from life cycle assessment methodology and carbon footprint methodology. In LCA there are midpoint and endpoint indicators, which can both be used to compare scenarios of waste collection. Pires et al. (2017) have used midpoint environmental impact categories to compare different packaging waste collection systems, as well as Larsen et al. (2010), namely, global warming, acidification, photochemical ozone formation, human toxicity, eutrophication, and ecotoxicity. In those environmental indicators based on LCA, they reflect the consumption of fuel and the consumption of raw materials to produce containers but normally exclude the raw materials and energy involved in the production of vehicles.

Some of the performance indicators can also be regarded as environmental indicators. Source separation rate, recycling rate, and the contamination rate of materials are indicators that reflect how materials are extending their life cycle, avoiding the extraction of raw materials. They can also be indicators of the degree of circular economy promoted by the waste collection system.

10.5 Social Indicators

Social indicators related to waste collection systems focus mostly on the response of citizens participating in the waste collection or in the recycling scheme. Indicators also applied in performance indicators like participation rate is also a social indicator, together with others proposed by Berg (1993), Dri et al. (2018), and Tai et al. (2011): willingness to participate, degree of satisfaction from users, accessible classified containers, and accessible classified vehicles.

Social indicators may also reflect working conditions. Waste collectors are exposed to hazardous conditions, including hazardous materials, biological pathogens, and sharp objects (HSE 2014). Statistics on injuries are mostly injuries per year or injuries per year per 100 full-time workers (Statista 2018). In the USA, between 2009 and 2015, waste collection has the highest workplace injuries, varying from 5.9 to 7.1 injuries and illness per 100 full-time workers (Statista 2018). Health and safety conditions of workers at waste management sector and waste collection subsector are relevant, however, there are few detailed studies by country or specific for waste collection to be helpful to compare between waste collection systems or compare countries.

10.6 Final Remarks

The way how waste management sector has operated, evaluated, and improved, their waste collection systems reflect the difficulty in getting data and treating data to provide useful information to make changes and improvements. Information and communication technologies can, nowadays, make the difference in the control of those systems by answering to the data gathering with reduced resources and capability to process information. Considering their variety, there is the need to focus always the indicators to the goal of the study.

The drivers to make monitoring and control of waste collection systems are, in general, the costs. The principal goal is, always, to make waste collection at a lowest cost. But the need to reach recycling targets and source separation of waste streams led waste managers to monitor waste collection systems in terms of the goals of the waste collection, the environmental impacts of the waste collection, and how the workers and citizens look for the waste collection system.

The most recent focus on the monitoring of waste management sector, including waste collection, is the standardization through norms like Eco-Management and Audit Scheme (EMAS), used by companies and other organizations to evaluate, report, and improve their environmental performance. To help EMAS implementation at waste sector, EMAS includes the Reference Documents on Best Environmental Management Practice for Waste Management Sector (Dri et al. 2018). Such document will support organizations showing how to measure/monitor the progress made and how to benchmark their performance (Schoenberger et al. 2014) and will allow the waste management sector to improve their vision on their how activity, pushing forward to the improvement of the sector in environmental and social aspects.

References

Accounts Commission (2000) Benchmarking refuse collection – a review of councils' refuse collection services. Audit Scotland, Edinburgh

Arendse L, Godfrey L (2001) Waste management indicators for national state of environment reporting Pretoria. Available via CSIR. http://www.unep.or.jp/ietc/kms/data/2010.pdf. Accessed 11 Dec 2015

Badran M, El-Haggar S (2006) Optimization of municipal solid waste management in port-said-Egypt. Waste Manag 26:534–545

Berg PEO (1993) Källsortering Teori, metod och implementering (source sorting theory, method and implementation). Dissertation, Chalmers Tekniska Högskola

Bertanza G, Ziliani E, Menoni L (2018) Techno-economic performance indicators of municipal solid waste collection strategies. Waste Manag 74:86–97

Bosch N, Pedraja F, Suárez-Pandiello J (2001) The efficiency of refuse collection services in Spanish municipalities: do non-controllable variables matter? Institut d'Economia de Barcelona, Centre de Recerca en Federalisme fiscal I Economia regional, Barcelona

Carrillo-Hermosilla J, del Río P, Könnölä T (2009) Eco-innovation: when sustainability and competitiveness shake hands. Palgrave Macmillan, Hampshire

Courcelle C, Kestemont MP, Tyteca D, Installe M (1998) Assessing the economic and environmental performance of municipal solid waste collection and sorting programs. Waste Manag Res 16:253–262

Dahlén L, Lagerkvist A (2010) Evaluation of recycling programmes in household waste collection systems. Waste Manag Res 28:577–586

Del Borghi A, Gallo M, Del Borghi M (2009) A survey of life cycle approaches in waste management. Int J Life Cycle Assess 14:597–610

Dri M, Canfora P, Antonopoulos IS, Gaudillat P (2018) Best environmental management practice for the waste management sector, learning from frontrunners – final draft. Office of the European Union, Luxembourg

European Environment Agency (EEA) (2005) EEA core set of indicators – guide. European Environmental Agency, Copenhagen

Faccio M, Persona A, Zanin G (2011) Waste collection multi objective model with real time traceability data. Waste Manag 31:2391–2405

Federico G, Rizzo G, Traverso M (2009) In itinere strategic environmental assessment of an integrated provincial waste system. Waste Manag Res 27:390–398

Fragkou M, Vicent T, Gabarrell X (2010) A general methodology for calculating the MSW management self-sufficiency indicator: application to the wider Barcelona area. Resour Conserv Recycl 54:390–399

Gallardo A, Bovea MD, Colomer FJ, Prades M, Carlos M (2010) Comparison of different collection systems for sorted household waste in Spain. Waste Manag 30:2430–2439

Gamberini R, Del Buono D, Lolli F, Rimini B (2013) Municipal solid waste management: identification and analysis of engineering indexes representing demand and costs generated in virtuous Italian communities. Waste Manag 33:2532–2540

García-Sánchez MI (2008) The performance of Spanish solid waste collection. Waste Manag Res 26:327–336

Greco G, Allegrini M, Del Lungo C, Savellini P, Gabellini L (2014) Drivers of solid waste collection costs empirical evidence from Italy. J Clean Prod 106:364–371

References

Hage O, Söderholm P (2008) An econometric analysis of regional differences in household waste collection: the case of plastic packaging waste in Sweden. Waste Manag 28:1720–1731

Hashimoto H, Ibaraki T, Imahori S, Yagiura M (2006) The vehicle routing problem with flexible time windows and traveling times. Discret Appl Math 154:2271–2290

Health and. Safety Executive (HSE) (2014) health and hazardous substances in waste and recycling. http://wwwhsegovuk/pubns/waste27htm. Accessed 28 Jan 2014

Karadimas N, Papatzelou K, Loumos V (2007) Optimal solid waste collection routes identified by the ant colony system algorithm. Waste Manag Res 25:139–147

Komilis DP (2008) Conceptual modeling to optimize the haul and transfer. Waste Manag 28:2355–2365

Larsen A, Vrgoc M, Christensen T, Lieberknecht P (2009) Diesel consumption in waste collection and transport and its environmental significance. Waste Manag Res 27:652–659

Larsen AW, Merrild H, Moller J, Christensen TH (2010) Waste collection systems for recyclables: an environmental and economic assessment for the municipality of Aarhus. Waste Manag 30:744–754

Lin CH (2008) A model using home appliance ownership data to evaluate recycling policy performance. Resour Conserv Recycl 52:1322–1328

Martinho G, Gomes A, Santos P, Ramos M, Cardoso J, Silveira A, Pires A (2017) A case study of packaging waste collection systems in Portugal part I: performance and operation analysis. Waste Manag 61:96–107

Moreira AR (2008) Municipal solid waste collection routes analysis and assessment of operational variables at routes productivity (in Portugusese: Análise de circuitos de recolha de RSU indiferenciados e avaliação da influência de variáveis operacionais na produtividade dos circuitos) Dissertation. Nova University Lisbon - Nova School of Science and Technology (FCT NOVA)

Nguyen T, Wilson B (2010) Fuel consumption estimation for kerbside municipal solid waste (MSW) collection activities. Waste Manag Res 28:289–297

Organisation for Economic Co-operation and Development (OECD) (2008) Key environmental indicators. OECD, Paris

Organisation for Economic Co-operation and Development (OECD) (2009) Eco-innovation in industry enabling green growth. OECD, Paris

Passarini F, Vassura I, Monti F, Morselli L, Villani B (2011) Indicators of waste management efficiency related to different territorial conditions. Waste Manag 31:785–792

Petersen CH, Berg PE (2004) Use of recycling station in Börlange, Sweden - volume weights and attitudes. Waste Manag 24:911–918

Peterson PJ, Granados A (2002) Towards sets of hazardous waste indicators: essential tools for modern industrial management. Environ Sci Pollut Res 9:204–214

Pires A, Sargedas J, Miguel M, Pina J, Martinho G (2017) A case study of packaging waste collection systems in Portugal–part II: environmental and economic analysis. Waste Manag 61:108–116

Rodrigues SSM (2016) Classification and benchmarking of municipal solid waste collection systems (in Portuguese: Classificação e benchmarking de sistemas de recolha de resíduos urbanos) Dissertation. Nova University Lisbon - Nova School of Science and Technology

Rodrigues S, Martinho G, Pires A (2016) Waste collection systems part B: benchmarking indicators. Benchmarking of the great Lisbon area, Portugal. J Clean Prod 139:230–241

Sanjeevi V, Shahabudeen P (2015) Development of performance indicators for municipal solid waste management (PIMS): a review. Waste Manag Res 33:1052–1065

Sanjeevi V, Shahabudeen P (2016) Optimal routing for efficient municipal solid waste transportation by using ArcGIS application in Chennai, India. Waste Manag Res 34:11–21

Schoenberger H, Canfora P, Dri M, Galvez-Martos JL, Styles D, Antonopoulos IS (2014) Development of the EMAS sectoral reference documents on best environmental management practice – learning from frontrunners, promoting best practice. Office of the European Union, Luxembourg

Sonesson U (2000) Modelling of waste collection - a general approach to calculate fuel consumption and time. Waste Manag Res 18:115–123

Sora M, González J (2014) Economic balance of door-to-door and road containers waste collection for local authorities and proposals for its optimization. Commissioned by the Association of Catalan Municipalities for door-to-door selective collection to Fundació ENT

Statista (2018) Recordable workplace injury and illness cases in the US waste industry from 2009 to 2015, by business (per 100 full time workers). https://wwwstatistacom/statistics/665031/us-waste-industry-worksplace-injury-and-illness-cases-by-business/. Accessed 22 Nov 2017

Tai J, Zhang W, Che Y, Feng D (2011) Municipal solid waste source-separated collection in China: a comparative analysis. Waste Manag 31:1673–1682

Teixeira CA, Avelino C, Ferreira F, Bentes I (2014a) Statistical analysis in MSW collection performance assessment. Waste Manag 34:1584–1594

Teixeira CA, Russo MM, Bentes I (2014b) Evaluation of operational, economic, and environmental performance of mixed and selective collection of municipal solid waste: Porto case study. Waste Manag Res 32:1210–1218

Viotti PP (2003) Genetic algorithms as a promising tool for optimisation of the MSW collection routes. Waste Manag Res 21:292–298

Waite R (1995) Household waste recycling. Earthscan Publicatons, London

Zaman A (2014) Identification of key assessment indicators of the zero waste management systems. Ecol Indic 36:682–693

Chapter 11
Assessment and Improvement

Abstract Today's environmental concerns are related to the population and its consumption of resources, which have led to significant ecological global changes, such as climate change and resources overexploitation. The solid waste management, in an integrated way, has been capable of influencing and contributing to the solution of such challenges. The purpose of this chapter is to discuss the assessment and improvement of the waste collection system by using life cycle thinking, with a sustainable perspective. Several methodologies such as life cycle assessment, carbon footprint, life cycle costing, and social life cycle assessment will be presented and discussed concerning its application to waste collection systems and contribution to the integrated waste management system.

Keywords LCA · Social LCA · LCC · Environmental impacts · Public participation · Behavior studies

11.1 Life Cycle Assessment and Carbon Footprint

The life cycle assessment is a process to (a) evaluate the environmental burdens associated with a product, process, or activity by identifying and quantifying the energy and materials used, wastes, and emissions released to the environment; (b) assess the impact of those energy and material uses and releases to the environment; and (c) identify and evaluate opportunities that lead to environmental improvements (Fava et al. 1991; Consoli et al. 1993). According to the International Organization for Standardization (ISO 14040 2006a), LCA addresses the environmental aspects and potential environmental impacts throughout a product's life cycle, from raw material acquisition through production, use, end-of-life treatment, recycling, and final disposal (i.e., cradle to grave). LCA is divided into four phases: goal and scope definition, inventory analysis, impact assessment, and interpretation. The goal and scope definition intends to define the purposes, specifications, and limits in the evaluation. The inventory analysis phase is responsible for the collection of data of the unit processes within the system and relating it to a functional unit. Impact assessment intends to make inventory information more understandable

---

© Springer International Publishing AG, part of Springer Nature 2019
A. Pires et al., *Sustainable Solid Waste Collection and Management*,
https://doi.org/10.1007/978-3-319-93200-2_11

through its translation into environmental impact categories. Final interpretation allows evaluating results obtained and comparing them with the initially defined goal (ISO 14040 2006a).

LCA applied to solid waste collection systems can serve two purposes: to evaluate the service provided (in terms of technology implemented) and find where more environmental impact is occurring and to evaluate which level of source separation (number of streams and quality of material source separated, recycling rate) should be promoted, to reach higher amounts of recyclables and higher quality of recyclables collected in such a way that could be beneficial to the environment. LCA has been applied to solid waste management since the 1990s to treatment and recovery technology processes, where the focus on specific collection and recycling schemes is being increasing more recently.

11.1.1 Goal and Scope Definition

For a waste collection system, the goal of an LCA study depends on the type of decision-making process: a microlevel decision, where the decision to be made will not impact the background system, and a meso−/macro-level, which can impact the background system. The micro-level decision is only devoted to the technical analysis and environmental inventory of the waste collection sector. Meso−/ macro-level is related to the analysis of strategies with large-scale to background sector (like the market for recyclables), which are related to studies on a national scale, with implications on national and international plans. To understand which type of LCA to perform, foreground and background systems must be defined. In a waste collection system LCA study, the foreground system is the waste collection system to be analyzed, where real data will be gathered related to the collection and transport and to waste container production and transport; the background system is generic data which is more related to the electricity grid, for example.

Functions of the System, the Functional Unit, and Reference Flow

At a glance, a waste collection system just performs one function: allows the temporary deposition of waste, its collection, and transport to a specific destination. However, the destination can also be included in the LCA, because waste collection can influence its destination. If packaging waste is source separated, it has recycling features; if organic waste is source separated, the production of a high-quality compost occurs. Even if the waste results from mix collection, it can also generate electric energy. When mixed waste collected is send for mechanical-biological treatment (by anaerobic digeston), or to incineration or even at the landfill, the biogas is generated and used to produce electric energy. The way how the system is defined will determine the number of functions of the system, and, if multifunctionality occurs, it has to be solved. A possible functional unit is the collection of a specific amount of waste generated in a period by a specific group of inhabitants, for example, the selective collection service of 1500 tons a month of MSW generated in an urban locality with a density of 5000 inhabitants/km^2 applied by Iriarte et al. (2009). Related to the functional unit is the reference flow for the normalization of input and output data (Chang and Pires 2015). In the case of Iriarte et al. (2009), the reference flow considered the theoretical recovery of 100% for the fractions: organic, paper, packaging, and glass present in the MSW. When comparing waste collection system for a specific waste stream, waste properties need to be studied, to ensure that the functional unit and the reference flow are the same. It is common that different waste collection vehicles have different waste compaction rates, changing the density of waste collected. Density is just one of the critical physical properties, but also moisture can be relevant for the collection of biodegradable waste or waste paper and waste cardboard.

System Boundaries

The waste collection system studied in the LCA can be:

- The service provided by a municipality or a private company.
- The number of recyclables collected by the collection system.

From a generic point of view, the waste collection system to be analyzed has, in the beginning, the stages of temporary deposition (containers) and waste collection vehicles. The frontiers of the waste collection also need to be addressed. The frontiers can be related to geographic locations, timescales, and technical components. Considering the technical components, LCA studies can be cradle-to-grave (starts with the extraction of materials, going through all life cycle), cradle-to-gate (from raw material extraction, going until the product leaves the factory), gate-to-gate (only regards a manufacturing process), and cradle-to-cradle (with a metabolic view where no waste exists) (Chang and Pires 2015). In waste collection systems, LCA can be only cradle-to-gate if the intention is to assess a particular waste collection technology without the use phase (like an environmental product declaration for waste bins), or the waste collection system itself can be a cradle-to-gate of the entire waste management system. A cradle-to-cradle applies to reverse logistic cases when the product is not waste or is for reuse. A typical waste collection LCA is a cradle-to-gate, or a streamlined LCA (also named screening and matrix LCA) because the assessment is only for a part of the life cycle. The definition is also applied when the LCA is not assessing all the environmental impact categories (Crawford 2011). The case study on Box 11.1 is an example of a streamlined LCA at recycling schemes.

> **Box 11.1 Comparison of Recycling Schemes in Portugal (Pires et al. 2017)**
>
> A comparative study based on LCA and economic analysis through indicators was conducted for three waste collection systems for packaging waste in a municipality in Portugal. The analysis compares the environmental impact of existing collection systems and the costs involved in the operation of those systems. For the LCA, Umberto 5.5 software package was used.
>
> Three waste recycling systems were compared: a curbside system, where all non-glass packaging waste is collected by the curbside bags; a drop-off system, where all packaging waste is collected by drop-off containers; and a mixed system, where glass is deposited at drop-off containers and lightweight packaging is deposited at drop-off and curbside systems. The LCA was used to analyze the environmental impacts, but only the collection system was assessed (the subsequent recycling was excluded). The results showed that the curbside system was less favorable economically and environmentally due to the more packaging and more fuel consumption per ton of waste, compared to drop-off collection system. Optimization of the curbside system is needed, through the use of reusable boxes and efficient collection routes (SEP 2018).

In a waste collection LCA, there is no need to include the environmental impacts resulting from the extraction of material and production of goods that originate waste, neither product reuse with the application of the "zero burden assumption." The zero burden assumes that waste brings no upstream environmental impacts into the waste collection system neither to the waste management system (Ekvall et al. 2007). Such can be applied because all product life cycle phases previous to the waste phase occur in the same way in the next waste collection alternatives.

To conduct an LCA is also needed to have in consideration aspects related to geographic boundaries and time horizons. Concerning geographic boundaries, the waste collection systems are typically local and regional, but also can be national or international, when materials collected are sent to recycling, which can be outside the borders of the country. Data referent to process far from the place where waste collection efficiently occurs can be hard to collect (Chang and Pires 2015). In the case of time horizon, the functional unit in a waste collection system can be dependent on the respective year or another time unit, which is the case of the amount of waste generated and collected, which is not constant through the years, with fluctuations during the year. There is the need to identify the time horizon of the analysis appropriately. When conducting an LCA for a waste collection system which the system is only the collection itself, there is the need to include equipment and infrastructure data. In this situation, the useful life of devices needs to fit into the functional unit period.

Another aspect to be defined during the goal and scope definition is the allocation procedure. According to ISO (2006b), allocation represents the portioning of input and output flows of a process or a system concerning the system under analysis and one or more other systems. Allocation applies in cases of multifunctionality and

11.1 Life Cycle Assessment and Carbon Footprint

when open recycling inside the system occurs (Ekvall and Tillman 1997). Multifunctionality is related to a multi-output and multi-input processes (or systems). A multi-output process occurs when a single system produces more than one product or only one product is processed inside the system and at least one product is generated and is used outside the system (what is called a coproduct) (Klöpffer and Grahl 2014). A multi-input process in waste management systems occurs when several waste streams are collected and treated, while LCA tries to isolate one of them (Tillman 2010). When a product is recycled not at the same product but in a different one, it is a case of open recycling (Tillman 2010). The way how to proceed to solve them is different if the LCA is of attributional type or consequential type, although there is no universal consensus. In general, the multi-output systems in waste management is usually solved by system expansion/substitution, whatever is an attributional or a consequential LCA, by ISO (2006b) and recommendations from EC-JRC-IES (2010). In a first step, system expansion is performed until all expanded systems produce the same quantities of the coproducts identified in the system, and in the next step, product outputs and inputs related with the coproducts are subtracted from all expanded systems (Bueno et al. 2015). In the case of multi-input, portioning made by physical or chemical classification is typically conducted (Meijer et al. 2017; Guinée et al. 2002). In the case of recycling allocation, "recycled content approach" (or cutoff approach) and the "end-of-life recycling approach" (or avoided burden approach) are used (Frischknecht 2010) (Fig. 11.1). The cutoff method considers the share of recycled material in the manufacture of the product, where the environmental impacts of recycled material were not attributed to the system under investigation because once recycled, they start a new life in a second product/process (Frischknecht 2010; Zampori and Dotelli 2014). In the avoided burden approach, the environmental impacts from the recycling include the system under investigation, avoiding the extraction of raw materials for the production of the product, and relating environmental impacts, crediting them to the product in the system in assessment (Frischknecht 2010; Zampori and Dotelli 2014).

According to Pelletier et al. (2015), the multifunctionality needs to be adequately justified, and the different approaches to solving multifunctionality mostly relate to the schools of LCA practitioners, which view the purpose of LCA in different ways:

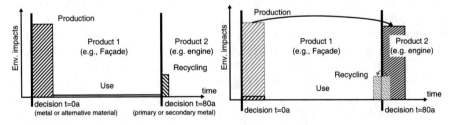

Fig. 11.1 Environmental impacts in the course of time during production, use, and end of life (recycling) of a long-living metal product. Left, recycled content approach; right end-of-life recycling approach. (Source: Frischknecht (2010))

Table 11.1 Alternative multifunctionality hierarchies consistent with competing for understanding of nature, purpose, and conditions necessary to LCA

| ISO 14044 | Consequential data modeling approach | Attributional data modeling approach | |
		Physical perspective	Socioeconomic perspective
Tier 1: Avoid allocation via subdivision or system expansion	Avoid allocation by subdivision or "system expansion + substitution"	Avoid allocation by subdivision or system expansion (reporting at level of all coproducts)	Avoid allocation by subdivision or system expansion (reporting at level of all coproducts)
Tier 2: Allocation based on an underlying physical relationship	NA	Avoid based on a relevant underlying physical relationship	Avoid based on a relevant underlying economic value of coproducts
Tier 3: Allocation based on some other reason	NA	NA	NA

Pelletier et al. (2015)

multifunctionality is to be solved by system expansion or by physical allocation or by economic allocation (representing allocation based on "some other relationship"). Whatever the school of the practitioner, Pelletier et al. (2015) suggest the alternative multifunctionality to help in making allocation consistent and more transparent to practitioners (Table 11.1).

11.1.2 Life Cycle Inventory

LCI represents the phase in the LCA where the collection and treatment of the data to perform the assessment occur. The steps of LCI include the data collection planning, collection itself, and validation of data. Concerning planning, there is the need to define the data to be collected regarding the type of LCA (attributional or consequential), type of the system (is data for the foreground or the background), and the LCA scale (is a full LCA or a streamlined). Depending on the type of LCA conducted – attributional or consequential – the type of information to be collected differs. For an attributional LCA, data to be collected is average or generic data that best represent the waste collection system. In consequential LCA, marginal data collection is related to operations during the life cycle that are affected by a change in the system under investigation (Ekvall and Weidema 2004). To develop the consequential analysis, scenario development and market forecasting can be applied. The one more used is the market forecasting, which only implies the knowledge of the existing market for outputs and inputs of the system, when the scenario development is critical to their application.

Concerning foreground and background data, the approaches to collect information are different. In the case of foreground data, or primary data, data collection intends to characterize as far as possible the system, being collected all data possible concerning inputs and outputs; background data, or secondary data, are related to information of secondary processes with no apparent influence on the core system (Chang and Pires 2015).

Choosing between a full and a streamlined LCA should be based on the goal and scope of the study and the time available to conduct the assessment, because LCA is time-consuming and expensive (Wang et al. 2016). Streamlined LCA occurs by (1) adjusting the system boundary (both foreground and background systems) and (2) limiting the inputs, outputs, and environmental impacts considered in the assessment. Previous full LCA studies can indicate areas which have low significance to the LCA results, allowing a justified streamlined LCA (Chang and Pires 2015).

LCI databases provide ready-made inventories to characterize waste collection systems. There are several databases which characterize several processes including waste collection and treatment processes. Most complete databases are the Ecoinvent (http://www.ecoinvent.org/), US Life Cycle Inventory Database (https://www.nrel. gov/lci/), and European Life Cycle Database (ELCD) (http://eplca.jrc.ec.europa.eu/ ELCD3/), just to name a few.

Documentation of data calculation for LCI occurs explicitly, where the explanation of all assumptions occurs. The validation of data and relating data to unit processes and functional unit is needed to ensure the quality of LCA (ISO 2006b). The validation of data during LCI should be made through mass balances, energy balances, and comparison with data from other sources, like emission factors for specific processes (Guinée et al. 2002).

11.1.3 Life Cycle Impact Assessment

The result of the LCI phase is the quantification of materials, energy, and substance flows which impact the environment. The LCIA intends to understand and evaluate the environmental impacts resulting from the system in the analysis, regarding the magnitude and the significance (ISO 2006b). The critical steps of the LCIA are (Curran 2006) selection and definition of impact categories, classification of substance flows with the selected impact category, and characterization of LCI impacts based on conversion factors scientifically based.

Complexity reduction of the conversion of inventory into impact categories occurs by the impact categories definition in midpoint and endpoint indicators. Midpoint indicators calculate the impact of LCI outputs through various environmental mechanisms with less uncertainty; endpoint indicators include the characterization factors to link midpoint indicators through additional environmental mechanisms, which incorporates greater uncertainty (Li and Khanal 2016) (Fig. 11.2).

In addition to the fundamental steps, other steps can be added to reach a more clear result, such as normalization, grouping, and weighting of impact categories, which will facilitate the comparison of LCA results and the interpretation phase (ISO 2006b). According to ISO (2006b) and Ashby (2009), normalization intends to

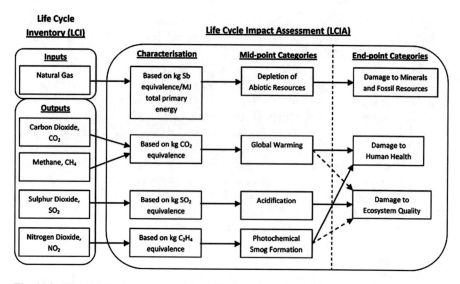

Fig. 11.2 Typical LCA framework linking LCI via midpoint categories to endpoint categories for selected damage types. Indicators can be formed from either category after normalization and optional weighting step. (Source: Rimos et al. (2014))

remove the units and reduce the data to a standard scale, grouping intends to sort and rank the impact categories if possible, and weighting of each impact category helps to understand which are the most critical impacts compared to the other category impacts. The result of these additional key steps is a value, an eco-indicator, which condensates all the information resulting from the LCA into one number. There is some criticism on the use of eco-indicators since there is no agreement on normalization and weighting factors and the value has no physical significance (Ashby 2009).

The LCIA is typically made by different methodologies, from their resulting indicators that could help to quantify and compare the environmental impact of the product or service. Several methodologies exist: CML (Guinée et al. 2002), Eco-Indicator 99 (EI'99) (Goedkoop and Spriensma 2000), Environmental Priority System 2000 (EPS 2000) (Steen 1999), EDIP (Hauschild and Potting 2005), IMPACT 2002+ (Jolliet et al. 2003), TRACI (Bare et al. 2003), USEtox 2.0 (Fantke et al. 2015), ReCiPe (Goedkoop et al. 2009), and ES'06 (Frischknecht et al. 2008). Choosing the LCIA system should address the following questions (Rosenbaum et al. 2018):

- Which impact categories do I need to cover and can I justify those that I am excluding?
- Which are the features of the region where the system in the analysis occurs?
- What kind of LCIA do I need, midpoint, endpoint, or both, and are the normalization steps also needed?
- Which elementary flows do I need to identify and know?
- Is there any information from organizations that could help me to choose?

- How can the LCIA results be interpreted and communicated?
- How well is the method scientifically supported?
- How proven is the method?
- Is there available data from my LCI to support the LCIA method?
- Is uncertainty an issue that needs to be quantified?

One possible strategy to choose an LCIA method can be the existence of LCA studies made to the waste collection system and the LCIA method used. Probably there is a method more frequently used, which can also be chosen, helping in the comparison of the results.

11.1.4 Interpretation

According to ISO (2006b), the interpretation is the last phase of the LCA, where results are summarized for conclusions and recommendation and help on decision-making, depending on the goals and scope defined at the beginning of the LCA. Due to its looping plus iterative procedure, the discussion conducted can dictate changes in previous decisions during the LCA like allocation rules, system boundaries, goal and scope features, data collected to perform the LCA, and environmental impact categories chosen, just to name a few of the possible consequences of interpretation phase (Chang and Pires 2015; ISO 2006b). Interpretation phase recommends a critical analysis done by an external entity (ISO 2006b).

Uncertainty and sensitivity analyses are conducted during the interpretation phase. Sensitivity analysis intends to understand how the model inputs influence the results; uncertainty analysis (also named propagation) aims to know quantitatively the overall uncertainty of results reached during LCA (Laurent et al. 2014). The most common method used to assess sensitivity is scenario analysis. These are one-factor-at-a-time (OFAT) methods with the intention to investigating the robustness of the results and finding the sensitive parameters that could influence LCA results and, in the last case, alter the recommendations to decision-makers (Laurent et al. 2014). Sensitivity analysis is performed by varying the inputs within a specific range and analyzing the impacts on the results, showing which are the results that must be regarded more carefully, and which assumptions must be justified and validated (Li and Khanal 2016). Uncertainty in waste collection system is generally related to the waste composition itself and the waste fraction distributions and chemical composition (e.g., water content, density). The system model used for the collection itself, the choice of a collection scheme, and the parameters dependent of the collection scheme, like fuel consumption, emissions, source-sorting efficiencies, and the transport distance, at least should be subjected to uncertainty analysis (Clavreul et al. 2012). Uncertainty analysis is usually conducted by Monte Carlo analysis, which consists in randomly sampling the probability distribution of each uncertain parameter in a large number of times, resulting in a frequency histogram and a probability distribution representing model results (Clavreul et al. 2012). When conducting an LCA comparing different waste management solutions, the

Monte Carlo simulation can indicate, for each solution, which is the probability that a specific result occurs (e.g., which is the probability of the result "incineration is better than anaerobic digestion" occur).

11.1.5 LCA Software

There are several on-market software to conduct an LCA study on waste collection, like Gabi, SimaPro, Team, and Umberto software. There is also more friendly software explicitly devoted to waste management, including solid waste collection, which can make the streamlined LCA easier. Those software/applications are IWM-2 (McDougall et al. 2001), WISARD/WRATE (Ecobilan 2004), EASEWASTE (Christensen et al. 2007), and ORWARE (Dalemo et al. 1997; Björklund et al. 1999). The development and use of waste LCA tools justify the need to deal with a reference flow composed of a mixture of materials (waste and its several waste streams); the LCA practitioner can evaluate more natural the influence of several parameters of the waste management scheme on the LCA results, making more accessible for the practitioner to track the impacts from heterogeneous waste streams and the impacts caused by each material (Clavreul et al. 2014).

No matter which is the software, the practitioner of an LCA to waste collection system must have in mind that capital goods may have significant importance on the LCA environmental impacts, not being adequate to exclude them. According to Brogaard and Christensen (2012), the impact of producing the capital goods for waste collection and transport – vehicles and containers – should not be neglected as the capital goods can be responsible for more than 85% of some of the environmental impact categories from all environmental impacts occurring for collection and transport waste (when a transport distance of 25 km was assumed).

11.1.6 Carbon Footprint

Common to most of these environmental impact systems is the one related to GHG emissions and climate change impact, i.e., when using LCA only to calculate the impact on climate change). GHG emissions impact can have several designations, where the most known is carbon footprint. Carbon footprint is also a subcomponent of ecological footprint, which is estimated by calculating the embodied life cycle energy plus GHG emissions associated with a specific system (Cifrian et al. 2013). According to EPLCA (2007), carbon footprint (also named as carbon profile) is the inventory of greenhouse gas emissions associated with a product, along with its life cycle, from the supply chain, use, and end-of-life.

Carbon footprint results from the indicator global warming potential (GWP), used in LCA. As defined by IPCC, the GWP reflects the relative effect of a GHG regarding the climate change, considering a fixed period (e.g., 100 years is GWP_{100}). GHG can have a different global warming impact, i.e., can contribute differently to the climate change, presenting different GWP_{100}. Carbon dioxide has 1, methane has 25, nitrous oxide has 298, HFCs have between 124 and 14,800, sulfur hexafluoride has 22,800, and PFCs have 7390–12,200 GWP_{100} (IPCC 2007).

Carbon footprint is a streamlined LCA, where the analysis is made only to the emissions that have a potential effect on climate change. Carbon footprint calculation uses databases for the background data, and for the foreground, it is necessary to collect real information as possible. When making a carbon footprint, it is necessary to have in mind that a possible "shifting of burdens" may occur, because other relevant environmental impacts are neglected (EPLCA 2007).

Several standards and norms could help in the development of a carbon footprint. The one specific for carbon footprint from ISO is ISO/TS 14067 (ISO 2013), which is based on the ISO norms for LCA (ISO 14040-14044 (ISO 2006a, b)), on standards for quantification, and on environmental labels and declarations for communication (ISO 14020, ISO 14024, and ISO 14025). Public initiatives to develop carbon footprint calculation methodologies also exist: the British Standards Institution Norm PAS 2050:2011 (BSI 2011), Protocol for the Quantification of Greenhouse Gases Emissions from Waste Management Activities (EpE 2010), Product Life Cycle Accounting and Reporting Standard of the GHG Protocol (WRI and WBCSD 2011), and US EPA Waste Reduction Model (WARM) (USEPA 2009), just to name a few.

Waste collection and transport contribute significantly to GHG emissions from municipal solid waste management system (Bernstad and la Cour Jansen 2012; Jaunich et al. 2016; Cleary 2009). The sources and magnitude of GHG depend on the type of collection and transport system in place: pneumatic systems (Teerioja et al. 2012; Punkkinen et al. 2012) or trucks (Fernández-Nava et al. 2014; Maimoun et al. 2013; Rose et al. 2013). In trucks' case, the fuel used by collection vehicles has a significant influence on the carbon footprint (López et al. 2009; Maimoun et al. 2013; Rose et al. 2013) and urban air quality (Fontaras et al. 2012; Sandhu et al. 2014). Authors have tried to adapt existing methodologies to calculate LCA or carbon footprint of waste collection systems, to reach more detailed inventories instead of just using average data. In Table 11.2 is presented a short review on assessment on waste collection systems made by LCA and carbon footprint.

11.2 Life Cycle Costing

Economic life cycle analysis, most known as life cycle costing (LCC), gives an economic perspective on the life cycle of the product or service. LCC involves three types of LCC assessments (Hunkeler et al. 2008): conventional, environmental, and societal. Conventional LCC represents standardized financial assessments, like accounting for marketed goods and services carried out typically by individual companies focusing on their direct costs. Environmental LCC includes the conventional LCC (also named financial LCC, where albeit costs from all stakeholders are included), to be in line with the system boundaries of the LCA (Rödger et al. 2018). The societal LCC further includes externality costs (i.e., it "internalizes"

Table 11.2 LCA and carbon footprint case studies on waste collection

LCA of waste collection systems		Carbon footprint of waste collection	
Source	Description	Source	Description
Punkkinen et al. (2012)	Comparing pneumatic and door-to-door collection systems	Pérez et al. (2017)	Developed a methodology for calculating the carbon footprint of waste collection vehicles
Rose et al. (2013)	Comparison of diesel and compressed natural gas-powered refuse collection vehicles	Maimoun et al. (2013)	Compare different fuels for waste collection vehicles
Pires et al. (2017)	Comparison of different packaging waste collection systems: curbside, drop-off, and mixed	Eriksson et al. (2015)	The carbon footprint of food waste management options

environmental and social impacts by assigning monetary values to the respective effects), by using accounting prices (Martinez-Sanchez et al. 2015; Rödger et al. 2018). The three types of LCC give the holistic view of the system, including LCA perspective and well societal concerns.

For each type of LCC, different costs may be considered (Table 11.3), such as internal costs, external costs, and social costs. Internal costs are referent to monetary costs occurring both inside and outside the waste management system in the analysis, being measured by market prices (Martinez-Sanchez et al. 2015). External costs occur outside the economic system, having no direct monetary value in the market, reflecting the impacts on third parties resulting from production and consumption (Martinez-Sanchez et al. 2015; Rödger et al. 2018). Social costs are the sum of internal and external costs, being defined as society's costs for managing waste (Porter 2002).

The results from the LCC are expressed in monetary terms per functional unit, for each of the life cycle phases, which can also be defined in the budget, transfers, and its sum (named convention LCC), and from the LCC result indicators that help to understand the assessment made. The societal LCC presents results from external costs, being divided into budget costs, externality costs, and societal costs (Martinez-Sanchez et al. 2015). Martinez-Sanchez et al. (2016), who have applied LCC to a food waste management system, have included in the LCC the indirect costs, related to the income effect associated with the marginal consumption and to indirect land use changes.

11.3 Social Life Cycle Assessment

To assess the social impact of a product, but also of services including waste collection system (also applicable to waste management system), a social life cycle can be conducted. Social impacts focus on aspects related to the well-being of humans (Yildiz-Geyhan et al. 2017). According to UNEP/SETAC (2009) and Benoît et al. (2010), social life cycle assessment (SLCA) is an assessment technique capable of evaluating the socioeconomic and societal impacts of products during their life cycle. Social impacts are consequences of behaviors/decisions,

11.3 Social Life Cycle Assessment

Table 11.3 Overview of costs incurred by waste agents and all members of society with regard to waste systems. Cost classes are (1) internal and external costs and (2) budget costs, externality costs, and transfers

	Internal costs	External costs	Social costs
Incurred by	Waste agents (e.g., waste generator and operators)	All the members of society (waste generators, waste management operators, and others)	Society
Budget cost	Bags, bins, capital goods, materials, and energy consumption, labor costs, material and energy sales		
Externalities cost		Time consumptions to source separate, health issues, disamenities, working environment issues	Sum of internal costs (excluding transfers) and external costs for society (i.e., waste generator, waste operator, and other agents)
Transfer	Fees, taxes, pecuniary externalities		Not applicable

Martinez-Sanchez et al. (2015)
Pecuniary externalities may be related to energy and material recovery within the waste system. These transfers represent financial losses occurring when existing facilities or industries outside the system boundary of the assessment have to operate below their design capacity as a result of the additional supply of energy and material resources offered by the waste system

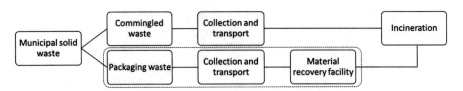

Fig. 11.3 Existing municipal solid waste collection system and system boundaries of the packaging waste collection system to be assessed in a hypothetical SLCA

socioeconomic processes, and capitals (human, social, and cultural), which can be either positive or negative (UNEP/SETAC 2009).

The methodology of SLCA follows the environmental LCA: goal and scope definition, inventory, impacts, and interpretation. In goal and scope definition, the critical aspect to have in mind is the functional unit to be defined, because it can be difficult to correlate a social impact with a process of a product or a service (Dreyer et al. 2006; Hauschild et al. 2008; Klöpffer 2008). For instance, Hosseinijou et al. (2014) indicated the social impacts would hardly be related to the functional unit (FU) of the product if the inventory data is based on semi-qualitative and qualitative data. Also, the frontiers of the system can also be challenging, although on assessing waste collection system, the task can be facilitated. In the case of Fig. 11.3, to assess a packaging waste collection system, only the packaging collection and transport are to be considered in the SLCA.

196   11 Assessment and Improvement

Another aspect to be defined in goal and scope definition is the impact categories to be assessed. There are 6 impact categories (human rights, working conditions, health and safety, cultural heritage, governance, socioeconomic repercussions) and 31 subcategories related with stakeholders' categories. In Table 11.4 are represented the stakeholders and a resume of the subcategories proposed by Benoît-Norris et al. (2011).

Since its beginnig in the 1990s, the SLCA has not been capable of being fully standardized, like what happens to LCA (Iofrida et al. 2018; Sureau et al. 2018). There is a difficulty in addressing social impacts into a physical flow of a product or of a service (Dreyer et al. 2006). Also, the SLCA published by UNEP/SETAC

Table 11.4 Five stakeholder categories in production system based on the UNEP's guideline for SLCA

Stakeholder categories	Subcategories
Worker	Freedom of association and collective bargaining
	Child labor
	Fair salary
	Working hours
	Forced labor
	Equal opportunities/discrimination
	Health and safety
	Social benefits/social security
Consumer	Health and safety
	Feedback mechanism
	Consumer privacy
	Transparency
	End-of-life responsibility
Local community	Access to material resources
	Access to immaterial resources
	Delocalization and migration
	Cultural heritage
	Safe and healthy living conditions
	Respect of indigenous rights
	Community engagement
	Local employment
	Secure living conditions
	Public commitments to sustainability issues
	Contribution to economic development
Society	Prevention and mitigation of armed conflicts
	Technology development
	Corruption
Value chain actors (excluding consumers)	Fair competition
	Promoting social responsibility
	Supplier relationships
	Respect of intellectual property rights

Source: Benoît-Norris et al. (2011)

(2009) lacks on specific impact assessment methodology, which has made practitioners apply different approaches (Chhipi-Shrestha et al. 2015) to solve it. Site-specific data collection is needed to characterize the foreground system in assessment, and those procedures are complicated to be implemented, requiring prioritization or cutoff criteria, as well a global social database to provide the rest of the data needed (Chhipi-Shrestha et al. 2015). All these issues make SLCA a not well-proven technique to assess the social impact of a product or a service, in this case, a waste collection system. The missing robustness of SLCA is even more problematic for a sector which is characterized by an informal sector in developing countries mostly, but also in developed countries, informal work may occur. In the study from Yildiz-Geyhan et al. (2017), the intention was to conduct an SLCA to packaging waste collection schemes, with scenarios of the formal and informal collection. The results showed that informal collection scenarios had socially fewer score than the formal scenarios in almost all impacts, but the best scenario was the ameliorate scenario, where the integration of formal and informal collection occurs.

11.4 Behavior Studies and Awareness Campaigns

The assessment of a waste collection system and, especially, of recycling schemes is mandatory to understand the effectiveness and what needs to be improved. The identification of residents' participation and acceptance of the recycling scheme is essential to the success of the scheme. Doing behavior studies and awareness campaigns will be helpful to reduce misjudgments that led to poor scheme design and performance, leading to high operational costs (Altaf and Hughes 1994; Jenkins et al. 2003). Although the behavior study is here presented as an assessment of waste collection or recycfling schemes, behavior studies are also made during the design, like is the case at Box 11.2, where public participation was considered in the design if the waste collection at the community of Didimoticho in Greece (Keramitsoglou and Tsagarakis 2013) (Box 11.2).

> **Box 11.2 Public Participation in Designing a Recycling Scheme in Didimoticho, Greece (Keramitsoglou and Tsagarakis 2013)**
> A public participation process was implemented in Greece, in a town where no source separation of waste exists. The process based on a structured questionnaire is divided into four parts: questions about residents' knowledge on recycling, including advantages and disadvantages; questions about nine recycling programs, intention to participate, and number of materials to be source separated; questions to assess attitudes to financial incentives; and questions on socioeconomic situation of respondents. A total of 343 validated answers were gathered. The result of this participative process was the strategy

(continued)

> **Box 11.2** (continued)
>
> for the introduction of recycling schemes in the town, where two stages were defined. In the first stage, drop-off systems should be provided for four waste flows, and simultaneously, domestic composting and hazardous waste recycling should be ensured. In the second stage, financial instruments and a curbside collection scheme in combination with the drop-off system should be explored. At the end of the two stages, evaluations must be conducted to verify if the strategy is successful or not.

When conducting the behavior studies, the intention is to understand which factors are affecting the participation of citizens in the collection system. Several factors have been studied so far by scientists (Miafodzyeva and Brandt 2013; Varotto and Spagnolli 2017):

- Socio-demographic factors: Include age, education level, income, gender, dwelling type, household size, home ownership, household type, employment status, and ethnicity.
- Psychological factors: Include information and knowledge, convenience/effort, social influence, responsibility, environmental attitudes, beliefs/perceptions of recycling consequences, specific recycling attitudes, motivation, recycling experience, behavior skills, the perception of the service provider, personality characteristics, emotion, and sense of community.
- Contextual factors: Include service, monetary incentives, the location of bins, characteristics of bins, and product characteristics.
- Other study-specific factors: Related to individual factors influencing the recycling behavior, including the share of immigrants in a community and significance of behavior habits and shopping behavior.

Understanding which are the factors affecting the recycling behavior will, consequently, determine the success of the source-separated collection system and influence the contents and shapes of the awareness campaign has to be elaborated. Awareness campaigns intend to communicate using strategies which could increase recycling behavior of citizens. Prompts, persuasive communication, verbal commitment, written contracts, and feedback, i.e., "the transmission of information about the effects of the behaviors of an individual or a group," are approaches reviewed and highlighted by Dupré and Meineri (2016).

11.5 Final Remarks

The assessment of any waste collection system requires a complete and robust definition of the system, to ensure that the goal of the assessment is correctly assessed. Most of the time, the missing data, the reduced time available, the budget constraints, and, most of all, the missing support of directors and managers to proceed with an accurate assessment of the system may lead to biased results that do not reflect the reality of the waste collection system. Efforts should be made, firstly, to make the assessment tools based on life cycle thinking available to this sector. Make available means cost affordable and scientifically understandable (concerning running the methods and result interpretation). However, such natural access (economically and technically) cannot make life cycle thinking models too simplified, in such way that it will not reflect the life cycle of the service provided – the collection of waste. The development of methodologies that could assess the sustainability of waste collection should be made at the light of Open Innovation 2.0 (Curley and Salmelin 2013). Practitioners, methodology developers, and academics should work together to make life cycle thinking methods, leading to the creation of wealth in the waste sector. When the waste sector reaches sustainable standards due to those assessment methods, the governments may establish such standards as the norms for an appropriate waste collection system operation. Citizens, in contact with those waste collection system, will demand all waste collection to perform in such sustainable way. The academia will force the entrance of the sustainable, holistic, and life cycle thinking in the waste-related course programs, repeating the quadruple helix cycle (academia, business, government, and citizens). In this new paradigm, waste collection and waste management sector can make the assessment and improvement of the sector a reality, preparing it for the challenges that will come in the future.

References

Altaf A, Hughes J (1994) Measuring the demand for improving urban sanitation services: results of a CV study in Ouagadougou, Burkina Faso. Urban Stud 31:19–30

Ashby MF (2009) Materials and the environment: eco-informed material choice. Elsevier, Oxford

Bare JC, Norris GA, Pennington DW (2003) TRACI, the tool for the reduction and assessment of chemical and other environmental impacts. J Ind Ecol 6:49–78

Benoît C, Norris GA, Valdivia S, Ciroth A, Moberg A, Bos U, Prakash S, Ugaya C, Beck T (2010) The guidelines for social life cycle assessment of products: just in time! Int J Life Cycle Assess 15:156–163

Benoît-Norris C, Vickery-Niederman G, Valdivia S, Franze J, Traverso M, Ciroth A, Mazijn B (2011) Introducing the UNEP/SETAC methodological sheets for subcategories of social LCA. Int J Life Cycle Assess 16:682–690

Bernstad A, la Cour Jansen J (2012) Review of comparative LCAs of food waste management systems – current status and potential improvements. Waste Manag 32:2439–2455

Björklund A, Dalemo M, Sonesson U (1999) Evaluating a municipal solid waste management plan using ORWARE. J Clean Prod 7:271–280

British Standards Institution (BSI) (2011) PAS 2050:2011 specification for the assessment of the life cycle greenhouse gas emissions of goods and services. BSI, London

Brogaard LK, Christensen TH (2012) Quantifying capital goods for collection and transport of waste. Waste Manag Res 30:1243–1250

Bueno G, Latasa I, Lozano PJ (2015) Comparative LCA of two approaches with different emphasis on energy or material recovery for a municipal solid waste management system in Gipuzkoa. Renew Sust Energ Rev 51:449–459

Chang NB, Pires A (2015) Sustainable solid waste management: a systems engineering approach, IEEE book series on systems science and engineering. Wiley–IEEE Press, Hoboken

Chhipi-Shrestha GK, Hewage K, Sadiq R (2015) 'Socializing' sustainability: a critical review on current development status of social life cycle impact assessment method. Clean Technol Environ Policy 17:579–596

Christensen TH, Bhander G, Lindvall H, Larsen AW, Fruergaard T, Damgaard A, Manfredi S, Boldrin A, Riber C, Hauschild M (2007) Experience with the use of LCA–modelling (EASEWASTE) in waste management. Waste Manag Res 25:257–262

Cifrian E, Andres A, Viguri JR (2013) Estimating monitoring indicators and the carbon footprint of municipal solid waste management in the region of Cantabria, Northern Spain. Waste Biomass Valoriz 4:271–285

Clavreul J, Guyonnet D, Christensen TH (2012) Quantifying uncertainty in LCA–modelling of waste management systems. Waste Manag 32:2482–2495

Clavreul J, Baumeister H, Christensen TH, Damgaard A (2014) An environmental assessment system for environmental technologies. Environ Model Softw 60:18–30

Cleary J (2009) Life cycle assessments of municipal solid waste management systems: a comparative analysis of selected peer–reviewed literature. Environ Int 35:1256–1266

Consoli F, Allen D, Boustead I, Fava J, Franklin W, Jensen AA, de Oude N, Parrish R, Perriman R, Postlethwaite D, Quay B, Séguin J, Vigon B (eds) (1993) Guidelines for life–cycle assessment: a 'code of practice'. SETAC, Brussels

Crawford RH (2011) Life cycle assessment in the built environment. Spon Press, New York

Curley M, Salmelin B (2013) Open innovation 20 – a new paradigm. https://ec.europa.eu/digital-single-market/en/news/open-innovation-20-%E2%80%93-new-paradigm-and-foundation-sustainable-europe. Accessed 10 Mar 2018

Curran MA (2006) Life cycle assessment: principles and practice. National Risk Management Research Laboratory – Office of Research and Development, USEPA, Cincinnati

Dalemo M, Sonesson U, Björklund A, Mingarini K, Frostell B, Nybrant T, Jönsson H, Sundqvist J-O, Thyselius L (1997) ORWARE – a simulation model for organic waste handling systems. Resour Conserv Recycl 21:17–37

Dreyer LC, Hauschild MZ, Schierbeck J (2006) A framework for social life cycle impact assessment. Int J Life Cycle Assess 11:88–97

Dupré M, Meineri S (2016) Increasing recycling through displaying feedback and social comparative feedback. J Environ Psychol 48:101–107

Ecobilan (2004) WISARD – waste integrated system for analysis of recovery and disposal. Ecobilan

Ekvall T, Tillman AM (1997) Open–loop recycling: criteria for allocation procedures. Int J Life Cycle Assess 2:155–162

Ekvall T, Weidema BP (2004) System boundaries and input data in consequential life cycle inventory analysis. Int J Life Cycle Assess 9:161–171

Ekvall T, Assefa G, Björklund A, Eriksson O, Finnveden G (2007) What life–cycle assessment does and does not do in assessments of waste management. Waste Manag 27:989–996

EpE (2010) Protocol for the quantification of greenhouse gases emissions from waste management activities. http://www.epe-asso.org/en/protocol-quantification-greenhouse-gases-emissions-waste-management-activities-version-5-october-2013/. Accessed 15 Jan 2018

Eriksson M, Strid I, Hansson PA (2015) Carbon footprint of food waste management options in the waste hierarchy – a Swedish case study. J Clean Prod 93:115–125

European Commission–Joint Research Centre–Institute for Environment and Sustainability (EC–JRC–IES) (2010) International reference life cycle data system (ILCD) handbook general guide for life cycle assessment – detailed guidance. European Commission–Joint Research Centre–Institute for Environment and Sustainability, Luxembourg

References

European Platform on Life Cycle Assessment (EPLCA) (2007) Carbon footprint – what it is and how to measure it. European Commission, Ispra

Fantke PE, Huijbregts MAJ, Margni M, Hauschild MZ, Jolliet O, Mckone TE, Rosenbaum RK, van De Meent D (2015) USEtox 20 user manual (version 2). http://usetox.org. Accessed 15 Jan 2018

Fava J, Dennison R, Jones B, Curran MA, Vigon B, Selke S, Barnum J (eds) (1991) A technical framework for life–cycle assessment. SETAC and SETAC Foundation for Environmental Education, Washington, DC

Fernández-Nava Y, del Río J, Rodríguez-Iglesias J, Castrillón L, Mara E (2014) Life cycle assessment of different municipal solid waste management options: a case study of Asturias (Spain). J Clean Prod 81:178–189

Fontaras G, Martini G, Manfredi U, Marotta A, Krasenbrink A, Maffioletti F, Terenghi R, Colombo M (2012) Assessment of on–road emissions of four Euro V diesel and CNG waste collection trucks for supporting air–quality improvement initiatives in the city of Milan. Sci Total Environ 426:65–72

Frischknecht R (2010) LCI modelling approaches applied on recycling of materials in view of environmental sustainability, risk perception and eco–efficiency. Int J Life Cycle Assess 15:666–671

Frischknecht R, Steiner R, Jungbluth N (2008) The ecological scarcity method – eco-factors 2006. Federal Office for the Environment (FOEN), Bern

Goedkoop M, Spriensma R (2000) The eco–indicator 99 – a damage–oriented method for life cycle impact assessment. https://www.pre-sustainability.com/download/EI99_annexe_v3.pdf. Accessed 15 Jan 2018

Goedkoop MJ, Heijungs R, Huijbregts M, De Schryver A, Struijs J, van Zelm R (2009) ReCiPe 2008, a life cycle impact assessment method which comprises harmonized category indicators at the midpoint and the endpoint level. http://www.lcia-recipenet/. Accessed 12 Nov 2012

Guinée JB, Gorree M, Heijungs R, Huppes G, Kleijn R, van Oers L, Wegener Sleeswijk A, Suh S, Udo de Haes A, de Buijn JA, van Duin R, Huijbregts MAJ (eds) (2002) Handbook on life cycle assessment: operational guide to the ISO standards. Kluwer Academic Publishers, Dordrecht

Hauschild MZ, Potting J (2005) Spatial differentiation in life cycle impacts assessment–the EDIP2003, Methodology environmental news no 80. Danish Ministry of the Environment, Environment Protection Agency, Copenhagen

Hauschild MZ, Dreyer LC, Jørgensen A (2008) Assessing social impacts in a life cycle perspective–lessons learned. CIRP Ann Manuf Technol 57:21–24

Hosseinijou SA, Mansour S, Shirazi MA (2014) Social life cycle assessment for material selection: a case study of building materials. Int J Life Cycle Assess 19:620–645

Hunkeler D, Lichtenvort K, Rebitzer G (eds) (2008) Environmental life cycle costing. SETAC, Pensacola, FL (US) in collaboration with CRC Press, Boca Raton

Intergovernmental Panel on Climate Change (IPCC) (2007) Fourth assessment report: climate change 2007 (AR4). IPCC. http://www.ipccch/publications_and_data/publications_and_data_reports–html. Accessed 10 Nov 2012

International Organization for Standardization (ISO) (2006a) ISO 14040 environmental management – life cycle assessment: principles and framework. ISO, Geneva

International Organization for Standardization (ISO) (2006b) ISO 14044 environmental management – life cycle assessment: requirements and guidelines. ISO, Geneva

International Organization for Standardization (ISO) (2013) ISO/TS 14067 carbon footprint of products–requirements and guidelines for quantification and communication. ISO, Geneva

Iofrida N, Strano A, Gulisano G, de Luca AI (2018) Why social life cycle assessment is struggling in development? Int J Life Cycle Assess 23:201–203

Iriarte A, Gabarrell X, Rieradevall J (2009) LCA of selective waste collection systems in dense urban areas. Waste Manag 29:903–914

Jaunich MK, Levis JW, DeCarolis JF, Gaston EV, Barlaz MA, Bartelt-Hunt SL, Jones EG, Hauser L, Jaikumar R (2016) Characterization of municipal solid waste collection operations. Resour Conserv Recycl 114:92–102

Jenkins RR, Martinez SA, Plamer K, Podolsky MJ (2003) The determinants of household recycling: a material–specific analysis of recycling program features and unit pricing. J Environ Econ Manag 45:294–318

Jolliet O, Margni M, Charles R, Humbert S, Payet J, Rebitzer G, Rosenbaum R (2003) IMPACT 2002+: a new life cycle impact assessment methodology. Int J Life Cycle Assess 8:324–330

Keramitsoglou KM, Tsagarakis KP (2013) Public participation in designing a recycling scheme in Didimoticho, Greece. Resour Conserv Recycl 70:55–67

Klöpffer W (2008) Life cycle sustainability assessment of products. Int J Life Cycle Assess 13:89–95

Klöpffer W, Grahl B (2014) Life cycle assessment (LCA): a guide to best practice. Wiley, Weinheim

Laurent A, Clavreul J, Bernstad A, Bakas I, Niero M, Gentil E, Christensen TH, Hauschild MZ (2014) Review of LCA studies of solid waste management systems – part II: methodological guidance for a better practice. Waste Manag 34:589–606

Li Y, Khanal SK (2016) Bioenergy: principles and applications. Wiley Blackwell, Hoboken

López JM, Gómez A, Aparicio F, Sánchez FJ (2009) Comparison of GHG emissions from diesel, biodiesel and natural gas refuse trucks of the City of Madrid. Appl Energy 86:610–615

Maimoun MA, Reinhart DR, Gammoh FT, Budh PM (2013) Emissions from US waste collection vehicles. Waste Manag 33:1079–1089

Martinez-Sanchez V, Kromann MA, Astrup TF (2015) Life cycle costing of waste management systems: overview, calculation principles and case studies. Waste Manag 36:343–355

Martinez-Sanchez V, Tonini D, Møller F, Astrup TP (2016) Life–cycle costing of food waste management in Denmark: importance of indirect effects. Environ Sci Technol 50:4513–4523

McDougall F, White P, Franke M, Hindle P (2001) Integrated solid waste management: a life cycle inventory. Blackwell Science Ltd, Oxford

Meijer J, Kasem N, Lewis K (2017) SM transparency report™/EPD framework part A – LCA calculation rules and report requirements. http://www.sustainablemindscom/files/transparency/SM_Part_A_LCA_calculation_rules_and_report_requirements_2017pdf. Accessed 15 Feb 2018

Miafodzyeva S, Brandt N (2013) Recycling behaviour among householders: synthesizing determinants via a meta–analysis. Waste Biomass Valoriz 4:221–235

Pelletier N, Ardente F, Brandão M, de Camillis C, Pennington D (2015) Rationales for and limitations of preferred solutions for multi–functionality problems in LCA: is increased consistency possible? Int J Life Cycle Assess 20:74–86

Pérez J, Lumbreras J, Rodríguez E, Vedrenne M (2017) A methodology for estimating the carbon footprint of waste collection vehicles under different scenarios: application to Madrid. Transp Res Part D 52:156–171

Pires A, Sargedas J, Miguel M, Pina J, Martinho G (2017) A case study of packaging waste collection systems in Portugal – part II: environmental and economic analysis. Waste Manag 61:108–116

Porter RC (2002) The economics of waste. Resources for the Future, Washington DC

Punkkinen H, Merta E, Teerioja N, Moliis K, Kuvaja E (2012) Environmental sustainability comparison of a hypothetical pneumatic waste collection system and a door–to–door system. Waste Manag 32:1775–1781

Rimos S, Hoadley AFA, Brennan DJ (2014) Environmental consequence analysis for resource depletion. Process Saf Environ Prot 92:849–861

Rödger J-M, Kær LL, Pagoropoulos A (2018) Life cycle costing: an introduction. In: Hauschild MZ, Rosenbaum RK, Olsen SI (eds) Life cycle assessment – theory and practice. Springer, Cham, pp 373–399

Rose L, Hussain M, Ahmed S, Malek K, Costanzo R, Kjeang E (2013) A comparative life cycle assessment of diesel and compressed natural gas powered refuse collection vehicles in a Canadian city. Energy Policy 52:453–461

References

Rosenbaum RK, Hauschild MZ, Boulay A-M, Fantke P, Laurent A, Núñez M, Vieira M (2018) Life cycle impact assessment. In: Hauschild M, Rosenbaum RK, Olsen S (eds) Life cycle assessment: theory and practice. Springer, Cham, pp 167–270

Sandhu GS, Frey HF, Bartelt-Hunt S, Jones E (2014) In–use measurement of the activity, fuel use, and emissions of front–loader refuse trucks. Atmos Environ 92:557–565

Science for Environment Policy (SEP) (2018) Kerbside waste–collection schemes may need optimisation, highlights Portuguese study. European Commission DG Environment News Alert Service, issue 504, 7 March 2018, edited by SCU, The University of the West of England, Bristol

Steen B (1999) A systematic approach to environmental priority strategies in product development (EPS). Available via Chalmers University of Technology, Technical Environmental Planning. http://www.cpmchalmersse/document/reports/99/1999_4pdf. Accessed 11 Nov 2012

Sureau S, Mazijn B, Garrido SR, Achten WMJ (2018) Social life–cycle assessment frameworks: a review of criteria and indicators proposed to assess social and socioeconomic impacts. Int J Life Cycle Assess 23:904–920

Teerioja N, Moliis K, Kuvaja E, Ollikainen M, Punkkinen H, Merta E (2012) Pneumatic vs door–to–door waste collection systems in existing urban areas: a comparison of economic performance. Waste Manag 32:1782–1791

Tillman AM (2010) Methodology for life cycle assessment. In: Sonesson U, Berlin J, Ziegler F (eds) Environmental assessment and management in the food industry. Woodhead Publishing, Cambridge, pp 59–82

UNEP/SETAC (2009) Guidelines for social life cycle assessment of products. United Nations Environment Program, Paris SETAC Life Cycle Initiative United Nations Environment Programme

United States Environmental Protection Agency (USEPA) (2009) Waste reduction model (WARM). https://www.epagov/warm. Accessed 12 Jan 2018

Varotto A, Spagnolli A (2017) Psychological strategies to promote household recycling: a systematic review with meta–analysis of validated field interventions. J Environ Psychol 51:168–188

Wang J, Zhuang H, Lin PC (2016) The environmental impact of distribution to retail channels: a case study on packaged beverages. Transp Res Part D: Transp Environ 43:17–27

World Resources Institute (WRI), World and Business Council for Sustainable Development (WBCSD) (2011) Product life cycle accounting and reporting standard. Word Resources Institute and World Business Council for Sustainable Development, USA

Yildiz-Geyhan E, Altun-Çiftçioğlu GA, Kadırgana MAN (2017) Social life cycle assessment of different packaging waste collection system. Resour Conserv Recycl 124:1–12

Zampori L, Dotelli G (2014) Design of a sustainable packaging in the food sector by applying LCA. Int J Life Cycle Assess 19:206–217

Part III
Sustainable Solid Waste Collection: Integrated Perspective

Chapter 12
Optimization in Waste Collection to Reach Sustainable Waste Management

Abstract This chapter focuses on one of the most used OR techniques which is optimization with linear programming modeling. This technique suits many problems faced when designing and operating a sustainable solid waste system. Single and multiple objective problems will be presented, and some OR special problems will be described with detail (the traveling salesman problem, the vehicle routing problem, the Chinese postman problem, the transportation problem, and the location problem). These problems appear in communal site collection and container collection, curbside collection, and location of containers or landfills, to name a few. Since in real-world problems, decision-makers pursue conflicting goals, strategies to deal with such issues are also presented. Several case studies are described, providing a deeper understanding of the applicability of such techniques.

Keywords Cost optimization · Environmental impact · Multi-objective programming · Postman problem · Single objective programming · Social impact · Vehicle routing problem

12.1 Introduction

Sustainable solid waste management involves a number of strategic, tactical, and operational decisions: the location and capacity of treatment sites and landfills, the selection of treatment technologies, waste flow allocation to processing facilities and landfills, service territory partitioning into districts, collection days' selection for each district and for each waste type, fleet composition, and routing and scheduling of collection vehicles (Ghiani et al. 2014). Operations research (OR) techniques, as modeling and optimization, can help decision-makers to attain significant cost savings and improve systems performance regarding environmental and social impact.

There are some reasons that favor the development of mathematical models. For instance, when building the model, some relations between different elements of the problem may be revealed, providing a deeper understanding of the system in study. Having a model, different scenarios can be tested and analyzed providing the planner

© Springer International Publishing AG, part of Springer Nature 2019
A. Pires et al., *Sustainable Solid Waste Collection and Management*,
https://doi.org/10.1007/978-3-319-93200-2_12

with better information on which to base his decisions. Often scenarios can not be experimented in "real life" due to the impact they may have on society or due to the cost of implementation of an "experimental" system.

One of the most used OR techniques is optimization, and mathematical programming models are part of the techniques when one has an objective that wants to maximize or minimize (Williams 2013). In this chapter, special types of mathematical programming models are analyzed concerning the adequacy to solve several problems faced when designing and operating a sustainable solid waste system. Single objective problems will be presented in Sect. 12.2. These have been applied to many of the solid waste management issues by means of some OR special problems: traveling salesman problem, the vehicle routing problem, the Chinese postman problem, the transportation problem, and the location problem. For instance, communal site collection and container collection problems are suitably modeled as the traveling salesman problem or the vehicle routing problem (Sects. 12.3.1 and 12.3.2); the curbside collection problems are modeled as the Chinese postman problem (Sect. 12.3.3); the waste volume to be sent from the depots to transfer stations is, in OR language, a transportation problem (Sect. 12.3.4); and the location of landfills or the placement of containers may be formulated as a facility location problem (Sect. 12.3.5).

Decision-makers often pursue conflicting goals, such as to locate facilities as close as possible to sources (to minimize transportation costs) and as far as possible from urban centers (to maximize distance). This enters a new area of mathematical programming which is the multi-objective optimization. In Sect. 12.4, the main differences between single and multiple objectives and some methods to solve the latter problems are addressed. Two case studies where these methods were successfully applied will be described to give some real examples where sustainable solid waste managers can see the real use of these techniques.

The first case addresses the collection and transportation system for supplying a waste-to-energy facility with solid waste from the municipalities and communities (Sect. 12.5). Case study 2 presents the retrofit of a waste of electric and electronic equipment (WEEE) recovery network, where economic and environmental minimization objectives are taken into consideration simultaneously (Sect. 12.6). Some conclusion and final remarks will end this chapter.

12.2 Single Objective Models

12.2.1 Linear Programming Model

Motivating Example 1

An SWM system has three intermediate facilities (F1, F2, F3) that send waste to two disposal facilities (D1, D2). The management needs to establish the amount to send to each disposal facility from each intermediate facility that minimizes the total

12.2 Single Objective Models

Table 12.1 Distance between intermediate facilities and disposal facilities (in km)

	F1	F2	F3
D1	20	30	60
D2	40	20	50

transportation cost. The distance between each facility is known (Table 12.1), and the cost is estimated as 0.2€ per km.

The disposal facilities, D1 and D2, have capacity limits and cannot receive more than 40 t and 60 t, respectively. There are 50 t to be sent from F1, 25 t from F2, and 30 t from F3.

To built the model, x_{11}, x_{12}, x_{21}, x_{22}, x_{31}, x_{32} are introduced as the variables representing the amounts sent from intermediate facilities 1, 2, and 3 to the disposal facilities 1 and 2 (e.g., x_{21} represents the amount sent from intermediate facility F2 to the disposal facility D1).

The total cost of transporting the waste is given by the expression:

$$0.2(20x_{11} + 40x_{12} + 30x_{21} + 20x_{22} + 60x_{31} + 50x_{32}) \tag{12.1}$$

The objective of the SWM system is to choose the values for $x_{11}, x_{12}, x_{21}, x_{22}, x_{31}, x_{32}$ so as to make the value of this expression as low as possible. This means that expression (12.1) is the objective function to be minimized (in this case). The capacities of the disposal facilities limit the values that x_{ij} can take. Since there are two disposal facilities, each has to receive at most the 40 t and 60 t, respectively. The expression (12.2) models disposal facility D1:

$$x_{11} + x_{21} + x_{31} \leq 40 \tag{12.2}$$

Being the expression for facility D2 given by:

$$x_{12} + x_{22} + x_{32} \leq 60 \tag{12.3}$$

Inequalities (12.2) and (12.3) are known as constraints. They restrict (constraint) the possible values variables x_{ij} can take. Similar constraints have to be written for the intermediate facilities since they have a maximum amount of waste to send. These are given by constraints (12.4) to (12.6):

$$x_{11} + x_{12} \leq 50 \tag{12.4}$$

$$x_{21} + x_{22} \leq 25 \tag{12.5}$$

$$x_{31} + x_{32} \leq 30 \tag{12.6}$$

Lastly, all variables are non-negative transportation amounts. This means that either one does not transport any waste from i to j or the amount must be a positive value (constraint (12.7)). This last constraint is named as domain constraint and assures that variable will assume values that make sense:

$$x_{11}, x_{12}, x_{21}, x_{22}, x_{31}, x_{32} \geq 0. \qquad (12.7)$$

The mathematical model that represents the problem faced by the SWM system is a Linear Programming model (LP) since all constraints and the objective function are linear expressions. In short, a LP model is characterized by:

- An objective function – A single linear expression to be maximized or minimized.
- Several linear constraints which must not exceed ($\leq$), must be at least ($\geq$), or must be exactly equal ($=$) a specific value.
- The domain constraint to get the variables fully characterized (as continuous variables).

Generically, a LP model is written in the form:

$$\min \sum_{i=1}^{n} c_i x_i$$

$$\text{s.t.} \sum_{i=1}^{n} A_{ij} x_i \leq b_j, \qquad \forall j = 1, \ldots, m$$

$$x_i \geq 0, \qquad \forall i = 1, \ldots, n$$

where c_i are the coefficients of the objective function, A_{ij} are the coefficients of the constraints, and b_j are the terms on the right side of the constraints which are all constants.

A LP may include all types of constraints (less-than-or-equal-to, greater-than-or-equal-to, and equal-to), two types of constraints, or only one kind. Only less-than-or-equal-to constraints are needed in the example of the SWM system. For more rigorous and technical details about linear programming models, refer to the text-books (Bertsimas and Tsitsiklis 1997; Vanderbei 2015).

12.2.2 Mixed-Integer Linear Programming

A huge amount of problems can be formulated using linear constraints and continuous and integer variables (Sioshansi and Conejo 2017). When a model has only continuous variables, it is a LP model; when it combines both continuous and integer variables, it is named as mixed-integer linear programming model (MILP). When a model is only composed of integer variables, it is often named as integer linear programming (ILP). A special case of integer variables are the binary variables. These only take two values, 0 and 1. They are extremely important to linearize some kind of constraints and/or to model "yes/no" decisions. Problems with only binary variables are called binary linear problems (BLP), and when combined with continuous variables, they are named as mixed binary linear problems. In this book, no

12.2 Single Objective Models

such distinction is made, existing only LP models (if only continuous variables are used) or MILP models (for the remaining cases).

Motivating Example 2

The SWM wants to determine which of the four drivers should be assigned to each of the four vehicles that collect waste. Each driver has his/her preference, and the company wants to maximize workers' total satisfaction. Therefore, each worker ($W1,\ldots,$ $W4$) was asked to classify each vehicle ($V1,\ldots,V4$) according to the satisfaction they have in driving it (Table 12.2).

For this problem one defines the variable x_{ij} has a binary variable: $x_{ij} = 1$ if driver V_i ($i = 1,\ldots,4$) is assigned to vehicle W_j ($j = 1,\ldots,4$); and 0, otherwise.

The objective function is then given by the following expression:

$$\max 20x_{11} + 12x_{12} + 15x_{13} + 12x_{14} + 8x_{21} + \ldots + 12x_{42} + 10x_{43} + 15x_{44}.$$

This problem has three sets of constraints. The first set assures that each driver is assigned to one vehicle:

$$x_{11} + x_{12} + x_{13} + x_{14} = 1$$
$$x_{21} + x_{22} + x_{23} + x_{24} = 1$$
$$x_{31} + x_{32} + x_{33} + x_{34} = 1$$
$$x_{41} + x_{42} + x_{43} + x_{44} = 1$$

The second set assures that each vehicle is only assigned to one driver:

$$x_{11} + x_{21} + x_{31} + x_{41} = 1$$
$$x_{12} + x_{22} + x_{32} + x_{42} = 1$$
$$x_{13} + x_{23} + x_{33} + x_{43} = 1$$
$$x_{14} + x_{24} + x_{34} + x_{44} = 1$$

The last constraint assure that all variables take only 0 or 1 values:

$$x_{ij} \in \{0, 1\}, \quad \forall i = 1, \ldots, 4 \; j = 1, \ldots, 4$$

Taking all elements together, the problem faced by the company is modeled by:

$$\max 20x_{11} + 12x_{12} + 15x_{13} + 12x_{14} + 8x_{21} + \ldots + 12x_{42} + 10x_{43}$$
$$+ 15x_{44} \tag{12.8}$$
$$\text{s.t.} \quad x_{11} + x_{12} + x_{13} + x_{14} = 1 \tag{12.9}$$

Table 12.2 Satisfaction of each driver regarding each vehicle (in satisfaction units)

	V1	V2	V3	V4
W1	20	12	15	12
W2	8	13	12	15
W3	10	15	7	16
W4	18	12	10	15

$$x_{21} + x_{22} + x_{23} + x_{24} = 1 \tag{12.10}$$

$$x_{31} + x_{32} + x_{33} + x_{34} = 1 \tag{12.11}$$

$$x_{41} + x_{42} + x_{43} + x_{44} = 1 \tag{12.12}$$

$$x_{11} + x_{21} + x_{31} + x_{41} = 1 \tag{12.13}$$

$$x_{12} + x_{22} + x_{32} + x_{42} = 1 \tag{12.14}$$

$$x_{13} + x_{23} + x_{33} + x_{43} = 1 \tag{12.15}$$

$$x_{14} + x_{24} + x_{34} + x_{44} = 1 \tag{12.16}$$

$$x_{ij} \in \{0, 1\}, \quad \forall i = 1, \ldots, 4 \ j = 1, \ldots, 4 \tag{12.17}$$

This problem is an example of an assignment problem.

Generically, an MILP model is written in the form:

$$\min \sum_{i=1}^{n} c_i x_i$$

$$\text{s.t.} \sum_{i=1}^{n} A_{ij} x_i \leq b_j, \qquad \forall j = 1, \ldots, m$$
$$x_i \geq 0, \qquad \text{for some } i = 1, \ldots, n$$
$$x_i \in \{0, 1\} \qquad \text{for the remaining } i = 1, \ldots, n$$

where c_i are the coefficients of the objective function, A_{ij} are the coefficients of the constraints, and b_j are the terms on the right side of the constraints which are all constants. As for LP models, MILP models may include all types of constraints (less-than-or-equal-to, greater-than-or-equal-to, and equal-to), two types of constraints, or only one kind. Only equal-to constraints are needed in the example of the SWM system.

12.2.3 Stochastic Programming

In real problems, data is often uncertain. The above models assume all data to be known in advance. However, that might not be the case. When facing uncertainty, one can design a stochastic model. These models are special cases of LP or MILP models where a finite number of scenarios is estimated and the objective function will minimize or maximize an expected value. Let us see a small example.

12.2 Single Objective Models

Table 12.3 Recyclable waste deposition amounts estimated by scenario and scenario occurrence probability and bin capacity

	Low	Average	High	Extremely high	Capacity per bin
Glass	110	130	150	200	3
Plastic/metal	60	80	120	150	1
Paper	100	150	170	210	2
Probability	0.1	0.5	0.35	0.05	–

Motivating Example 3

Suppose one needs to decide on the number of recyclable bins that should be made available in a new neighborhood. The amounts of waste that will be disposed of are unknown. However, four scenarios have been estimated based on the data from similar neighborhoods (Table 12.3). The decision-maker wants not only to know the minimum number of bins per recyclable material to make available but also to minimize the excess and shortfall of waste install capacity. Notice that the number of bins is a decision that has to be made before one knows the amount that will be disposed of.

The decision concerning how many bins to make available in the neighborhood is not scenario dependent (called first-stage decisions). But the excess and the shortfall capacities are scenario-dependent decisions (called second-stage decisions). Let us then define the variable x_i as the number of bins from recyclable waste type i, $i = 1$ (glass), 2 (plastic), 3 (paper) to be made available in the neighborhood, and y_{is} and z_{is} are the excess and shortfall capacity variables for recyclable waste type i in scenario s, $s = 1$ (low), 2 (average), 3 (high), 4 (ext. high). Additionally, consider the parameters v_i as the capacity of each type of bin, p_s as the probability of each scenario, and b_{is} as the estimated amount of waste type i generated in scenario s.

$$\min \sum_{i=1}^{3} v_i x_i + \sum_{s=1}^{4} p_s \left(\sum_{i=1}^{3} y_{is} + z_{is} \right) \tag{12.18}$$

$$\text{s.t. } v_i x_i - y_{is} + z_{is} = b_{is}, \quad \forall i = 1, \ldots, 3 \ s = 1, \ldots, 4 \tag{12.19}$$

$$\begin{aligned} x_i \geq 0 \text{ and integer } \forall i = 1, \ldots, 3 \\ y_{is}, z_{is} \geq 0 \quad \forall i = 1, \ldots, 3 \ s = 1, \ldots, 4 \end{aligned} \tag{12.20}$$

Expression (12.18) models the objective of minimizing the number of bins together with the excess and shortfall capacity. Notice the number of bins is modeled by their corresponding capacity (since all terms of the expression have to have the same unit). The second term of this expression models the expected value of excess and shortfall capacity. Equations (12.19) assure that the installed capacity minus the excess capacity plus the shortfall capacity meets the estimated waste generation volume for each waste type in each scenario. Lastly, constraints (12.20) define variable domain.

The model proposed in the motivating example is named as two-stage stochastic programming model. One should note that extra information has been incorporated into this model. While in the (simple) MILP model, data is in the form of a single point, in the (stochastic) MILP, data is represented by a series of estimates, each weighted by a probability. In fact, it is the MILP model where the objective function optimizes an expected value. The solution of these problems is not the optimal value to any scenario (this solution can be computed considering each scenario as an independent problem), but it reduces the risk of taking the decision before knowing all the data. As pointed out by Williams (2013), the solution of stochastic models may be seen as "keeping one's options open" or not "putting all one's eggs in one basket."

A multistage scenario context may appear when successive scenarios occur over time. In this case from each scenario in the second stage, there will be several other scenarios in a third stage, and so on and so forth. A scenario tree will be built which easily becomes very large, leading to intractable models. For more rigorous and technical details about stochastic linear programming models, refer to Birge and Louveaux (1997) and Greenberg and Morrison (2008).

12.2.4 Nonlinear Programming

A problem to be modeled using the mathematical programming techniques presented above has to have some characteristics. For instance, a resource used by an activity has to be proportional to the level of the activity. Another assumption is that the total use of a resource by some activities corresponds to the sum of the uses by each activity. If some of these conditions do not apply, one can no longer formulate the problem as a linear programming model. A nonlinear programming model (NLP) can be defined with three characteristics:

(i) The objective function is a nonlinear function of real values:

$$g : \mathbb{R}^n \to \mathbb{R}$$

(ii) The constraints are nonlinear expressions of the form:

$$f_i(x_1, \ldots, x_n) \leq 0, \quad \forall i = 1, \ldots, p$$
$$h_j(x_1, \ldots, x_n) = 0, \quad \forall j = 1, \ldots, q$$

(iii) The decision variables are continuous:

$$(x_1, \ldots, x_n) \in \mathbb{R}^n$$

where p and q are the number of inequality and equality constraints, respectively.

In its generic form, a NLP model can be defined as:

$$\min g(x)$$
$$\text{s.t.} f_i(x) \leq 0, \qquad \forall i = 1, \ldots, p$$
$$h_j(x) = 0, \qquad \forall j = 1, \ldots, q$$

This kind of problems presents several issues that make their solution very demanding. Whenever one has the possibility to approximate a nonlinear model with a linear model, the resulting solution (from the linear model) will provide insightful information to the decision-maker. In this book nonlinear models are not addressed. The reader is referred to Bazaraa et al. (2013) and Luenberger and Ye (2016).

12.2.5 Solving a Linear Programming Model

Software to solve mathematical programming problems is often composed of two different pieces: a modeling language and a solver (Sioshansi and Conejo 2017). The former allows the definition of the model using a human-readable format. The user defines the specific data, variables, constraints, and the objective function. One of the first mathematical programming languages is GAMS (Bisschop and Meeraus 1982). Many others have appeared since then: AMPL (Fourer et al. 2002) and OPL (IBM 2017b), among others. Commonly, these programming languages include features that allow the user to combine it with other software as spreadsheets, databases, and text files, which considerably facilitates the handling of large amounts of data. The solver receives a file, in machine language, that has been created by the mathematical programming language. The optimization problem is, in fact, solved by the solver. The solution file is then sent back to the modeling language software to be translated back to a human-readable format. Two state-of-the-art solvers are CPLEX (IBM 2017a) and GUROBI (Gurobi 2016) which solve LP, MILP, and a few types of NLP models. For NLP, one can find other state-of-the-art solvers as BARON (Tawarmalani and Sahinidis 2005), MINOS (Murtagh and Saunders 1983), and CONOPT (Drud 1994).

12.3 Some Special Problems

The first publication regarding waste collection was presented by Beltrami and Bodin (1974). The vehicle routing problem was the mixed-integer linear programming model used to solve the collection activities of the New York City Department of Sanitation. In this work, waste to be collected was located in points in the plane as in any communal site collection and container collection context. In curbside collection, on the other hand, every house needs to be visited. Consequently, the number of spots to visit in communal site collection and container collection is significantly lower than the number of customers served in curbside collection. Therefore, communal site collection and container collection problems are more suited to being modeled as a variant of the traveling salesman problem (TSP), while curbside collection problems are more suited to being modeled as a variant of the Chinese postman problem (CPP). The TSP identifies the least cost route of a single vehicle that includes every node in the network and then returns to the starting node. The CPP identifies the least cost route of a single vehicle that includes every arc in the network (Beliën et al. 2014). When several vehicles are used, one no longer models the problem as a TSP but as a vehicle routing problem (VRP).

Another common issue arising in sustainable solid waste management is the location of facilities. High costs associated with property acquisition and facility construction make facility location or relocation projects long-term investments. For instance, building a new treatment or disposal facility may take 1–4 years, while the operating life of a facility is estimated to be around 15–30 years (Ghiani et al. 2014). Thus, decision-makers must select sites that will not simply perform well according to the current system state but that will continue to be of value for the facility lifetime, even as environmental factors change, populations shift, and market trends evolve. Finding robust facility locations is thus a difficult task, demanding that decision-makers account for uncertain future events (Owen and Daskin 1998).

The classical MILP formulation of these special problems that are so common in sustainable solid waste management will be presented below. To illustrate some of them, a brief description of works where such models have been used is provided.

12.3.1 Traveling Salesman Problem

The traveling salesman problem (TSP) is a very well-known optimization problem. Since its early studies, the problem has been defined as follows: a traveling salesman wishes to visit exactly once each of a list of m cities and then return to the home city; knowing that the cost of traveling between cities i and j is c_{ij}, what is the minimum cost route the traveling salesman can take? Its mathematical structure is a graph where each city is denoted by a point (or node) and lines are drawn connecting every two nodes (called arcs or edges). Associated with every line is a distance (or cost, or any other metric that suits the context). If the direction in which an edge of the graph is traversed matters, the TSP is formulated differently. This aspect is particularly relevant to waste collection problems. Therefore, two cases must be one distinguishes the asymmetric (where $c_{ij} \neq c_{ji}$) and the symmetric (where $c_{ij} = c_{ji}$) TSP.

The asymmetric TSP can be formulated as:

12.3 Some Special Problems

$$\min \sum_{i=1}^{m} \sum_{j=1}^{m} c_{ij} x_{ij} \tag{12.21}$$

$$\text{s.t.} \quad \sum_{i=1}^{m} x_{ij} = 1, \quad j = 1, \dots, m \tag{12.22}$$

$$\sum_{j=1}^{m} x_{ij} = 1, \quad i = 1, \dots, m \tag{12.23}$$

$$\sum_{i \in S} \sum_{j \in S} x_{ij} \leq |S| - 1, \quad S \subset \{1, \dots, m\}, 2 \leq |S| \leq m - 2 \tag{12.24}$$

$$x_{ij} \in \{0, 1\}, \quad i, j = 1, \dots, m \tag{12.25}$$

where $x_{ij} = 1$ if arc (i, j) belongs to the solution, and 0 otherwise; and $S \neq \varnothing$ is a subset of the nodes $1, \dots, m$. Constraints (12.22) and (12.23) impose that if a node is visited, then there is exactly one arc leaving and one entering the node, respectively, while (12.24) are subtour elimination constraints (SECs) and impose that no partial circuit exists (Roberti and Toth 2012).

The symmetric TSP can be formulated as:

$$\min \sum_{i < j} c_{ij} x_{ij} \tag{12.26}$$

$$\text{s.t.} \quad \sum_{i < k} x_{ik} + \sum_{k < j} x_{kj} = 2, \quad k = 1, \dots, m \tag{12.27}$$

$$\sum_{i \in S} \sum_{j \in S} x_{ij} \leq |S| - 1, \quad S \subset \{1, \dots, m\}, 3 \leq |S| \leq m - 3 \tag{12.28}$$

$$x_{ij} \in \{0, 1\}, \quad i, j = 1, \dots, m \tag{12.29}$$

This symmetric formulation is a particular case of the asymmetric one. Interestingly, the algorithms developed for the asymmetric TSP do not (in general) perform well when applied to the symmetric TSP (Laporte 2010). An innumerous amount of algorithms has been proposed over the years to solve these problems (Hoffman et al. 2013). Concorde is today the best available solver for the symmetric TSP and is freely available at www.tsp.gatech.edu. A detailed description of Concorde can be found in the book by Applegate et al. (2006).

Blazquez et al. (2012) address the problem of low-cost PM_{10} (particulate matter with aerodynamic diameter < 10 μm) street sweeping route. To do so, only a subset of the streets of the urban area to be swept is selected for sweeping, based on their PM10 emission factor values. Subsequently, a low-cost route that visits each street in the set is computed. Unlike related problems of waste collection where streets must be visited once (Chinese or rural postman problem, respectively), in this case, the sweeping vehicle route must visit each selected street exactly as many times as its number of street sides, since the vehicle can sweep only one street side at a time. Additionally, the route must comply with traffic flow and turn constraints. This arc routing problem was transformed into a node routing problem by representing the street network topology by a directed graph G, where the nodes represent the street sides to be swept. Arcs connecting nodes in graph G correspond to the shortest routes between the street sides to be swept that do not contain high-priority streets. With this transformation, the problem can be solved by applying any known solution to the asymmetric traveling salesman problem. The proposed method was applied to the northeast area of the municipality of Santiago (Chile). Results show that the proposed methodology achieved up to 37% savings in kilometers traveled by the sweeping vehicle when compared to the solution obtained by solving the TSP problem with geographic information systems (GIS) tools.

12.3.2 Vehicle Routing Problem

The vehicle routing problem (VRP) is defined as the problem of determining the least cost delivery routes from a depot to a set of customers, subject to some constraints (Laporte 2009). This is one of the central problems in waste collection management and must be frequently solved by planners. In real-world applications, there are several variants of the problem since one may encounter diverse operating policies and constraints.

Formally, the VRP may be defined as follows. Let $G = (V, A)$ be a directed graph where $V = \{0, \ldots, n\}$ is the node set and $A = \{(i,j) : i,j \in V, i \neq j\}$ is the arc set. Node 0 represents the depot, whereas the remaining vertices correspond to collection points. A fleet of m identical vehicles of capacity Q is based at one single depot. Each collection point i has a non-negative volume to be collected q_i. A cost matrix c_{ij} is defined on A. In the classical VRP and for simplicity, travel costs, distances, and travel times are assumed to be equivalent. The VRP involves the determination of m vehicle routes such that each route starts and ends at the depot, each collection point is visited once, and only once, by one vehicle, the total collection volume of a route does not exceed Q, and the total length of a route does not exceed a fixed limit

12.3 Some Special Problems

L. When $m = 1$ and $Q = \infty$, the VRP "becomes" the traveling salesman problem. According to Laporte (2007), the VRP is significantly more difficult to solve than a TSP of the same size. Different formulations model the symmetric and the asymmetric VRPs.

Let x_{ij} be an integer variable representing the number of times arc (i,j) appears in the solution, and let $E = \{(i,j) : i, j \in V, i < j\}$ be the set of undirected arcs. The symmetric VRP problem may then be formulated as:

$$\min \sum_{(i,j) \in E} c_{ij} x_{ij} \tag{12.30}$$

$$\text{s.t.} \sum_{j \in V \setminus \{0\}} x_{0j} = 2m \tag{12.31}$$

$$\sum_{i<k} x_{ik} + \sum_{k<j} x_{kj} = 2, \quad k = 1, \ldots, m \tag{12.32}$$

$$\sum_{\substack{i \in S, j \notin S \\ \text{or } i \notin S, j \in S}} x_{ij} \leq \left\lceil \sum_{i \in S} \frac{q_i}{Q} \right\rceil, \quad S \subset \{1, \ldots, m\} \tag{12.33}$$

$$x_{0j} \in \{0, 1, 2\}, \quad j = 1, \ldots, m \tag{12.34}$$

$$x_{ij} \in \{0, 1\}, \quad i, j = 1, \ldots, m \tag{12.35}$$

In this formulation, the objective function minimizes the total routing cost (Eq. (12.30)). Constraints (12.31) define the number of arcs entering or leaving node 0. Note that the right-hand side can be a constant if m is known a priori. Constraints (12.32) ensure that two arcs are incident to each node (one arc corresponds to the arrival of the vehicle to the collection site; the other models the vehicle leaving the site). Constraints (12.33) prevent the formation of subtours by forcing any subset of customers to be connected to the depot. These constraints also assure that capacity constraints are meet (the notation [b] stands for the smallest integer grated or equal to b).

The multi-depot vehicle routing problem (MDVRP) is a generalization of the vehicle routing problem (VRP) in which, beyond the definition of vehicle routes, it is necessary to determine from which depot collection nodes are to be visited (Ramos et al. 2014). The MDVRP simultaneously establishes the service areas of each depot and the associated vehicle routes. The vehicle routes are defined such that (1) each route starts and ends at the same depot, (2) each collection node is visited exactly once by a vehicle, (3) the total collected volume of each route does not exceed the vehicle capacity, and (4) the total duration of each route (including travel and service times) does not exceed a preset limit. The best solution is typically one that

220   12 Optimization in Waste Collection to Reach Sustainable Waste Management

minimizes the total routing cost (distance, time, among others). One possible formulation for this problem can be found in Chap. 14.

12.3.3 Chinese Postman Problem

In general, any problem that requires that all edges of a graph (streets, etc.) be served at least once while traveling the shortest total distance overall is a Chinese postman problem (CPP). An example of such a problem could be a postman who must visit each house along each street in a neighborhood, or a door-to-door waste collection system where a single vehicle performs the route, or even a street sweeping system. While in TSP all collection points are discrete nodes, in the CPP one needs to visit all arcs.

In traversing a postman route, one must be able to leave every node that the route visits. In a directed graph (a graph where arcs are oriented), this means that the number of arcs leading into a given node must equal the number of arcs directed out of that node. Since the original graph may not satisfy this condition, additional copies of some arcs must be added to bring this about. This operation is called balancing and below is outlined an optimal procedure to perform it.

Let $G = (V, A)$ be a directed graph. For every node $i \in V$, let s_i be the number of arcs entering node i minus the number of arcs leaving node i. Any node with $s_i \neq 0$ is an imbalance node and therefore needs to be balanced in order to solve the postman problem. Let S be the set of nodes i with $s_i > 0$ and T the set of nodes i with $s_i < 0$. Assume l_{ij} denotes the length of the shortest path from i to j and x_{ij} is the number of additional copies of arc (i, j) to be added to G. The direct Chinese postman problem can be formulated as (Benavent et al. 2000):

$$\min \sum_{i \in S} \sum_{j \in T} l_{ij} x_{ij} \tag{12.36}$$

$$\text{s.t.} \sum_{j \in T} x_{ij} = s_i, \quad i \in S \tag{12.37}$$

$$\sum_{i \in S} x_{ij} = -s_j, \quad j \in T \tag{12.38}$$

$$x_{ij} \geq 0, \quad i \in S, j \in T \tag{12.39}$$

This model is the linear programming formulation of the transportation problem which is well known for always having an integer optimal solution when the constraints right-hand sides are integer.

12.3 Some Special Problems

Curbside collection of municipal solid waste (MSW) represents a significant fraction of waste management cost. Filipiak et al. (2009) applied an algorithm for the Chinese postman problem to evaluate the truck routes and the optimum sequence of each of these vehicles needed to collect the generated waste in the township of Millburn, USA. With a population of 20,000 in approximately 7000 households, this township divides the service area into three sections and collects waste twice a week from curbsides of single-family and two-family units using up to four trucks. Then with the aid of commercial software (graph Magics software), the optimum routes for each of the trucks were obtained. Results from this optimization study were then compared with the current waste collection practice. It is shown in this case study that the average total route length of four trucks that collected solid waste was more than 2 mi shorter than the real case truck. Any of the four trucks had better results than the real case. Based on the case study results, improvements to the current MSW collection procedure were proposed to the township of Millburn, N.J.

A more general problem is the Windy Postman Problem (WPP) where one aims at finding the minimum cost route traversing all arcs of G at least once. In this problem, the cost of traversing an arc depends on the direction of travel. Therefore, let $G = (V, E)$ be an undirected graph where for each arc $(i, j) \in E$, there are two associated non-negative costs c_{ij} and c_{ji} corresponding to the two different directions: from i to j and from j to i, respectively. One possible formulation for such problem is given by Benavent et al. (2000) where δ_i is the set of all arc incident in node i:

$$\min \sum_{(i,j) \in E} \left(c_{ij} x_{ij} + c_{ji} x_{ji} \right) \tag{12.40}$$

$$\text{s.t.} \quad x_{ij} + x_{ji} \geq 1, \quad (i, j) \in E \tag{12.41}$$

$$\sum_{(i,j) \in \delta(i)} \left(x_{ij} - x_{ji} \right) = 0, \quad i \in V \tag{12.42}$$

$$x_{ij}, x_{ji} \geq 0 \text{ and integer} \tag{12.43}$$

The objective function (12.40) minimizes the total cost. Constraints (12.41) assure that every arc is traversed at least once regardless of the direction, and constrains (12.42) ensure that if node i has a way in arc, it must also have a way out arc. In the two previous models, all arcs have to be traversed at least once. However, in some problems only a subset of arcs needs to be traversed. This is known as the rural postman problem. Both CCP and WPP are particular cases of a larger class of models: the arc routing problem.

Braier et al. (2017) apply an integer programming model to optimize the routes of a recyclable waste collection system servicing Morón, a large municipality outside Buenos Aires, Argentina. The truck routing problem posed by the system is a particular case of the generalized directed open rural postman problem given particular conditions in the Argentinean traffic regulation. These conditions included the prohibition on left turns at traffic lights and U-turns (the exception for U-turns in cul-de-sacs), the inclusion of perimeter bands for detours outside the serviced area, and street directionality (one-way and two-way). The route solutions generated by the proposed methodology perform significantly better than the previously used, manually designed routes. The main improvement is the 100% coverage of blocks within the municipality with the model solutions, whereas with the manual routes, as much as 16% of the blocks went unserviced. The proposed routes were adopted by the municipality in 2014, and the national government is planning to introduce the methodology elsewhere in the country.

12.3.4 Transportation Problem

The transportation problem is concerned with finding the minimum cost of transporting a single product from a given number of sources (e.g., depots) to a given number of destinations (e.g., transfer stations). At each source there are s_i units of product to be shipped to some destination, $i = 1, \ldots, m$. Each destination has available capacity to receive d_j units of product, $j = 1, \ldots, n$.

Let x_{ij} be the number of units shipped from source i to destination j, and let c_{ij} be the cost of transporting one unit of product from i to j. The linear formulation of the transportation problem is given by Eqs. (12.44) to (12.47) (Hillier and Lieberman 2014):

$$\min \sum_{i=1}^{m} \sum_{j=1}^{n} c_{ij} x_{ij} \tag{12.44}$$

$$\text{s.t.} \quad \sum_{i=1}^{m} x_{ij} = d_j, \quad j = 1, \ldots, n \tag{12.45}$$

$$\sum_{j=1}^{m} x_{ij} = s_i, \quad i = 1, \ldots, m \tag{12.46}$$

12.3 Some Special Problems   223

$$x_{ij} \geq 0, \quad i = 1, \ldots, m; j = 1, \ldots, n \qquad (12.47)$$

This formulation assumes that

$$\sum_{i=1}^{m} s_i = \sum_{j=1}^{n} d_j.$$

If in the real problem this assumption does not hold, it can be reformulated considering either a dummy destination (dummy source) to take up the extra amount there is to be shipped (to be received).

12.3.5 Location Problem

Facility location problem has been studied for long, and several models have been proposed to this date. The simplest location model considers only the trade-off between fixed operating and variable delivery cost. This problem is concerned with finding the (undermined number of) facilities to be opened/used among a set of possible locations that minimize the total cost. This cost accounts not only for the variable cost of serving a set of customers but also the fixed cost of opening the facilities. Let z_{ij} be the fraction of customer zone j's demand satisfied by the facility located at location i and y_i a binary variable that assumes a value of 1 if a facility is to be established at location i. Concerning costs, assume f_i as the fixed cost of establishing a facility at location i and c_{ij} the total distribution cost for supplying all of customers zone site j's demand by the facility at location i. Verter (2011) proposed model (12.48) to (12.51) as a formulation for the uncapacitated facility location problem:

$$\min \sum_{i=1}^{m} \sum_{j=1}^{n} c_{ij} z_{ij} + \sum_{j=1}^{n} f_j y_j \qquad (12.48)$$

$$\text{s.t.} \quad \sum_{j=1}^{n} z_{ij} = 1, \quad i = 1, \ldots, m \qquad (12.49)$$

$$z_{ij} \leq y_j \quad i = 1, \ldots, m; j = 1, \ldots, n \qquad (12.50)$$

$$0 \leq z_{ij} \leq 1, y_j \in \{0, 1\} \quad i = 1, \ldots, m; j = 1, \ldots, n \qquad (12.51)$$

The objective function (12.48) represents the total fixed and variable costs. Equations (12.49) guarantee that the demand at each customer zone is satisfied. Constraints (12.50) ensure that customer demand can be satisfied only from the locations where a facility is established, i.e., if $y_i = 1$, and in such case, the firm incurs the associated fixed costs (the second term of the objective function).

Badran and El-Haggar (2006) propose a mixed-integer programming model for the location of collection stations in Port Said to minimize the municipal solid waste management system total cost. The municipal solid waste includes both the hazardous and the construction and demolition waste. The best locations for collection stations are selected from a given set of candidate locations so that system cost is minimized. Given the lack of detailed data, several scenarios were studied to analyze the best answer to three questions: (i) should the construction and demolition waste and the hazardous waste flows be firstly sent to the composting plant and then to the landfill or directly from the collection station to the landfill? (ii) should the daily capacity of collection stations be 10 ton, 15 ton, or a mix of the two? (iii) should the waste generated from each source be sent only in the collection stations available in that source or not? The least cost solution presents mixed capacities to the collection stations with no limitation on the waste flow from the source to the collection station. The objective function of the best solution has a total value of 10,122 LE/day (US$1716/day). The total number of collection stations is 29, where 27 of them have a 15 tons daily capacity, and two have a daily capacity of 10 tons. In this scenario, the capacity of the composting plant is fully utilized. The flow from the composting plant to the landfill is 92.84 tons per day.

Many location models minimize some function of the distances between facilities. This is appropriate when locating service facilities. However, if one is locating an obnoxious facility, such as a waste bin, a landfill, or an incineration facility reactor, closeness is undesirable. In such instances, a model which maximizes some function of distance may be more appropriate (Erkut and Neuman 1989).

12.4 Multiple Objectives

In current real-world problems, the decision-maker (DM) has often to face more than one criteria (cost, service quality, equity, etc.) which often are conflicting when searching for the optimal solutions (Antunes et al. 2016). Such problems enter new areas of operations research where multiple criteria need to be simultaneously accounted for when finding a solution.

Under the designation of multiple criteria approaches, two distinct areas emerge in the specialized literature:

- Decision-making with multiple attributes.
- Decision support with multiple objectives.

Areas differ in the way alternatives are defined. The former one refers traditionally to the selection, ranking, or categorization problems of a finite set of alternatives

12.4 Multiple Objectives

(these are known a priori). This lies within what is named as multi-criteria decision-making which is addressed in Chap. 13. The latter concerns problems in which the alternatives are defined by a set of constraints. These problems enter a new area of mathematical programming: multi-objective programming (MOP) where multiple objective functions are explicitly considered.

The concept of optimal solution is, in MOP, known as Pareto optimal solution (also known as efficient, nondominated, or noninferior solution). A solution is said to be nondominated if there is no other feasible solution that simultaneously improves all the objective function values (Steuer 1986). Notice that in MOP, improving an objective implies deteriorating, at least, one of the other objective function values.

The MOP methods have been traditionally divided into three categories, according to the process of modeling the decision-maker preferences:

- An a priori modeling of preferences is made.
- No articulation of preferences is made (generating methods).
- Progressive articulation of preferences (interactive methods).

In the a priori methods, the aggregation of preferences is made before any computation has been performed; consequently, the problem is first transformed into a single objective problem (e.g., selecting one objective function to be optimized considering the remaining objectives as constraints, optimizing a weighted sum of the objective functions, or minimizing a distance function to a reference point using different metrics). These methods do not allow the DM to have an active role in the decision-making process which often leads to lesser receptiveness regarding the proposed solutions. In the second category of methods (generating methods), all nondominated solutions are generated and presented to the decision-maker. These methods present two major drawbacks when applied to real-world problems: the computational burden required for computing the entire set of nondominated solutions is too high; proposing hundreds or thousands of solutions to a decision-maker is not useful for the exploitation of results in practice. Lastly, interactive methods enable a step-by-step articulation of the decision-maker's preferences (e.g., STEM (Benayoun et al. 1971), interval criterion weights (Steuer 1986). These methods imply the interaction between a computation and a dialogue phase. After each computation phase, one (or several) nondominated solution(s) is (are) proposed to the DM. He/she reacts providing the necessary information to start a new computation phase or deciding to stop the procedure. Generically, a MOLP model is written in the form:

$$\min f_1 = \sum_{i=1}^{n} c_{i1} x_i$$

$$\cdots$$

$$\min f_k = \sum_{i=1}^{n} c_{ik} x_i$$

$$\text{s.t.} \sum_{i=1}^{n} A_{ij} x_i \leq b_j, \qquad \forall j = 1, \ldots, m$$
$$x_i \geq 0, \qquad \forall i = 1, \ldots, n$$

When optimizing a single objective function, the feasible region, in the decision space $x \in S$, is mapped onto $\mathbb{R}$. In the multi-objective context, the decision space is mapped onto a k-dimensional space $Z = \{z = f(x) \in \mathbb{R}^k : x \in S\}$, named "objective function space" or "criteria space." In this latter space, each solution $x \in S$ is represented by a vector $z = (z_1, z_2, \ldots, z_k) = f(x) = (f_1(x), f_2(x), \ldots, f_k(x))$. Given that one faces conflicting objective functions, no feasible solution $x \in S$ simultaneously optimizes all objective functions. Therefore, in multi-objective optimization, decision-makers seek for "good compromise solutions." These "good" solutions are called efficient, nondominated, or Pareto optimal solutions. The projection of the Pareto optimal set under the objective functions is called Pareto front (Barth et al. 2004).

12.4.1 Lexicographic Method

Considerer the MOLP model proposed above. Suppose the decision-maker ranks the objectives according to his preference such that $f_1(x)$ is of higher importance than $f_2(x)$, $f_2(x)$ is of higher importance than $f_3(x)$, and so on.

First, one minimizes $f_1(x)$ subjected to $x \in S$ and determines an optimal solution x^* with $f_1(x^*) = z_1$. Next one solves the problem of minimizing $f_2(x)$ subjected to $x \in S$ and $f_1(x) \leq z_1 + \delta_1$ where δ_1 is a positive deviation from the optimal value of the most important objective function accepted by the decision-maker. The remaining objective functions are sequentially optimized adding in each step a new constraint to the feasible region. In general, in the ith iteration, one solves:

$$\min_{x \in S} \{f_i(x) : f_j(x) \leq z_j + \delta_j, j = 1, \ldots, i - 1\}.$$

Further details can be found in Khorram et al. (2010).

> Erkut et al. (2008) propose a mixed-integer multiple objective linear programming model, which helps to solve the location-allocation problem of municipal solid waste management facilities in the Central Macedonia region in North Greece. Five objectives are considered: minimize the greenhouse effect, minimize the amount of final disposal, maximize the amount of energy recovery, maximize the amount of material recovery, and minimize the total

(continued)

opening, transportation, and processing costs. The multi-objective problem is formulated as a lexicographic minimax problem to find a fair nondominated solution, a solution with all normalized objectives as equal as possible. How to replace the original lexicographic minimax problem with the lexicographic minimum problem is discussed. The model is applied to compare and contrast the prefectural and regional planning for MSW management. Computational experiments with data from Central Macedonia show that the gains achieved by moving from a prefectural to a regional plan are minimal since the waste flow between prefectures is small. However, when assuming all objective functions as equally important, the regional plan generated is superior to the prefectural plan only on the total cost objective. Notice, that this decision being so data dependent, there may be other instances where regional plans dominate prefectural plans by a wider margin.

12.4.2 Weighted Sum Method

One of the most frequent methods to deal with multi-objective problem is to define a new objective function as the weighted sum of the multiple objective (function (12.52)). The weights should reflect the decision-maker's preferences regarding each objective. This method transforms a multi-objective problem into a single objective one:

$$\min f(x) = w_1 \cdot f_1(x) + w_2 \cdot f_2(x) + \ldots + w_k \cdot f_k(x) \qquad (12.52)$$

$$\text{s.t.} \quad \sum_{i=1}^{n} A_{ij} x_i \leq b_j, \qquad \forall j = 1, \ldots, m$$
$$x_i \geq 0, \qquad \forall i = 1, \ldots, n$$

with $w_i > 0$, $i = 1, \ldots k$ and $w_1 + w_2 + \ldots + w_k = 1$.

For further details, refer to textbook of Antunes et al. (2016).

12.4.3 Distance Minimization to the Ideal Point

If a efficient solution closer to the ideal situation is to be found, the distance minimization to the ideal point is the adequate method used. The ideal solution z_I is, in fact, a reference point, and one way to compute it is to solve each objective function individually.[1] Consequently, $z_I = \left(z_1^*, z_2^*, \ldots, z_k^*\right)$. Too often, this reference point represents an unattainable outcome, and therefore one may wish to find the feasible outcome that is closer to the ideal. The compromise solution found by this method is always an efficient solution.

Two distance functions allow one to define a multiple objective problem into a (mixed integer) linear programming model: *Manhattan* and *Chebyshev* distances. The latter is the largest individual absolute difference of any pair of coordinates between two points and is useful in minimax settings. The model that minimizes the *Manhattan* distance to the ideal point is as follows:

$$\min f(x) = |f_1(x) - z_1^*| + |f_2(x) - z_2^*| + \ldots + |f_k(x) - z_k^*| \qquad (12.53)$$

$$\text{s.t.} \sum_{i=1}^{n} A_{ij} x_i \le b_j, \qquad \forall j = 1, \ldots, m$$
$$x_i \ge 0, \qquad \forall i = 1, \ldots, n$$

Since $f_i(x) - z_i^* \ge 0, i = 1, \ldots, k$, then function (12.53) can be transformed into the linear function (12.54):

$$\min f(x) = \left(f_1(x) - z_1^*\right) + \left(f_2(x) - z_2^*\right) + \ldots + \left(f_k(x) - z_k^*\right). \qquad (12.54)$$

The model that minimizes the *Chebyshev* distance to the ideal point is as follows:

$$\min \max_{i=1,\ldots k} |f_i(x) - z_i^*| \qquad (12.55)$$

$$\text{s.t.} \sum_{i=1}^{n} A_{ij} x_i \le b_j, \qquad \forall j = 1, \ldots, m$$
$$x_i \ge 0, \qquad \forall i = 1, \ldots, n$$

The linearization of the above model is given by model (12.56). For further details, refer to Eiselt and Sandblom (2007) and Antunes et al. (2016).

$$\min v \qquad (12.56)$$

$$\text{s.t.} \sum_{i=1}^{n} A_{ij} x_i \le b_j, \qquad \forall j = 1, \ldots, m$$
$$f_i(x) - z_i^* \le v, \qquad \forall i = 1, \ldots, k$$
$$x_i \ge 0, \qquad \forall i = 1, \ldots, n$$

---

[1] The reference point can be any point that properly reflects the decision-maker's preferences.

12.4 Multiple Objectives

12.4.4 ε-Constraint Method

The last method presented is the ε-constraint method where the most important objective $f_j(x)$ is selected to be minimized, while the remaining objectives are converted into inequality constraints. The multiple objective programming problem defined above is redefined as:

$$\min f_j = \sum_{i=1}^{n} c_{ij} x_i$$

$$\text{s.t.} \sum_{i=1}^{n} A_{ij} x_i \leq b_j, \qquad \forall j = 1, \ldots, m$$

$$\sum_{i=1}^{n} c_{ip} x_i \leq \varepsilon_p, \qquad \forall p = 1, \ldots j - 1, j + 1, \ldots, k$$

$$x_i \geq 0, \qquad \forall i = 1, \ldots, n$$

Varying the ε_p values, it is possible to reach all nondominated solution (Pet-Armacost et al. 2013).

According to Mavrotas (2009), the ε-constraint method presents three aspects needing careful attention: (1) the calculation of the range of the objective functions over the efficient set; (2) the guarantee of efficiency of the obtained solution, and (3) the increased solution time for problems with more than two objective functions. In this work, the author proposes an improved version of the ε-constraint method to address those aspects, the so-called augmented ε-constraint method. To overcome the first issue, the lexicographic optimization approach is applied over every objective function in order to compute the range of the objective functions (the payoff table) over the efficient set. The lexicographic approach will ensure the Pareto optimality by optimizing a first objective function and then, among the possible alternative optima, optimizing a second objective function, and so on. If the range of the objective functions is obtained only by individual optimization, it is not guaranteed that the solutions obtained are Pareto optimal solutions since alternative optima may be presented (weak efficiency). To overcome the second issue, Mavrotas (2009) suggests the objective functions set as constraints are transformed into equalities (instead of inequalities as in the traditional method) by incorporating slack or surplus non-negative variables. These new variables are then used as a second term in the objective function penalizing it if their value differs from zero. This strategy forces the model to produce only efficient solutions. When dealing with three objective functions, a total of $(m + 1) \cdot (n + 1)$ runs are performed to obtain the Pareto front, where m and n are the equal intervals dividing the range of each objective function. Lastly, to decrease the number of runs, the algorithm initiates with the more relaxed version of the constrained objective function and gradually restricts the bounds. When the problem becomes infeasible, it means that from that point below, only infeasible models will be obtained. Therefore, the algorithm proceeds to the next grid point.

12.4.5 Iterative Methods

If the decision-maker is unable to give enough information about the relative importance of the objective functions, methods such as weighted sum and lexicographic are not adequate. Several methods have been defined to overcome such drawback. These are traditionally iterative methods where the facilitator and the decision-maker discuss the adequacy of the solutions provided by the methods. When the decision-maker is happy with one solution, the iterative process ends.

One example of such a method is the STEM method proposed by Benayoun et al. (1971). This iterative exploration procedure finds the best compromise among objectives (compromise solution) after a certain number of cycles. Each cycle has a calculation phase, and a decision-making phase, i.e., a conversation between the facilitator and the decision-maker. During the decision-making phase, the decision-maker examines the results of the calculations and identifies the satisfactory and unsatisfactory objectives and also indicates which objectives in the current solution can be decreased to achieve an improvement in the unsatisfactory objectives. A new constraint is added to the model; a new compromise solution is determined. The procedure ends when the decision-maker is satisfied with the solution.

Other methods such as the Zionts and Wallenius method, the TRIMAP, and Pareto race method have been successively applied to real-world problems (Antunes et al. 2016). However, one cannot say one of them is superior to all the others. In fact, some methods may suit different DMs and problems better than the others. One must always keep in mind that for an iterative method to be successfully applied, the DM must be available and willing to actively participate in the solution process and direct it according to her/his preferences (Miettinen et al. 2008).

12.5 Case Study 1: Integrated Assessment of a New Waste-to-Energy Facility in Central Greece in the Context of Regional Perspectives

Perkoulidis et al. (2010) examined a centralized waste-to-energy (WtE) plant sited in Thiva (Viotia, Greece) to serve municipalities as a potential treatment facility. The supplying of the candidate facility was assessed through the evaluation of different transfer schemes considering local conditions and economic criteria. In particular, the direct transportation of MSW from the producers to the WtE facility with refuse collection vehicles (RCV) or via a transfer station that was at the time inexistent. Moreover, six different cases regarding the energy recovery of the candidate WtE facility and the relevant energy efficiency, as EC defines it, were assessed to examine how WtE could meet the requirements of the new 2008/98/EC Directive on waste (EC 2008).

The transfer station possible locations were evaluated and identified from the solution of facility location model previously developed by the authors so that an efficient supplying chain between the waste producers and the waste-to-energy facility could be put in place. Four potential scenarios for the regional waste management were assessed via the multi-criteria decision-making method and ELECTRE III method (see Chap. 13) using four criteria: total cost, biodegradable municipal waste diversion from landfill, energy recovery, and greenhouse gas emissions.

The study demonstrated that a waste management scenario based on a waste-to-energy plant with an adjacent landfill for disposal of the residues would be the best performing option for the region, depending however on the priorities of the decision-makers. Also, it was also showed that efficient planning is necessary and the simultaneous operation of sanitary landfills and a WtE should be avoided.

A sensitivity analysis was performed to evaluate the effects of increased recycling rate, on the calorific value of treated municipal solid waste and the gate fee of the candidate plant. It showed that increased recycling efforts would not diminish the potential for incineration with energy recovery from waste and neither would have adverse impacts on the gate fee of the waste-to-energy plant.

12.6 Case Study 2: A Recovery Network for WEEE – A Sustainable Design

Furtado et al. (2011) and Gomes and Barbosa-Povoa (2014) studied the redesign of Amb3e recovery network in order to minimize both the costs and the environmental impacts. Amb3e is an association responsible for the management of waste of electric and electronic equipment (WEEE) recovery network This network involves different activities: people drop off products at gathering points (GP); the unsorted products are sent to collection centers (CC), sorted, and then sent to proper treatment facilities (TF). Products are sorted according to five operational flows defined by Amb3e: the flow A encompasses all the big equipment, the flow B includes cooling and refrigeration equipment, the flow C has all the small equipment, the flow D has lighting equipment, and the flow E has televisions and cathodic ray tubes. In 2012, flow A was the more representative one, with 35.5% of all products collected, followed by flow B (25.9%), flow E (13.8%), flow C (13.4%), and, finally, flow D (1.4%).

The network comprised 315 gathering points, but given the strategic nature of this work and without losing accuracy, these were grouped into 278 GPs, one for each municipality of mainland Portugal. There were also eight treatment facilities spread over mainland Portugal. The objective was to know the best location for the collection centers which could be located in all Portuguese municipalities. Amb3e outsources all product transportation to the treatment facilities.

In short, the problem can be defined as follows: given a superstructure composed by all possible locations of the entities in the network, the distance between all the pairs of entities, the estimated WEEE volume to be collected at each generation source, the recovery target set by legislation, the sorting criteria to be performed at the CCs, the initial stock levels at the entities, the maximum storage capacity, maximum and minimum processing capacities, upper and lower limits for flows between all pair of entities, and all the costs and environmental impacts involved in the network, determine the locations of CCs, the flows between GPs and CCs and between CCs and TFs, the storage volumes at CCs and TFs, and the processed and disposed volumes at each TF, that minimizes the total cost and the global environmental impact of the network.

The developed MILP model accounts for all the abovementioned data. A two-time scale allows the simultaneous modeling of strategic and tactical decisions such as the location of CC (strategic) and the planning decisions related with collection, sorting, storage, processing, and transportation volumes (tactical).

Environmental aspects were modeled through an LCA approach where environmental impacts were associated to transportation (a distance-dependent parameter that accounts for gas consumption and emission to the atmosphere and heavy metal emissions to land and water); processing activities at TF's facilities (accounts for all processing activities in terms of energy consumption and gas emissions); disposal activities (linked to the proper disposal of products that cannot be recycled, this might represent a benefit for the environment function since products may be incinerated and energy produced); CC installation (accounts for land use impact and all end-of-life activities related to the (future) closed down of such units); storage of material (emissions caused by stored material and land used); and lastly no product collection (reflecting the harm caused by products that are improperly disposed, either by being left at the nature or being landfilled with organic waste). All values were calculated according to SimaPro (Goedkoop et al. (2004)).

Different analyses were performed: (1) the total cost of the network was minimized, comparing the current scenario with the optimized one; (2) a similar analysis was performed with the minimization of the environmental impacts; and (3) a multi-objective approach was followed to define an approximation of the Pareto front. All results were provided by GAMS/CPLEX (build 23.3).

12.6.1 The Current Network

For the current network, all the 278 possible locations for the collection centers were assumed to be installed. There were also 278 GPs, one for each municipality, and 8 TFs that existed in mainland Portugal. For this network the total cost scenario generated a result of about 59,000 thousand euros. The major contribution comes from processing costs (66%), followed by the transportation costs (17.8%), CCs opening costs (7.1%), the compensation fees to CCs (6.8%), and storage costs (2.3%). The global environmental impact of the current scenario is about 4000 points. The largest part comes from to CCs installation representing a total of 62.1%, followed by TFs operation (14.6%), processing (12.0%), transportation (9.3%), disposal (1.4%), and finally storage impacts (0.6%). These results set the baseline comparison for the optimized networks presented next.

12.6.2 The Optimal Network

The optimal number of CC opened was optimized considering each objective functions independently, leading to different results. In addition to the 278 GPs and the 8 TFs, the total cost minimization scenario proposes the opening of 20 CCs. This represents a total cost of about 55,500 euros, which leads to 6% decrease regarding the current network cost (Table 12.4). The cost structure is similar to the current one, except on the costs of opening CC that represents only 0.5% of the total against 6.8% of the previous one.

Table 12.4 Environmental impacts and total cost of both optimal network structures

Optimal solutions	Cost (10^3€)	Environmental impact (10^3 pts)	No. of CC
Cost minimization	55,500	1800	20
Environmental impact minimization	63,600	1690	15

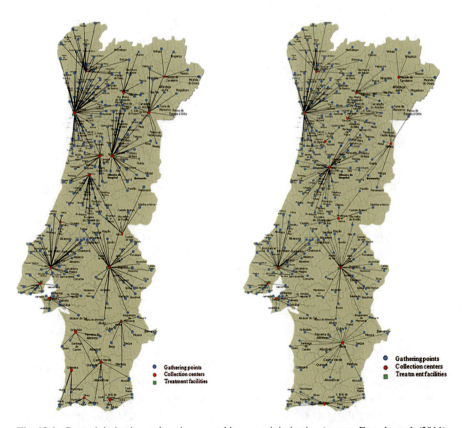

Fig. 12.1 Cost minimization and environmental impact minimization (source: Furtado et al. (2011))

The environmental impact minimization scenario proposes only 15 CCs be opened, five less than the optimal cost scenario. This led to a global environmental impact of about 1690 points, representing a massive reduction from the 4000 points of the current structure. This reduction was mainly due to the reduction of CCs opened. Figure 12.1 shows all connections between CC and TF in the two optimal solutions.

12.6.3 Scenario Comparison

A comparison of all the results above was made to have a better perception of what happens in each scenario. In the total cost scenario, the compensation fee, the processing, the transportation, and the storage costs are virtually the same whether the current or the optimized network is considered. This happens because these costs are directly linked to the collection of WEEE that is performed. So, being both scenarios modeled for the same country and the same collection rates, it is only normal that those costs are almost the same. So, it is safe to assume that the variable that makes a difference is the number of CCs opened. Table 12.4 shows that a 7% reduction in the total environmental impact (1800 vs. 1690) corresponds to a 14% increase in total cost (55,550 vs. 63,600). This cost increase is caused by the reduction of the number of CC from 20 to 15 and consequently the increase in transportation costs.

12.6.4 Multi-Objective Analysis

To simultaneously analyze both objectives, a multi-objective methodology was applied. The optimal values obtained previously (Table 12.4) were assumed as approximations to the two extreme points of the Pareto front. The ε-constraint method (section 12.4.4) was applied to compute other efficient points. The economic objective function was chosen to be minimized, while the environmental objective function was set as a constraint. The efficient solutions were obtained by varying the limit imposed according to the constraint (12.57):

$$F_e = \varepsilon \cdot I_{max}. \tag{12.57}$$

where F_e is the equation of the environmental objective function and $\varepsilon \in [0, 1]$ and I_{max} is the total environmental impact of the lowest-cost network.

Initially, it was intended to apply the multi-objective methodology to mainland Portugal. However, given the model dimensions, it was computationally impossible to realize such study. Thus, the analysis was performed on a smaller geographical area, the municipalities of Lisbon and Vale do Tejo (LVT), which is the Portuguese region with largest number of municipalities (53 of the 278 municipalities in mainland Portugal) and is the location of four of the eight treatment and recovery units.

12.7 Final Remarks

Table 12.5 Total cost and environmental impacts of network structures for LVT area

Optimal solutions	Cost (10^3€)	Environmental impact (10^3 pts)	No. of CC
Cost minimization	19,534	669	6
Environmental impact minimization	22,190	608	4

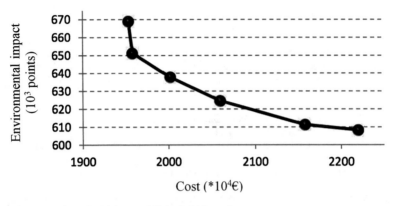

Fig. 12.2 Pareto front for Lisbon and Vale do Tejo region

Table 12.5 presents the cost and impact values obtained as extreme points for the Pareto front related to LVT. The minimization of the environmental function leads to a 10% reduction of the environmental impact of the minimum cost network. This reduction is traded off by a 13% increase in the total cost of the logistics network.

Figure 12.2 shows an approximation to the Pareto front. These points were computed varying ε between 0.9 and 1, with increments of 0.02. Its negative exponential form suggests that benefits regarding environmental impacts become smaller as the total cost network reaches its minimum value. This is to mean that as the network becomes greener, the investments need to be larger to have an effective environmental impact benefit.

12.7 Final Remarks

In this chapter, mathematical programming models were presented and illustrated by several works addressing solid waste management issues. In fact, the existing literature focusing on such topics is extremely rich where contributions aim at finding the optimal solution for reduced problems of the more complex ones. Nonetheless, models generate insights which lead to better decisions and improve thinking skills by breaking problems down into components. They make assumptions explicit and may be seen as a laboratory in which decision-makers can experiment and learn by testing the implications of alternative courses of action.

They are a helpful tool in developing not only a recommended decision, but they also in building the rationale behind the why a given decision is preferred.

By no means, the intention was to be exhaustive in this chapter. The focus was on what are the most common models and methods used in the literature concerning sustainable waste management.

References

Antunes CH, Alves MJ, Clímaco J (2016) Multiobjective linear and integer programming. EURO advanced tutorials on operational research. Springer, Schweiz

Applegate DL, Bixby RE, Chvaátal V et al (2006) The Traveling Salesman Problem: a computational study. Princeton series in applied mathematics Princeton. Princeton University Press, Princeton

Badran MF, El-Haggar SM (2006) Optimization of municipal solid waste management in Port Said–Egypt. Waste Manag 26(5):534–545

Barth M, Scora G, Younglove T (2004) Modal emissions model for heavy-duty diesel vehicles. Transp Res Rec 1880:10–20

Bazaraa MS, Sherali HD, Shetty CM (2013) Nonlinear programming: theory and algorithms. John Wiley & Sons, New York

Beliën J, De Boeck L, Van Ackere J (2014) Municipal solid waste collection and management problems: a literature review. Transp Sci 48:78–102

Beltrami EJ, Bodin LD (1974) Networks and vehicle routing for municipal waste collection. Networks 4(1):65–94

Benavent E, Corberán A, Sanchis JM (2000) Linear programming based methods for solving arc routing problems. In: Dror M (ed) Arc routing: theory, solutions and applications. Springer, Boston, pp 231–275

Benayoun R, De Montgolfier J, Tergny J et al (1971) Linear programming with multiple objective functions: step method (STEM). Math Program 1(1):366–375

Bertsimas D, Tsitsiklis JN (1997) Introduction to linear optimization. Athena Scientific, Belmont

Birge JR, Louveaux F (1997) Introduction to stochastic programming. Springer, Berlin

Bisschop J, Meeraus A (1982) On the development of a general algebraic modeling system in a strategic planning environment. In: Goffin JL, Rousseau JM (eds) Applications, mathematical programming studies, vol 20. Springer, Berlin/Heidelberg, pp 1–29

Blazquez CA, Beghelli A, Meneses VP (2012) A novel methodology for determining low-cost fine particulate matter street sweeping routes. JAPCA J Air Waste Manage Assoc 62(2):242–251

Braier G, Durán G, Marenco J et al (2017) An integer programming approach to a real-world recyclable waste collection problem in Argentina. Waste Manag Res 35(5):525–533

Drud AS (1994) CONOPT – a large-scale GRG code. ORSA J Comput 6(2):207–216

European Parliament, Council (2008) Directive 2008/98/EC of the European Parliament and of the Council of 19 November 2008 on waste and repealing certain directives. Off J Eur Union L312:3–30

Eiselt HA, Sandblom CL (2007) Linear programming and its applications. Springer Science & Business Media, Berlin/Heidelberg

Erkut E, Neuman S (1989) Analytical models for locating undesirable facilities. Eur J Oper Res 40 (3):275–291

Erkut E, Karagiannidis A, Perkoulidis G, Tjandra SA (2008) A multicriteria facility location model for municipal solid waste management in North Greece. Eur J Oper Res 187:1402–1421

Filipiak KA, Abdel-Malek L, Hsieh HN et al (2009) Optimization of municipal solid waste collection system: case study. Pract Period Hazard Toxic Radioact Waste Manage 13 (3):210–216

References 237

Fourer R, Gay DM, Kernighan BW (2002) AMPL: a modeling language for mathematical programming. Duxbury Press

Furtado P, Gomes MI, Barbosa-Povoa AP (2011) Design of an electric and electronic equipment recovery network in Portugal-Costs vs. Sustainability. In: Pistikopoulos N, Georgiadis MC, Kokossis AC (eds) Computer aided chemical engineering, vol 29. Elsevier, Amsterdam, pp 1200–1204

Ghiani G, Laganà D, Manni E, Musmanno R, Vigo D, (2014) Operations research in solid waste management: A survey of strategic and tactical issues. Comput. Oper. Res. 44, 22–32

Goedkoop M, Oele M, Leijting J, Ponsioen T, Meijer E (2004) SimaPro 6-Introduction to LCA with SimaPro. *Amersfoort, the Netherlands: PRé Consultants*

Gomes MI, Barbosa-Povoa AP (2014) Projecto de uma rede logística para a recolha de equipamentos elétricos e eletrónicos. In: Oliveira RC, Ferreira JS (eds) Investigação operacional em ação: casos de aplicação. Imprensa da Universidade de Coimbra, chapter 11.

Greenberg H, Morrison T (2008) Chapter 14: Robust optimization. In: Ravindran AR (ed) Operations research and management science. CRC Press, Boca Raton

Gurobi Optimization, Inc. (2016) Gurobi optimizer reference manual. http://www.gurobi.com

Hillier FS, Lieberman GJ (2014) Introduction to operations research. McGraw Hill. Inc., New York

Hoffman KL, Padberg M, Rinaldi G (2013) Traveling salesman problem. In: Encyclopedia of operations research and management science. Springer US, Berlin, pp 1573–1578

IBM ILOG CPLEX Optimization Studio (2017a) CPLEX User's Manual Version 12 Release 7. www.cplex.com

IBM ILOG CPLEX Optimization Studio (2017b) OPL Language User's Manual Version 12 Release 7. https://www.ibm.com/support/knowledgecenter/en/SSSA5P

Khorram E, Zarepisheh M, Ghaznavi-Ghosoni BA (2010) Sensitivity analysis on the priority of the objective functions in lexicographic multiple objective linear programs. Eur J Oper Res 207 (3):1162–1168

Laporte G (2007) What you should know about the vehicle routing problem. Nav Res Logist 54:811–819

Laporte G (2009) Fifty years of vehicle routing. Transp Sci 43:408–416

Laporte G (2010) A concise guide to the traveling salesman problem. J Oper Res Soc 61(1):35–40

Luenberger DG, Ye Y (2016) Linear and nonlinear programming. In: Price CC (ed) International series in operations research & management science, vol 228. Springer International Publishing, Cham

Mavrotas G (2009) Effective implementation of the epsilon-constraint method in multi-objective mathematical programming problems. Appl Math Comput 213:455–465

Miettinen K, Ruiz F, Wierzbicki AP (2008) Multiobjective optimization: introduction to multiobjective optimization: interactive approaches. In: Branke J, Deb K, Miettinen K, Słowiński R (eds) Lecture notes in computer science, vol 5252. Springer, Berlin/Heidelberg, pp 27–57

Murtagh BA, Saunders MA (1983) MINOS 5.4 User's Guide. Report SOL 83-20R, Systems Optimization Laboratory

Owen SH, Daskin MS (1998) Strategic facility location: a review. Eur J Oper Res 111:423–447

Perkoulidis G, Papageorgiou A, Karagiannidis A, Kalogirou S, (2010) Integrated assessment of a new Waste-to-Energy facility in Central Greece in the context of regional perspectives. Waste Manag. 30, 1395–1406

Pet-Armacost J, Mollaghasemi M, Armacost RL (2013) Interactive multiple objective mathematical programming. In: Gass S, Fu M (eds) Encyclopedia of operations research and management science. Springer US, Boston, pp 784–792

Ramos TRP, Gomes MI, Barbosa-Póvoa AP (2014) Economic and environmental concerns in planning recyclable waste collection systems. Transp Res Part E Logist Transp Rev 62:34–54

Roberti R, Toth P (2012) Models and algorithms for the asymmetric traveling salesman problem: an experimental comparison. EURO J Transp Logist 1(1–2):113–133

Sioshansi R, Conejo AJ (2017) Optimization in engineering: models and algorithms, vol 120. Springer, Cham

Steuer R (1986) Multiple criteria optimization: theory computation and application. Wiley, New York

Tawarmalani M, Sahinidis NV (2005) A polyhedral branch-and-cut approach to global optimization. Math Program 103(2):225–249

Vanderbei RJ (2015) Linear programming. Springer, Heidelberg

Verter V (2011) Uncapacitated and capacitated facility location problems. In: Eiselt HA, Marianov V (eds) Foundations of location analysis. Springer, Boston, pp 25–37

Williams HP (2013) Model building in mathematical programming. Wiley, UK

Chapter 13
Multi-criteria Decision-Making in Waste Collection to Reach Sustainable Waste Management

Abstract Multi-criteria decision analysis (MCDA) is concerned with theory and methodology that can deal with complex problems encountered in sustainable waste management. It provides methodologies to support decision-makers when selecting the best compromise among a set of alternative characterized by different and conflicting criteria. The top five methods used in solid waste management (SAW, AHP, TOPSIS, PROMETHEE, ELECTRE) are described with detail. The existence of multiple stakeholders and MCDA software is also addressed. This chapter ends with five case studies providing an overview of how these methods have been supporting decision-making in SWM.r

Keywords AHP · ELECTRE · PROMETHEE · SAW · Stakeholders · Sustainable decisions · TOPSIS · Waste management

13.1 Introduction

Multi-criteria decision-making (MCDM) is concerned with theory and methodology that can deal with complex problems encountered in business, engineering, and other areas of human activity (Achillas et al. 2013). MCDM includes two complementary areas: mathematics-based multiple objective programming (MOP) – addressed in Chap. 12 – and decision-maker-driven multi-criteria decision analysis (MCDA). In a nutshell, MCDA methods aim to compare or rank any set of alternatives based on the criteria adopted, whereas MOP techniques are focused on determining the set of optimal alternatives according to the criteria considered (Goulart-Coelho et al. 2017).

MCDA is a sub-discipline of operations research that explicitly considers multiple criteria (e.g., points of view, goals) in decision-making environments (Achillas et al. 2013). A complex problem is characterized by noncomparable and conflicting criteria or objectives such as cost, performance, reliability, safety, productivity, and affordability. In the presence of multiple criteria, a unique optimal decision for the problem does not exist, but rather many decisions are suitable (Wiecek et al. 2008). A very large number of MCDA methods have been

© Springer International Publishing AG, part of Springer Nature 2019
A. Pires et al., *Sustainable Solid Waste Collection and Management*,
https://doi.org/10.1007/978-3-319-93200-2_13

239

proposed over the years to help in selecting the best compromise alternatives rather than taking decisions based only on personal thoughts, views, or experiences.

Sustainable solid waste management is a complex process which includes waste collection routes, transfer station locations, treatment strategy, treatment plant location, and energy recovery. When designing and managing sustainable waste systems, decision-makers should define local and regional goals on all or some of these activities and then plan a strategy accordingly. MCDA provides methodologies to help decision-makers in selecting the best compromise among alternatives through (i) the organization of the different elements into a hierarchical structure, (i) the apprehension of the relationships between components of the problem, and (iii) the encouragement of the communication among stakeholders (Malczewski 2006). Moreover, MCDA methodologies provide decision-makers with a powerful tool toward convincing the public over the optimal waste management strategy since these techniques attempt to make (subjective) decision-making process as transparent and explicit as possible (Belton and Stewart 2001; Hajkowicz and Higgins 2008).

Guitouni and Martel (1998) categorized MCDA methods in three groups based on the way they model the decision-maker(s)'s preferences: single synthesizing, outranking, and interactive approaches (Guitouni and Martel 1998). The first approach builds functions that aggregate alternatives scores and then maximizes the final score. The outranking methods model decision-maker(s)s preferences making use of binary comparisons of alternatives in each criterion (Martel 1999). Lastly, methods with interactive approaches present trade-off solutions and dialogues with the decision-makers to reach a conclusion.

This chapter unfolds as follows. The generic methodology of multi-criteria decision analysis methods is presented together with the basic concepts and terminology (Sect. 13.2). Section 13.3 presents five of the most known MCDA models that have been extensively applied to decision-making in SWM; this section ends with a brief comparison of the models. Uncertainty is a major issue in decision-making. It may be related to decision-maker values and judgments and/or related to imperfect knowledge concerning consequences of actions. Sensitivity analysis is discussed in Sect. 13.4, and it is a powerful and well-studied tool that provides valuable information concerning the decision robustness. Given the models' complexity, software is an essential tool when dealing with real-world challenges; a brief review of some tailored design software is presented in Sect. 13.5. Dealing with multiple stakeholders in decision-making is the topic address in Sect. 13.6. Lastly, five case studies are presented to provide with an overview of how have been these methods of help in SWM and what kind of decisions have been addressed.

13.2 Generic Multi-Criteria Analysis Methodology

An MCDA method aims to rank or score a finite number of decision options based on a set of evaluation criteria. The MCDA model can be represented by a performance table M of n alternatives (decision options) and m criteria, where each entrance x_{ij} reflects the score for alternative i with respect to criterion j:

$$M = \begin{bmatrix} x_{11} & \cdots & x_{1m} \\ \vdots & \ddots & \vdots \\ x_{n1} & \cdots & x_{nm} \end{bmatrix}$$

x_{ij} may be quantitative or qualitative data. The importance of each criterion is given in a one dimensional weights vector W containing m weights, where w_j denotes the weight assigned to the jth criterion, $W = (w_1, \ldots, w_m)$. The MCDA methods aim to:

- Define the function $r_i = f(X, W)$, $R = (r_1, \ldots, r_n)$ and provide a rank order of the alternatives.
- Define the function $u_i = g(X, W)$, $U = (u_1, \ldots, u_n)$ and compute a utility score for each alternative.

The utility score may be viewed as a measure of the overall benefit or worth of an alternative relative to the other alternatives in study. Values r_i and u_i may then be used by the decision-maker to:

- Select a single alternative.
- Select a subset of alternatives.
- Determine an ordering of all alternatives.

Under some conditions, such as strict dominance and nondiscrimination, criteria should be removed from an MCDA model. Criteria are in "strict dominance" when one alternative is outperformed by another in all criteria. When criteria do not provide differentiation among the alternatives (all have the same performance evaluation), one says they are nondiscriminating (Hajkowicz and Higgins 2008).

The application of MCDA methods usually includes several steps, which are depicted in Fig. 13.1:

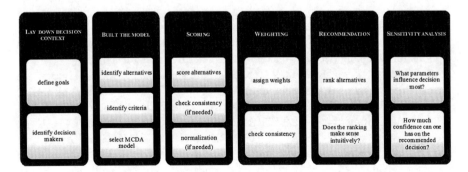

Fig. 13.1 MCDA decision-making steps

1. Establish decision context – where goals are defined and the decision-makers selected.
2. Build the model:

   (a) Identify criteria: define and describe all criteria that will be use to evaluate the alternatives; depending on the context, they might be clustered into a hierarchy. There are no strategies to assess the optimal number of criteria. In some cases, the type of MCDA or the capacity of the decision-makers to deal information put a limit on (or favor a certain) the number of criteria. When in doubt about excluding a criterion, it is always prudent to take it into account in the analysis, since some MCDA methods allow a criterion to be dropped in a later stage.
   (b) Identify alternatives.
   (c) Select the MCDA method.

3. Scoring – this may require decision-maker preference inputs:

   (a) Score the alternatives for each criterion (x_{ij}).
   (b) Check consistency of scores.
   (c) Score normalization: transform scores into commensurate units; this step is needed for some of the MCDA methods.

4. Weighting – this step heavily dependents on decision-maker preferences:

   (a) Assign weights to each criterion.
   (b) Check consistency of weights.

5. Recommendation:

   (a) Rank the alternatives based on scores and weights.
   (b) Analyse if the rank makes sense intuitively.

6. Perform sensitivity analysis (weights, performance measures, etc.).

13.3 Multi-Criteria Decision Aid Methods

The number of MCDA methods has increased rapidly over the past decades (for a comprehensive review, see Figueira et al. (2016b)). They provide almost unlimited options for modeling decision-makers preferences to attain a result.

A recent work reviewed of 221 papers published since 1980 in solid waste management where MCDA techniques were applied. Goulart-Coelho et al. (2017) determined the top five methodologies: Analytic Hierarchy Process (42%), Simple Additive Weighting (24%), ELECTRE methodology (10%), and PROMETHEE and TOPSIS (both with about 7%). ELECTRE and PROMETHEE are outranking methods, where preferences are based on comparisons between pairs of alternatives to assess if an alternative is at least as good as another (Cinelli et al. 2014).

13.3 Multi-Criteria Decision Aid Methods

ELECTRE and PROMETHEE were placed fourth and sixth, respectively, as the most used methods mentioned in the revised works.

13.3.1 Simple Additive Weighting (SAW)

Simple Additive Weighting is one of the simplest and widely applied techniques of MCDA. In weighted average, for each criterion, the scores are transformed onto a normalized scale (commonly 0 to 1, where 0 represents the worst performance and 1 the best one), multiplied by weights, and summed to attain overall score. The selection (or rank) of alternatives should be made according to the score determined as a weighted average of the normalized values:

$$u_i = \sum_{j=1}^{m} w_j x_{ij},$$

where $w_1 + \ldots + w_m = 1, 0 < w_j < 1$.

Simple weighted averaging models are compensatory methods; this is to mean that good values for an alternative on one criterion can offset poor values on another criterion (Adelman 2013).

The examples of such methods are SMART (simple multi-attribute rating method), MAUT (multiple attribute utility theory), and MAVT (multiple attribute value theory). In SMART, weights are derived using direct numerical ratio judgments of the relative importance of attributes – swing weighting method. First the decision-makers rank the order of the criterion in decreasing order of importance and assign an arbitrary value to the most important criteria. Then they judge how much less important each of the remaining criteria is in relation to the most important and attribute a corresponding value. Finally, the ratio weights are normalized (Wang and Yang 1998). This way of computing weights allows the weight on a criterion to reflect both the range of difference of the alternatives and how much that difference matters to the decision-makers. Both MAVT and MAUT methods assume there is a value (utility)-based function representing the decision-maker preferences. Such functions need to be assessed and then are used to aggregate and rank the alternatives (Guitouni and Martel 1998).

13.3.2 Analytic Hierarchy Process (AHP)

One of the most popular MCDA methods is AHP, proposed by Saaty (1980). AHP hierarchically decomposes the decision problem and, making use of pairwise comparison, computes a numerical value allowing for a final raking of alternatives. To measure criteria of relative weights using AHP, decision-makers are asked to make a series of pairwise comparison judgments on a five-point ratio scale [1, 3, 5, 7, 9]. The same approach is used to access relative importance (weights) between the alternatives within each criterion. The computation of relative weights involves three steps. Considering the pairwise comparison matrix, the sum of each column is first calculated. Then, the matrix is normalized by dividing each element in the matrix by the corresponding column total (the new matrix is called the normalized pairwise comparison matrix). Lastly, the average of the elements in each row of the normalized matrix is made by dividing the sum of normalized scores for each row by the number of criteria. A consistency test should be performed to access the quality of the comparison matrix. Only consistent matrices should be used to compute the relative weights. Consistency means that the decision-maker is coherent when making pairwise comparisons (Taha 2003).

Despite wide applications of the AHP in a variety of domains and at different levels of the decision hierarchy, the AHP has been criticized from several viewpoints. Perez (1995) shows that the method should be handled with care since in AHP the introduction of a new alternative may (or may not) change the alternatives ranking (rank reversal).

13.3.3 TOPSIS

Technique for Order of Preference by Similarity to Ideal Solution, TOPSIS, looks for alternatives with the shortest distance from the ideal alternative and the farthest from the most disadvantage option (Hwang and Yoon 1981). The ideal alternative is a hypothetical alternative that comprises the most desirable outcomes for the evaluation criteria. The nadir represents another hypothetical alternative comprising the least desirable outcomes for evaluation criteria.

The set of alternatives is compared by (1) identifying weights for each criterion; (2) normalizing scores in each criterion; (3) calculating a distance, D^+, between each alternative and the ideal alternative (best on each criterion) and a distance, D^-, to the nadir alternative (worst) across the weighted criteria, using one of several possible distance measures (e.g., Euclidean distance); (4) calculating the index of similarities; and finally, (5) ranking the alternatives in decreasing order with the calculated index. The index of similarities is calculated as the ratio between the distance (separation) from the nadir alternative and the sum distance from the ideal and nadir alternatives:

$$\frac{D^-}{D^+ + D^-}.$$

The alternative that is closest to the ideal point, and at the same time farthest from the nadir, is the best alternative under this decision rule (Nyerges and Jankowski 2009). Benefits of TOPSIS include that the only judgments required to the decision-maker are weights, while the ratio depends on the weights and the range of alternatives themselves.

13.3.4 PROMETHEE

Preference Ranking Organization Method for Enrichment, PROMETHEE, is a methodology for MCDA problems that was developed by J.P. Brans during the early 1980s. They are part of the outranking MCDA family and are based on a set of prerequisites (Brans and De Smet 2016): (i) the extent of difference between the performance of two alternatives must be accounted for; (ii) the scales of the criteria are irrelevant as comparisons are performed on a pairwise base; (iii) three cases are possible: alternative a is preferred to alternative b, alternative a and alternative b are indifferent, and alternative a and alternative b are incomparable; (iv) the methods should be easily understandable by the decision-makers; and (v) weights must be assigned in a flexible manner.

The assessment procedure requires information between and within the criteria. Concerning the information between the criteria, this is expressed as the relative importance among them and consists of weights that are independent from the measurement scales.

To obtain information within the criteria, a preference function for each criterion, expressing the difference in performance of alternative a over alternative b, must be identified adopting, as a result, the pairwise comparison approach. Six different shapes for the preference function have been defined, and the identification of the appropriate ones is a task of the analyst who has to question the DM in a structured manner (Brans and De Smet 2016).

Once the preference functions for all the criteria and the weights (w_i) of the criteria are identified, a global preference index indicating the degree of preference of a over b can then be calculated as the weighted average. Subsequently, two parameters, leaving and entering outranking flows, must be calculated, indicating the outranking power and weaknesses of each alternative over the other, respectively. Lastly, the leaving and entering flows can be combined, resulting in the net outranking flow that provides the performance of each alternative (Cinelli et al. 2014).

13.3.5 ELECTRE

Elimination and choice expressing the reality, ELECTRE is, as PROMETHEE, a methodological approach belonging to outranking MCDA family. As an outranking approach, it aims to assess whether alternative a is at least as good as (in other words it outranks) b. The first method was developed by Bernard Roy in the 1960s. Preferences are structured on four elementary binary relations: indifference, preference, weak preference, and incomparability (Figueira et al. 2016a). To identify the outranking relations, concordance and discordance indexes are employed that refer to the cases where the criteria of alternative a are the same or better than those of b (aSb) and to the cases where criteria of a are not as good as those of b (bSa), respectively. ELECTRE methods were developed in order to account for heterogeneous criteria whose aggregation in a common scale is difficult, to prevent compensation behavior and to account for differences in terms of preferences, leading in this way to the introduction of thresholds. Several ELECTRE methods have been developed to solve different decision problems: choice (ELECTRE I, Iv, and IS), ranking (ELECTRE II, III, and IV), and sorting (ELECTRE TRI, TRI C, and TRI nC) (Figueira et al. 2016a).

13.3.6 Comparison of MCDA Methods

Table 13.1 synthesizes the five most applied MCDA methods in SWM studies. Advantages and disadvantages are also presented to ease the understanding of the limitation each method presents. Notice there is no "best" method; all have pros and cons. Since there is no general rule for the choice of a specific MCDA method for a waste stream, the only criterion considered seems to be the question whether the decision-maker necessitates or not a ranking of the alternatives (Achillas et al. 2013). Undoubtedly, the decision on the MCDA method to be employed is also influenced by the authors' previous experience and the availability of adequate software.

13.4 Sensitivity Analysis

Sensitivity analysis is an ex post way to analyze how robust the model is to possible uncertainties (Mustajoki and Marttunen 2017). One type of sensitivity analysis is investigating the impact of parameter values in weights and performances on MCDA outcomes. One can assess such impact on the overall value after weighting and aggregating have been performed. In the literature, this process is also sometimes named as "robustness analysis". While these two terms have different meanings, similar methods can be used to reveal the impact of uncertainty in the estimated values. Among them, the most frequently used methods for sensitivity analysis are deterministic and probabilistic sensitivity analyses. Next is presented the description of the deterministic process. For details concerning the probabilistic approach, refer to Marsh et al. (2017).

Deterministic sensitivity analysis, one parameter, either a criterion weight or an alternative score, is varied at a time, and its impact on the alternatives ranking is observed. If the rank order of alternatives remains unchanged, the decision can be seen as a robust one. Otherwise, one can measure how much the parameter can be changed (by increasing or decreasing its value) without affecting the rank of the alternatives. Deterministic sensitivity analysis is also a valuable tool to investigate the influence of criterion weights on the final ranking. As previously, one should vary the criterion weights one by one and analyze the changes in the overall values of the alternatives.

13.4 Sensitivity Analysis

Table 13.1 Comparison of MCDA methods

MCDA method	Description	Advantages	Disadvantages
SAW	Value-based method	Easy to use and well understandable	Normalization is required
	Use utility as measurement	Applicable when exact and total information is collected	Compensation between good scores on some criteria and bad scores on other criteria can occur
		Well-proven technique	
		Good performance when compared with more sophisticated methods	
AHP	Use of value-based, compensatory, and pairwise comparison approach	Applicable when exact and total information is collected	Compensation between good scores on some criteria and bad scores on other criteria can occur
		Incorporates qualitative and quantitative criteria	
	Use of hierarchical structure to decompose a complex problem	Decision problem can be fragmented into its smallest elements, making evidence of each criterion applied	Implementation is quite inconvenient due to complexity
		Generation of inconsistency index to assure decision-makers	Complex computation is required
			Time-consuming due to pairwise comparisons
			Normalization is required
TOPSIS	Use of value-based compensatory method	Easy to implement and understandable principle	Normalization is required
	Measure distances of alternatives to the ideal and nadir solution	Provision of a well-structured analytical framework for alternatives ranking	
	Alternative ranking according to the closeness to the ideal solution	Consideration of both the positive and negative ideal (nadir) solutions	
		Applicable when exact and total information is collected	
		Allows the use of fuzzy numbers to deal with uncertainty problems	
PROMETHEE	Use of outranking method, pairwise comparison, and compensatory method	Applicable even when there is missing information	Time-consuming without using specific software

(continued)

Table 13.1 (continued)

MCDA method	Description	Advantages	Disadvantages
	Use of positive and negative preference flows for each alternative in the valued outranking		When using many criteria, it becomes difficult for decision-maker to obtain a clear view of the problem
	Applicable even when simple and efficient information is needed		
	Generation of ranking in relation to decision weights		
ELECTRE	Use of outranking method, pairwise comparison, and compensatory method	Applicable even when there are incomparable alternatives	Time-consuming without using specific software due to complex computational procedure
		Applicable even when incorporation of uncertainties is required	
	Use of indirect method that ranks alternatives using pairwise comparison	Applicable for quantitative and qualitative attributes	May or may not provide the preferred alternative

Source: Based on Pires et al. (2011)

Tornado graph is a very helpful deterministic sensitivity analysis visualization graph. Such graphs show the impact on model's final result of fixed changes (e.g., -15%, -2.5% and $+2.5\%$, $+15\%$) in one model parameters at a time (Fig. 13.2). It is particularly valuable for showing which parameter has the greatest influence on the final outcome. A tornado graph is a horizontal bar chart where each bar represents one parameter and the length of the bar represents the variation of the model outcome when the parameters vary from its lowest value to its highest. Only one parameter is analyzed at a time, while the remaining parameters are set to their initial values. The name of the chart comes from the way the different bars are placed: the longest bar (corresponds to the most influential parameter) is placed on the top of the chart, the second most influential parameter bar is placed second, and so on (Clemen and Reilly 2013). For instance, the vertical axis of Fig. 13.2 represents three primary cost and revenue variables with uncertainty limit of $\pm40\%$, respectively, Yadav et al. (2018). Uncertainty limit of $\pm40\%$ means that the value of final score is examined for reduction and increment of 40% in the values of cost and revenue parameters. This limit is chosen based on the capacities of existing MSW processing facilities. In the figure MSW generation ($\left[G_i^L, G_i^U\right]$) and recycled fraction of MSW ($\left[\alpha_i^L, \alpha_i^U\right]$) are the most sensitive cost and revenue parameters as 40% reduction in the mid-values reduces final score value to 124,060 and 40% reduction in the values increases final score value to 234,973.

13.4 Sensitivity Analysis

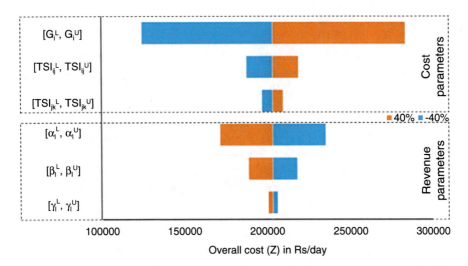

Fig. 13.2 Tornado diagram example. (Source: Yadav et al. (2018))

Goulart-Coelho et al. (2017) in their review found out that among the articles that performed a sensitivity analysis the main objectives were: to evaluate the impact of weight changes on the final ranking (72%); to study the sensitivity related to aggregation, normalization, and thresholds assumed (20%); to assess the effect on the data, such as criteria values, and waste quantities and composition (18%); and to investigate the impact of the criteria selected (7%). The heavy dependency of criteria weights on definition regarding decision-maker's judgments might justify the prevalence of such studies. Nevertheless, the impact on the final ranking due to the criteria adopted, the input data, the aggregation method, the normalization process, and the thresholds values can be as significant as, or even more significant than, the one related to weights.

Gómez-Delgado and Tarantola (2006), using variance-based methods based on Monte Carlo simulation, performed an extensive sensitivity analysis of the criteria used to determine the best location for a hazardous waste landfill site in Madrid, Spain. They concluded that only 3, out of 11, criteria jointly account for more the 97% of the output variance in the multi-criteria spatial decision model. These criteria were: slope (46%), sensitivity of groundwater system to pollution (28%), and type of soil (23%). The analysis performed on weights showed that all the weights remained very far from the factors that represented the major source of uncertainty. Vafaei et al. (2016) study the importance of selecting suitable normalization techniques when using a MCDA method such as TOPSIS. With an illustrative example, the authors concluded that without considering a proper normalization technique as well as appropriate representation of criteria (cost or benefit), the decision results might prove highly inaccurate.

13.5 MCDA Software

Various multi-criteria software tools or decision support systems (DSS) have been developed to support the application of MCDA methods in practice (see Weistroffer and Li (2016) for an exhaustive review). These tools not only give computational methods implementation but also usually provide various ways to support other phases of the process, such as model building and results analysis (Soltani et al. 2015). Graphical user interfaces are an essential features of MCDA software since they determine how comfortable the user is in working with the DSS, and the level of comfort influences how frequently the model is used and how believable its results are (MacDonald 1996). Moreover, by visualizing the process and the results, it facilitates the illustrative, transparent, and understandable realization of a decision-making analysis (Reichert et al. 2013).

Software support is provided for all the MCDA methods presented above, although the features of each of them are different. Five different software will be presented below to provide a brief insight of the existing spectrum concerning decision-making software.[1]

DecideIT software[2] employs a SAW-based method and aims at being a decision tool that handles imprecision. It distinguishes itself from other methods within this approach by refraining from precise numerical inputs (although this is also possible) and builds on various degrees of imprecise statements including comparisons to meet conditions of uncertainty. It provides the modeling of decision trees, scenarios and criteria are categorized, and criteria weights, as well as scenario performances, can be defined with various options of imprecise statements (Buchholz et al. 2009).

SANNA 2009[3] is an Excel freeware add-in for multi-criteria decision support of problems up to 180 alternatives and 50 criteria. It allows the estimation of weights using several methods (e.g., pairwise comparisons) and incorporates methods such as SAW, TOPSIS, two ELECTRE methodologies (I and III), and PROMETHEE, among others (Weistroffer and Li 2016).

HIPRE 3C[4] is a software family that allows individual and group decision support. It integrates AHP and SMART, which can be run separately or combined in one model. The visual interface is customizable to ease the process of structuring, prioritization, and analysis of complex decision problems (Weistroffer and Li 2016).

ELECTRE III–IV[5] aggregate partial preferences into a fuzzy or several non-fuzzy outranking relations. The non-fuzzy ones are useful when criteria cannot be

---

[1]In http://www.cs.put.poznan.pl/ewgmcda/index.php/software and http://www.mcdmsociety.org/content/software-related-mcdm, one can find a brief description and the link to several available software.

[2]https://preference.nu/products/decideit/

[3]http://nb.vse.cz/~jablon/sanna.htm

[4]http://www.hipre.hut.fi

[5]http://www.lamsade.dauphine.fr/spip.php?article240&lang=en

[6]http://www.promethee-gaia.net/softwareF.html

weighted. A "distillation" procedure proposes two complete pre-ordering of the alternatives. The intersection of these pre-orderings shows the most consistent global preferences among alternatives (Weistroffer and Li 2016).

Visual PROMETHEE,[6] also known as Decision Lab 2000, is an interactive decision support system based on two outranking methods PROMETHEE and GAIA (Brans and De Smet 2016), suitable for individual and group decision-making. Sensitivity analyses can be performed using techniques such as multi-scenario comparisons and intervals of stability, among others. The embedded methodology requires a much smaller number of comparisons from the decision-maker than the AHP method; it allows the definition of measurement scale particular to the decision-maker. The PROMap GIS feature connects Visual PROMETHEE with Google Maps.

13.6 Dealing with Multiple Stakeholders in the Decision-Making Process

Soltani et al. (2015) performed a literature review to investigate how MCDA methods helped to address decision-making in sustainable solid waste management when multiple stakeholders (as government, municipalities, industries, experts, and/or general public) are involved. Authors pointed out four characteristics as being significant in this context:

- Extent of stakeholders' involvement in the decision-making process (i.e., stakeholders can choose criteria of concern, rank criteria based on their importance, and/or evaluate the performance of alternatives in each criterion).
- Stakeholder groups (i.e., local governments, municipalities, public or residents, experts, and other nongovernmental organizations or industry).
- Hierarchy of stakeholders (i.e., some stakeholders may have priority or veto power in decision-making).
- Relationship among stakeholders (i.e., competition, coalition, or both); this category is mainly looking for solutions toward conflicts among stakeholders.

Although MCDA is effective for decision-making with multiple stakeholders, these should first agree on some elements including "set of alternative options, set of criteria, scores to be attributed to each of these criteria for each of those options, weights to apply to criteria, ranking method to be used to compare options" (Van den Hove 2006). Conflict often rises when stakeholders, having diverse interests, express different priorities over criteria of decision-making. Different views among the decision-makers may be due to three main causes (Roy 1989; Belton and Pictet 1997):

- Uncertainty: caused by the inherent limitations in apprehending and representing the problem and context and in the precision of appropriate data. This occurs in any modeling activity.

- Conflict: about problem definition, criteria, and/or alternative definition, which results from the different values or priorities of the decision-makers; if the decision-makers reach the point of not speaking to each other, then methods such as mediation or litigation have to be used.
- Misunderstanding: a consequence of different perspectives and partial information.

Belton and Pictet (1997) proposed a framework for contextualizing MCDA in group decision support, which focuses on a range of procedures to come up to an answer to the question: to which extent should decisions be shared among the stakeholder? The authors propose three procedures defining the way in which the views of individuals are brought together with the aim of achieving a group decision:

- *Sharing* aims to obtain a common element by consensus, through a discussion of the views and the negotiation of an agreement; it addresses the differences and tries to reduce them by explicitly discussing their cause.
- *Aggregating* aims to obtain a common element by compromise, through a vote or calculation of a representative value; it acknowledges the differences and tries to reduce them without explicitly discussing their cause.
- *Comparing* aims to obtain an individual element (to reach an eventual consensus based on negotiation of independent individual results); it acknowledges the differences without necessarily trying to reduce them.

The authors argue that for the sake of consistency alternatives, criteria and weight should be common to all decision-makers. The role of the facilitator in the process of decision-maker is considerably different at each perspective. While in sharing, the facilitator has to be constantly aware of the process; in the aggregating and comparing, a lesser commitment is needed since, in the extreme, the facilitator will interact with each decision-maker separately. Other issues, related to the procedure adopted to reach a decision, are also pointed out: cyclicity of the process and shared understanding, level of comparison, basis for comparing individual results, use of time (sharing is much more time-consuming, but the expectation of a consensual outcome is also higher), face-to-face meetings or computer-supported collaborations, choice of participants, difference in expertise, and ownership and commitment to the decision.

13.7 MCDA Case Studies

Most studies on sustainable waste management strategies have mainly focused on two stages: the waste treatment strategy and the location of treatment plant because of the magnitude of their ecological and financial impacts (Soltani et al. 2015). Nonetheless, other very noteworthy works have been published. In this section, five case studies addressing different issues are presented. The first case addresses waste management in a refugee camp. Authors study different alternatives to deal with the particular reality of such a context (people live under very poor conditions, and the consumption of food coming from humanitarian aids causes the generation of much solid waste). The second case study is also related with waste management and has two aspects that should be pointed out: one of the alternatives is "business as usual" meaning that the authors consider the scenario of not changing the current status quo. This is a real relevant aspect since, very often, studies lack access to the quality of the current waste management with regard to the new alternatives. The other significant aspect is the way very similar final scores in the top rank alternatives are handled. The third case study concerns waste management strategies. The results proceeded by two methods are compared: a single indicator (the ecological footprint) and MCDA method with six criteria (one of them is the ecological footprint). The fourth, and the last, case study addressing waste management tackles the retrofit of a management system to comply with European Union recycling targets. One of the interesting aspects of this work is how the authors have dealt with uncertainty issues. The last case study proposed a single score index to evaluate the performance of two distinct selective collection schemes.

13.7.1 Multi-criteria Decision Analysis for Waste Management in Saharawi Refugee Camps

Garfì et al. (2009) applied AHP to compare and rank different waste management solutions in Saharawi refugee camps (Algeria). Given the particular conditions in which environmental and social aspects need to be taken into account, four different waste collection and management alternatives were scored according to the 39 criteria. The selected alternatives were S1 (waste collection by using three tipper trucks, disposal, and burning in an open area), S2 (waste collection by using seven dumpers and disposal in a landfill), S3 (waste collection by using seven dumpers, three tipper trucks, and disposal in a landfill), and S4 (waste collection by using three tipper trucks and disposal in a landfill). The criteria were divided into general criteria for human development projects (technical, social, environmental, and economic criteria for humanitarian and environmental projects) and specific criteria for waste management (technical objectives specifically applied in waste management).

The results show that the second and the third solutions (seven dumpers and disposal in a landfill and seven dumpers, three tipper trucks, and disposal in a landfill) were the better alternatives for waste management. In fact, what made them more sustainable were, according to the authors, (i) the preference for dumpers instead of trucks in terms of greater suitability and appropriateness of small-scale technologies and lower environmental impacts; (ii) the appropriateness of a waste disposal system (landfill instead of burning) that improves environmental impacts and avoids health risks; and (iii) the increasing of local staff employment. Regarding the importance of criteria on alternatives selection, some interesting insights were observed: health, living conditions, and income and employment criteria played a fundamental role in the alternative comparison; they were considered to be more important than environmental impacts and natural resources consumption.

13.7.2 Multi-Criteria Analysis as a Tool for Sustainability Assessment of a Waste Management Model

Milutinović et al. (2014) propose a model for assessing the sustainability of waste management which assists decision-makers in the selection of waste management scenarios with energy and resource recovery. The proposed model can also be applied to compare various waste treatment scenarios regarding their sustainability performance. Based on the AHP methodology, an iterative procedure is proposed to rank the alternatives (or scenarios, as called by the authors) so that their ranking scores are sufficiently different to distinguish the alternatives. The study of waste management sustainability of Nis City, Serbia, is presented as an illustrative case. The authors compared four scenarios:

1. Business as usual (all waste is landfilled with the exception of a fraction of metal and glass waste that is recycled).
2. Aerobic process (most of organic waste is composted; a fraction of plastic glass, paper, and metal waste is recycled; and the remaining waste is landfilled).
3. Incineration (glass and metal waste is recycled; residual waste is sent to a cogeneration plant).
4. Anaerobic digestion (glass, metal, and plastic waste is recycled; other waste is sent to anaerobic digestion plant).

These scenarios were scored according to the three economic criteria (investment, operational costs, and revenues), three environmental criteria (GHG, SO_2 emissions, and the volume reduction fraction), and two social criteria (job creation and public acceptance). The results lead to a tie in the first ranking position between scenarios 2 and 4. To better differentiate both alternative, three additional criteria were added to the model and the new ranking assessed. The new criteria were NOx emission, energy consumption, and fuel cost. This time, scenario 2 presented a slightly better final score than scenario 4 (33% vs. 31.6%) and therefore was ranked first. The other two scenarios maintained their ranking positions. To investigate whether an increase on the number of criteria would support and intensify the difference between scenarios 2 and 4, a new procedure iteration was performed considering now the previous 11 criteria plus 3 additional ones: VOC emissions (with relevance in composting and anaerobic digestion), recycling rate (with relevance regarding recycling), and heavy metals (Pb) released in water. This final ranking showed that the difference between scenarios 2 and

13.7 MCDA Case Studies

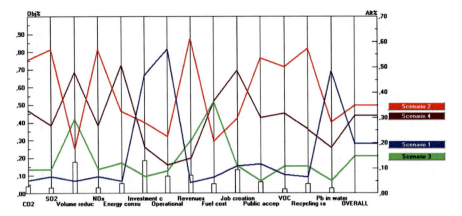

Fig. 13.3 Scenario ranking with 14 criteria. (From Milutinović et al. 2014)

4 increased once again and presented a difference between scores of 2.8% in favor of scenario 2, and there was no change in ranking scenarios. Figure 13.3 shows the 14 criteria and the corresponding score of each alternative (colored lines), criteria weights (white bars), and the overall ranking score.

13.7.3 Ranking Municipal Solid Waste Treatment Alternatives Based on Ecological Footprint and Multi-criteria Analysis

Herva and Roca (2013) address the selection of municipal solid waste treatment techniques with the support of PROMETHEE, GAIA, and AHP multi-criteria decision methodologies. Four different waste treatment alternatives are considered (landfilling with energy recovery, incineration with energy recovery, biological treatment of the organic fraction with energy recovery from the refuse-derived fuel, and thermal plasma gasification) and classified regarding only the environmental point of view. Although with a single point of view, the classification is made considering six different criteria: ecological footprint (EF), water consumption, air emissions of organic compounds and of dusts, water emissions of suspended solids, and occupied landfill volume. The EF is used both independently and together with the other five criteria, since in this work authors also wanted to compare the results yielded by the two methods proposed (EF vs. MCDA).

Regarding the MCDA approach, AHP is implemented in standard software tools (as Excel and MATLAB) and is employed to establish the criteria weights. The ranking of the four alternatives is performed with PROMETHEE model through the software Decision Lab 2000. The context under which the weight values were defined is not made explicitly in the paper. But since legal thresholds existed for the majority of the criteria selected, especially associated with the incineration of waste, a sensitivity analysis was performed to determine the impact on the final ranking if different weights were used. For each criterion, an interval is computed indicating within which bounds the weight of a criterion can be modified without affecting the final ranking of the alternatives. The analysis of the six intervals proved the proposed ranking to be robust.

The final ranking of waste treatment alternatives was then established as follows, from best to worst: thermal plasma gasification; biological treatment of the organic fraction with energy recovery from the refuse-derived fuel; incineration; and, lastly, landfilling with energy recovery.

Authors pointed out that the EF proved to be a good screening indicator although it did not provide a comprehensive measure of environmental impacts associated with the waste treatment options considered. Besides, the combined application of AHP and PROMETHEE/GAIA as MCA methodology was found to be a suitable way, not very complex at user level, to integrate the information provided by a set of environmental criteria and to aid decision-making. Moreover, the ranking obtained was in agreement with the general hierarchy recommended by legislation, prioritizing treatment techniques that allow for energy or materials recovery.

13.7.4 An AHP-Based Fuzzy Interval TOPSIS Assessment for Sustainable Expansion of the Solid Waste Management System in Setúbal Peninsula, Portugal

Pires et al. (2011) applied some MCDA methodologies to investigate the 18 managerial alternatives so that the sustainability of solid waste management in Setúbal region, Portugal, could be improved. One major concern in this study is modeling of uncertainty issues inherent to waste management which can seriously affect the compliance of European Union directives' targets and the choice of the best waste management solution. Among uncertainty sources, authors point out to uncertainties from model parameters, type of models, inherent process uncertainties, uncertainties due to lack of knowledge about a specific process or processes, or uncertainties embedded in decision-making. To face this additional challenge, an improved TOPSIS method was chosen to screen and rank the alternatives; this method uses an interval-valued fuzzy (IVF) method to model uncertainty. In short, one may say that instead of using a single value to score each alternative in each criterion, an interval is used to capture the impreciseness of the values. AHP method was the choice to determine the criteria weights.

The proposed criteria were selected considering the requirements of the new waste management philosophy brought by Thematic Strategy on the Prevention and Recycling of Waste (Council and European Parliament 2008). Therefore, one finds among criteria technical (landfill space saving), environmental (abiotic depletion, acidification, eutrophication, global warming potential, human toxicity, photochemical oxidation, and gross energy requirement), economical (investment and operational costs and operational revenues), and social aspects (economic sufficiency, fees, and odor).

The management alternatives are waste collection and separation of the three packaging materials through bin systems. In detail, alternative 0 refers to the predicted change that will take place in the Setúbal region SWM system so that national targets are reach to comply with European Directives. The remaining alternatives were designed to examine some special options for complying with the Landfill Directive. For example, alternative 1 emphasizes the inclusion of aerobic mechanical biological treatment facility (MBT), alternative 4 implies the use of anaerobic digestion (AD) in MBT, and alternative 6 examines the specific case of using biodegradable solid wastes anaerobic digestion line. In general, alternatives 0, 3, and 5 are options for a suite of intermediate processing. Separation of high calorific fraction of waste for refuse-derived fuel production was also considered in two options being defined for collecting the high calorific fraction from MRF refuse and from AD MBT separation.

Authors concluded the best solution for the system in study would be to implement of anaerobic digestion MBT and anaerobic digestion plant of biodegradable municipal waste followed by the RDF production and alternative 5 is the best option. If, however, criteria weights were equally important, the decision would turn out different. Therefore, it is suggested that other methods other than AHP should be applied access criteria weights.

13.7.5 Assessment Strategies for Municipal Selective Waste Collection Schemes

Ferreira et al. (2017) proposes a performance index that brings together a set of performance indicators highlighting collection trends using the simple additive weighting method. The study focuses on the urban reality of Oporto Municipality, Portugal, where two distinct selective collection schemes are in operation (manual rear-loading vehicles handling street-side containers with open lids and crane-loading vehicles handling drop-off and underground containers). The collection is performed for each selective waste type (light packaging, paper and cardboard, and glass) by nine circuits with drop-off and three with street-side containers. Each collection vehicle is assigned to a team of three workers from Monday to Saturday, 8 h per day. The disposal site is located 5 km outside the municipal limits.

The performance index is defined for each type of collection scheme and is able to aggregate, in a single value, the contributions of the selected indicators chosen to access performance of the two collection schemes: effective collection distance (in kilometers per tons), effective collection time (in hours per tons), and effective fuel consumption (in liters per tons). As the indicators are expressed in different units and scales, a previous normalization was provided to convert indicators into comparable values. In particular, authors used an inverse min-max normalization to convert indicators into values between 0 and 1. Therefore the normalized value of indicator j of collection scheme s and collected material i is given by

$$N_{ij}^s = \frac{I_{ij}^{max} - I_{ij}^s}{I_{ij}^{max} - I_{ij}^{min}}$$

where I_{ij}^s denotes the sample median value of indicator j of collection scheme s and collected material i and I_{ij}^{max} and I_{ij}^{min} denote, respectively, the maximum and minimum values observed for indicator j and collected material i considering all collection schemes. Formally, the performance index for collection scheme s, I_s, is given by

$$I_s = \sum_{i=1}^{n} \sum_{j=1}^{m} w_i^s \cdot \alpha_j \cdot N_{ij}^s$$

where the coefficient α_j weights the relative importance of indicator j in the overall service assessment (may assume values according to experts' criteria) and w_i^s reflects the collection proportions that differ among collected materials due to their specific density characteristics. Additional constraints are imposed on the weights coefficients reflecting the assumptions underlying the SAW method:

$$0 \le \alpha_{ij} \le 1 \text{ and } \sum_{j=1}^{m} \alpha_j = 1$$
$$0 \le w_i^s \le 1 \text{ and } \sum_{i=1}^{n} w_i^s = 1.$$

When applied to the two collection scheme in operation and given the α weights provided by the stakeholders, the index showed that street-side collection performed better than drop-off containers collection (0.8 vs. 0.76). Notice, the municipality stakeholders consider the effective collection time as the main critical factor to take into account. In fact, faster collection schemes reduce the probability of undesirable traffic congestion, which should be strongly avoided in urban high-density areas. No sensitive analysis was made to α weights.

13.8 Final Remarks

This chapter brings an overview of the multi-criteria decision analysis methods most applied to sustainable solid waste management problems. Simple additive weighing (SAW), AHP, TOPSIS, PROMETHEE, and ELECTRE methods are described and compared. Nowadays, these methods are being extensively used in the waste management context. However, few themes have been addressed. The largest majority of studies have focused on facility location (e.g., landfill site selection) or management strategy. And AHP ranks first among the applied MCDA models (Goulart-Coelho et al. 2017).

Uncertainty aspects common in SWM decision-making may be addressed with sensitivity analysis. Therefore, there is a section dedicated to sensitivity analysis as well as a case study showing a different approach to deal with this topic. The presence of different decision-makers is also addressed.

Several case studies concerning waste management strategies and site locations are presented to provide deeper insights concerning the used o MCDA methods in SWM. Some of the described works integrate more than one MCDA method to reach the final decision. One case study shows how MCDA methods can be of use to the definition of single value indexes.

References

Achillas C, Moussiopoulos N, Karagiannidis A et al (2013) The use of multi-criteria decision analysis to tackle waste management problems: a literature review. Waste Manag Res 31:115–129

Adelman L (2013) Choice theory. In: Gass S, Fu M (eds) Encyclopedia of operations research and management science. Springer US, pp 164–168

Belton V, Pictet J (1997) A framework for group decision using a mcda model: sharing, aggregating or comparing individual information? J Decis Syst 6:283–303

Brans JP, De Smet Y (2016) PROMETHEE methods. In: Figueira J, Greco S, Ehrgott M (eds) Multi criteria decision analysis: state of the art surveys. Springer, New York, pp 187–219

Buchholz T, Rametsteiner E, Volk TA, Luzadis VA (2009) Multi criteria analysis for bioenergy systems assessments. Energy Policy 37(2):484–495

Cinelli M, Coles SR, Kirwan K (2014) Analysis of the potentials of multi criteria decision analysis methods to conduct sustainability assessment. Ecol Indic 46:138–148

Clemen RT, Reilly T (2013) Making hard decisions with decision tools. Cengage Learning, Mason

Council, European Parliament (2008) Directive 2008/98/EC of the European Parliament and of the Council of 19 November 2008 on waste and repealing certain. Directives Off J Eur Commun, L312/3

Ferreira F, Avelino C, Bentes I et al (2017) Assessment strategies for municipal selective waste collection schemes. Waste Manag 59:3–13

Figueira JR, Mousseau V, Roy B (2016a) Electre methods. In: Figueira J, Greco S, Ehrgott M (eds) Multi criteria decision analysis: state of the art surveys. Springer, New York, pp 155–185

Figueira JR, Ehrgott M, Greco S (eds) (2016b) Multiple criteria decision analysis: state of the art surveys. Springer, New York

Garfi M, Tondelli S, Bonoli A (2009) Multi-criteria decision analysis for waste management in Saharawi refugee camps. Waste Manag 29:2729–2739

Goulart-Coelho LM, Lange LC, Coelho HM (2017) Multi-criteria decision making to support waste management: a critical review of current practices and methods. Waste Manag Res 35:3–28

Gómez-Delgado M, Tarantola S (2006) GLOBAL sensitivity analysis, GIS and multi-criteria evaluation for a sustainable planning of a hazardous waste disposal site in Spain. International Journal of Geographical Information Science, 20(4), 449–466

Guitouni A, Martel JM (1998) Tentative guidelines to help choosing an appropriate MCDA method. Eur J Oper Res 109:501–521

Hajkowicz S, Higgins A (2008) A comparison of multiple criteria analysis techniques for water resource management. Eur J Oper Res 184:255–265

Herva M, Roca E (2013) Ranking municipal solid waste treatment alternatives based on ecological footprint and multi-criteria analysis. Ecol Indic 25:77–84

Hwang C.-L. and Yoon K (1981) Multiple Attribute Decision Making: Methods and Applications. Springer-Verlag, Berlin

MacDonald ML (1996) A multi-attribute spatial decision support system for solid waste planning. Comput Environ Urban Syst 20:1–17

Malczewski J (2006) GIS-based multicriteria decision analysis: A survey of the literature. International Journal of Geographical Information Science, 20:703–726

Marsh K, Goetghebeur M, Thokala P et al (eds) (2017) Multi-criteria decision analysis to support healthcare decisions. Springer, US

Milutinović B, Stefanović G, Dassisti M et al (2014) Multi-criteria analysis as a tool for sustainability assessment of a waste management model. Energy 74:190–201

Mustajoki J, Marttunen M (2017) Comparison of multi-criteria decision analytical software for supporting environmental planning processes. Environ Model Softw 93:78–91

Nyerges TL, Jankowski P (2009) Regional and urban GIS: a decision support approach. Guilford Press, UK

Pérez J (1995) Some comments on Saaty's AHP. Manag Sci 41:1091–1095

Pires A, Chang NB, Martinho G (2011) An AHP-based fuzzy interval TOPSIS assessment for sustainable expansion of the solid waste management system in Setúbal peninsula, Portugal. Resour Conserv Recycl 56:7–21

Reichert P, Schuwirth N, Langhans S (2013) Constructing, evaluating and visualizing value and utility functions for decision support. Environ Model Softw 46:283–291

Roy B (1989) Main sources of inaccurate determination, uncertainty and imprecision in decision models. Math Comput Model 12:1245–1254

Saaty TL (1980) The analytic hierarchy process. McGraw-Hill, New York

Soltani A, Hewage K, Reza B, Sadiq R (2015) Multiple stakeholders in multi-criteria decision-making in the context of municipal solid waste management: a review. Waste Manag 35:318–328

Taha HA (2003) Operations research: an introduction. Prentice-Hall of India Private Limited, US

Vafaei N, Ribeiro RA, Camarinha-Matos LM (2016) Normalization techniques for multi-criteria decision making: analytical hierarchy process case study. In: Camarinha-Matos LM, Falcão AJ, Vafaei N, Najdi S (eds) Doctoral conference on computing, electrical and industrial systems. Springer, Cham, pp 261–269

Van Den Hove S (2006) Between consensus and compromise: acknowledging the negotiation dimension in participatory approaches. Land use policy, 23(1):10–17

Wang M, Yang J (1998) A multi-criterion experimental comparison of three multi-attribute weight measurement methods. J Multi-Criteria Decis Anal 7:340–350

Weistroffer HR, Li Y (2016) Multiple criteria decision analysis software. In: Figueira J, Greco S, Ehrgott M (eds) Multi criteria decision analysis: state of the art surveys. Springer, New York, pp 1301–1341

Wiecek M, Ehrgott M, Fadel G, Figueira JR (2008) Multiple criteria decision making for engineering. Omega 36:337–339

Yadav V, Karmakar S, Dikshit AK, Bhurjee AK (2018) Interval-valued facility location model: an appraisal of municipal solid waste management system. J Clean Prod 171:250–263

Chapter 14
A Sustainable Reverse Logistics System:
A Retrofit Case

Abstract This chapter presents a real case study of a recyclable waste collection system aiming at redesigning service areas and associated vehicle collection routes to support a sustainable operation. Not only economic objectives are to be considered, but also one should account for environmental and social aspects. The economic dimension is modeled through traveling distance that directly influences the global cost. The environmental one is modeled throughout the calculations of the CO_2 emissions. Finally, the social aspect is considered by aiming to define a balanced solution regarding working hours among drivers. A multi-objective solution approach based on mixed-integer linear programming models is developed and applied to real data.

Keywords Carbon dioxide emissions · Global cost · Multi-objective programming · Routing problem · Working hours

14.1 Introduction

Waste collection systems usually plan their operations according to administrative territorial boundaries (e.g. municipalities, county, district...). Even when managing two adjacent municipalities, operations are plan independently. The company studied in this chapter is no exception. All operations have been managed under a municipality perspective, i.e., the service areas of each depot and the collection routes have been defined taking into account the municipalities' boundaries. This approach has proved to be very costly and motivated the restructure of the company's tactical and operational planning decisions. Moreover, the company aims to foster the system's sustainability by integrating economic, social, and environmental objectives in the new plan.

This company responsible for the recyclable collection system covering 19 rural municipalities with a total area of 7000 km^2. It involves 1522 glass bins, 1238 paper bins, and 1205 plastic/metal bins spread over 207 sites (see Fig. 14.1). A collection site is assumed to correspond to an area instead of an individual container to reduce the problem size. Due to the proximity of the bins within an urban area (an average distance of 500 m), it is realistic to assume the containers to collect within this site as

---

© Springer International Publishing AG, part of Springer Nature 2019   261
A. Pires et al., *Sustainable Solid Waste Collection and Management*,
https://doi.org/10.1007/978-3-319-93200-2_14

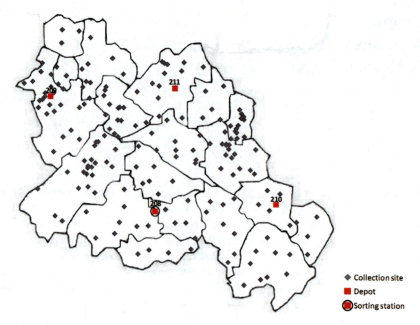

Fig. 14.1 Collection sites and depot locations

a single node. The number of containers at each site is known in advance. The company operates four depots and a vehicle fleet of eight vehicles. One of the depots acts also as a sorting station (depot 208). The remaining three depots are only transfer stations where the recyclable waste is consolidated and afterward transferred to the sorting station.

The company provided a dataset with historical data concerning all routes performed over a year. For each route, this available data contains the day, the collected recyclable material type and the corresponding number of containers, the traveled distance (in kilometers), the route duration, and the total collected weight. To estimate the collected amounts at each site and the corresponding collection frequency, the daily collected weight average per container was estimated. It took into account the time interval between two consecutive collections sites and the average collected amount per container in each route.

The three recyclable materials present different collection frequencies. Glass has to be collected every 6 weeks, plastic/metal every 3 weeks, and paper every 2 weeks. Therefore, a 6-week planning horizon is assumed. The materials are collected in separated routes since the vehicle fleet has no compartments. Taking into account the materials' densities and vehicles' maximum capacities, it was considered that vehicles can load a maximum of 8500 kg of glass, 3000 kg of paper, and 1000 kg of plastic/metal. For the outbound transportation (from the depots to the sorting station), larger vehicles are used, and their weight capacities are, under the same assumption, 12,000 kg for glass, 5000 kg for paper, and 3000 for plastic/metal.

All collection routes start at a depot, visit several sites collecting a single type of material, and return to a depot to unload. Multiple trips per day, as well as inter-depot routes (routes starting and ending at different depots), are allowed. However, by the end of a working day, all vehicles have to return to their depot of origin. Collection is performed 5 days a week, 8 h per day. The new plan should consider a vehicle route planning for a 6-week period that is to be repeated every 6 weeks. To avoid containers' overflow, managers should set a minimal and a maximum interval between two consecutive collections when defining route scheduling for each material.

14.2 Sustainability Objectives

The economic objective accunts, only for the variable costs of the system, since the fixed costs are associated with strategic decisions that have already been taken, (as number of depots, number of vehicles, and number of drivers), and cannot be changed. Hence, the variable costs are mainly related to the distance traveled by vehicles when collecting containers and transporting waste from depots to the sorting station. This includes fuel consumption and maintenance of the vehicle. Such costs depend linearly on the distance traveled, and thus the economic objective function is assessed by the total distance traveled. This includes the inbound distance (from the collection sites to the depots) and the outbound distance (from the depots to the sorting stations), to which adds the possible distance covered by empty vehicles between depots (heavily penalized). Currently, the total distance traveled is about 270,000 km per year.

On the environmental objective, and since transportation is this system's main activity, the greenhouse gas emissions (like CO_2, CH_4, HFCs, NO_x) are generated, in particular CO_2, which negatively impact the environment. The function is defined as the total CO_2 emitted by all vehicles in the system: each collection route performed and the round-trips between depots and the sorting station. Notice that since these last vehicles travel empty when returning to the depots, different CO_2 values are assumed for each direction. It was estimated that a total of 340,000 kg of CO_2 are emitted per year.

Lastly, the social objective promotes equity among human resources, in this case, the drivers. In the current plan, drivers' schedules are imbalanced with some drivers operating larger number of routes than others. From the historical data, a maximum of 220 and a minimum and 100 driving hours, are observed in a 6-week horizon. The company wants to put into practice a new operation scheme which will account for this organizational issue. Hence, the social objective is modeled as the minimization of the maximum working hours among all drivers in the planning horizon. This metric has a twofold contribution toward social sustainability. First, it promotes equity among drivers, enabling balanced workloads since all drivers are assigned to collection activities with similar number of hours (see Fig. 14.2 for an illustrative example). Second, with the minimization of the maximum working hours, drivers may be released to perform tasks other than just collection, as sorting activities, participation in recycling awareness campaigns, or training. This latter activity helps to improve the career development and promotes versatility among the human resources.

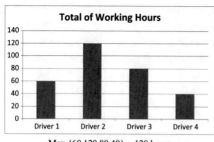
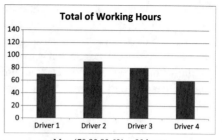

Fig. 14.2 Example of the effect of minimizing the maximum working hours

14.3 Modeling and Solution Approach

This case study involves the definition and scheduling of vehicle routes in multiple depot system, where inter-depot routes and multiple trips per vehicle are allowed. It is modeled as a multi-depot periodic vehicle routing problem with inter-depot routes (MDPVRPI). This model allows for the simultaneous selection of a set of visiting days for each client, the definition of the service areas of each depot, and of the multiple routes to be performed in each day of the planning horizon (see Annex A for the full model formulation). The MDPVRPI combines three problems: a multi-depot vehicle routing problem (MDVRP), a periodic vehicle routing problem (PVRP), and a vehicle routing problem with multiple use of vehicles (VRPMU). While the MDVRP considers a planning horizon of a single time unit, the PVRP considers a planning horizon with several time units, since it assumes customers to have different delivery (or collection) patterns. In this problem, a customer specifies a service frequency and a set of allowable delivery patterns, and the company has to decide on which day the delivery will occur. In the VRPMU, a vehicle can perform several routes during a working day and/or the planning horizon. The multiple uses of vehicles appear when the fleet is either small or the working day period is larger than the average route duration (see Petch and Salhi (2003), Oliveira and Vieira (2007), Azi et al. (2010), and Rieck and Zimmermann (2010)).

In the classical MDPVRP, all routes have to start and end in the same depot (closed routes). Whereas, in the MDPVRP with inter-depot routes (MDPVRPI), vehicles can renew their capacity in any depot in order to continue delivering or collecting materials without being forced to return to their home depot before the end of the working day. Hence, routes can start and finish at different depots enabling a vehicle rotation composed by inter-depot routes. The different routes concepts are illustrated at Fig. 14.3. The difference between an open and an inter-depot route is that in the latter a rotation has to be defined in order to get the vehicle back to its home depot. One defines a rotation as a set of inter-depot routes that can be performed consecutively until the home depot is reached.

A solution approach is developed to solve the case study as multi-objective MDPVRPI. Since the problem is modeled with the set partitioning formulation, a set of a large number of feasible routes has to be generated, and then the most

14.3 Modeling and Solution Approach

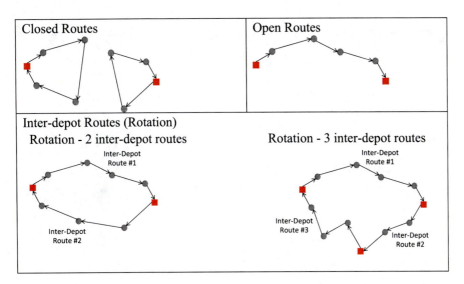

Fig. 14.3 Illustration of closed, open, and inter-depot routes

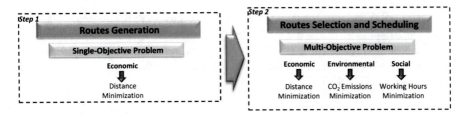

Fig. 14.4 Solution approach overview

adequate ones are selected from that pool. Therefore, the solution approach involves a first step to generate the routes and a second step where the multi-objective problem is solved (see Fig. 14.4). As the goal is to obtain a solution where costs are balanced with environmental and social concerns, the set of routes is defined considering only the economic objective. However, when selecting and scheduling the routes, at step 2, the three objectives are taken into account by solving the multi-objective MDPVRPI with the augmented ε-constraint method (see book Sect. 12.3.4). With such approach, an approximation to the Pareto front is obtained, which can be used by the decision-maker to evaluate trade-offs and to select the most adequate solution to put into practice.

The goal of step 1 is then to build the set of feasible routes required by the multi-objective MDPRVPI formulation. Generating all the feasible routes is however intractable (Laporte 2007), so only a subset will be defined. Accounting for the characteristics of the problem addressed, a diverse set of closed and inter-depot routes are generated representing alternative solutions to collect all sites. To build only closed ones, a MDVRP is solved – procedure 1. To build closed and inter-depot

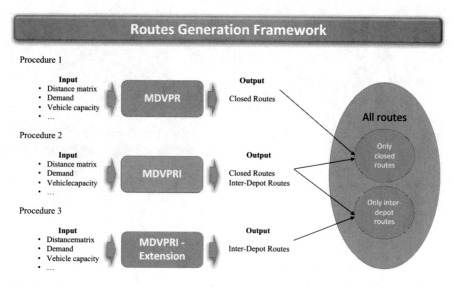

Fig. 14.5 Routes generation procedures

routes, a MDVRPI is solved – procedure 2. To build only inter-depot routes, a MDVRPI Extension is solved – procedure 3 (see Annexes B.1.1, B.1.2, and B.1.3 for all the details). Therefore, the set of all routes is fed by three independent procedures modeling the three alternative solutions to collect waste from all collection sites (see Fig. 14.5).

14.4 Results and Analysis

The solution approach proposed is applied to the described case study in order to define a sustainable plan for the recyclable waste collection in 19 Portuguese municipalities. It was implemented in GAMS 23.7 and solved through the CPLEX Optimizer 12.3.0, on an Intel Xeon CPU X5680 @ 3.33 GHz.

14.4.1 Routes Generation

Three procedures were applied to generate a set of diverse closed and inter-depot collection routes for each of the three recyclable materials. The number of routes provided by each procedure for each material is shown in Table 14.1.

The mixture of plastic and metal, which is assumed as a single material, requires more collection routes than the other two materials. This mixture has a lower density

14.4 Results and Analysis

Table 14.1 Number of routes defined per procedure and recyclable material

	Glass	Paper	Plastic/metal
Procedure 1			
Closed routes	39	42	66
Procedure 2			
Closed routes	37	41	64
Inter-depot routes	9	6	9
Procedure 3			
Inter-depot routes	38	40	62

when compared to the other two materials, and thus the vehicle weight capacity for plastic metal is smaller for the same vehicle volume capacity.

14.4.2 Sustainable Collection System

Step 2 of Figure 14.4 selects routes from set K while considering the number of available vehicles (eight in total) and where they are based. It also takes into accounts the planning horizon of 6 weeks (i.e., 30 working days) and observes the interval between collections. Then step 2 the multi-objective problem is solved by applying the augmented ε-constraint method Marieloas (2009) to define an approximation to the Pareto front. The proceedure ends with the application of a compromise solution method to compute a sustainable solution for the case study (see Annex B.2).

The payoff table generated by the lexicographic method (see section 12.4.1) is shown in Table 14.2. When minimizing the total distance (economic objective), a solution with 27,261 km is obtained. This solution emits 34,982 kg of CO_2, and the maximum number of hours among the eight vehicles is 200 h. When minimizing the CO_2 emissions (environmental objective), a solution with 34,747 kg of CO_2 is achieved. It implies less 0.7% of CO_2 emissions and more 0.3% kilometers when compared to the economic solution. The number of working hours remains unaltered. When minimizing the maximum number of working hours in the planning horizon (social objective), a solution with a maximum of 165 h is obtained. This solution implies a total of 30,118 km (about 11% more than in the economic solution) and 38,042 kg of CO_2 (about 10% more than in the environmental solution).

Figure 14.6 shows the total hours each driver has to work (social concern) in the collection activity for each of the three optimal plans: economic, environmental, and social. Both economic and environmental optimal plans are quite unbalanced, with differences between the maximum and minimum working hours of 102 and 120 h, respectively. On the contrary, the social optimal plan presents a totally balanced plan, where all drivers work the same number of hours in collection activities (165 h).

Table 14.2 Payoff table obtained with the lexicographic optimization of the objective functions

		Optimized objective function		
		Economic (km)	Environmental (kg)	Social (h)
Optimal solution of the objective	Economic	27,261	34,982	200
	Environmental	27,337	34,747	200
	Social	30,118	38,042	165

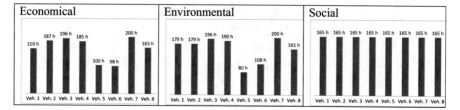

Fig. 14.6 Number of working hours per vehicle in the three solutions

The daily schedule for each vehicle with the assignment of all the routes to be operated in each day is the output of the procedure step 2. Vehicle 7's daily schedule is given in Fig. 14.7. Each day has the number and type of routes to be performed (*Pl* stands for plastic/metal, *Gl* for glass, and *Pa* for paper) and the total duration (in minutes). For example, in day 1 of the economic plan, the vehicle has to perform route #56 to collect plastic/metal and afterward route #250 to collect paper. The total duration (including unloading activities) is 461 min. Route #250 is performed three times during the planning horizon in line with the collection frequency set for the material paper (days 1, 12, and 22). The interval between consecutives visits respects the minimum and maximum interval allowed for this material (9 and 11 days, respectively).

Comparing both schedules (Fig. 14.7a, b), fewer routes are performed by vehicle 7 in the social solution (44 against 52 routes in the economic solution). On the one hand, in the "economic schedule," routes are to be performed every day, while in the "social schedule" there is one day (day 6) where no routes are assigned. In the "social schedule," the driver of vehicle 7 works 165 h in collection activities, while in the "economic schedule," he/she works 200 h. To reduce 35 working hours from vehicle 7, the scheduled hours for the remaining vehicles have to increase. This can be achieved with the reconfiguration of depot service areas. As an illustrative example, the service areas for the material glass for the three solutions are shown in Fig. 14.8. In the social solution, the number of collection sites assigned to depot 208 (114 sites) is the lowest when among the three solutions (128 sites in the economic solution and 136 in the environmental solution). In opposition, the number of sites assigned to depot 209 is the largest (46 sites in the social solution against 32 and 26 in the economic and environmental solutions, respectively). Depot 209 (where vehicles 5 and 6 are based) is the one with less working hours in the economic and

14.4 Results and Analysis

(a) Economical plan
$\sum$ Hours = 200h

1	2	3	4	5
#56 (Pl); #250 (Pa) 461 m	#54 (Pl); #249 (Pa) 448 m	#253 (Pa); #254 (Pa) 480 m	#59 (Pl); #63 (Pl) 467 m	#251 (Pa) 346 m
6	**7**	**8**	**9**	**10**
#61 (Pl); #62 (Pl) 428 m	#252 (Pa) 295 m	#248 (Pa) 299 m	#255 (Pa); #267 (Pa) 387 m	#57 (Pl); #58 (Pl) 471 m
11	**12**	**13**	**14**	**15**
#55(Pl);#60(Pl);#81(Pl) 425 m	#49 (Pl); #250 (Pa) 478 m	#249 (Pa); #254 (Pa) 467 m	#253 (Pa) 295 m	#397 (Gl) 249 m
16	**17**	**18**	**19**	**20**
#251 (Pa) 346 m	#56 (Pl); #252 (Pa) 461 m	#54 (Pl); #248 (Pa) 465 m	#255 (Pa); #267 (Pa) 387 m	#59 (Pl); #63 (Pl) 467 m
21	**22**	**23**	**24**	**25**
#61 (Pl) 258 m	#62 (Pl); #250 (Pa) 465 m	#249 (Pa) 282 m	#253 (Pa); #254 (Pa) 480 m	#57 (Pl); #58 (Pl) 471 m
26	**27**	**28**	**29**	**30**
#55(Pl);#60(Pl);#81(Pl) 425 m	#251 (Pa) 346 m	#49 (Pl); #252 (Pa) 478 m	#248 (Pa) 299 m	#255 (Pa); #267 (Pa) 387 m

(b) Social plan
$\sum$ Hours = 165 h

1	2	3	4	5
#250 (Pa) 295 m	#249 (Pa); #254 (Pa) 467 m	#62 (Pl); #253 (Pa) 465 m	#63 (Pl) 165 m	#391 (Gl) 292 m
6	**7**	**8**	**9**	**10**
	#49 (Pl); #54 (Pl) 349 m	#248 (Pa); #267 (Pa) 378 m	#255 (Pa) 308 m	#56 (Pl); #81 (Pl) 214 m
11	**12**	**13**	**14**	**15**
#413 (Gl) 337 m	#250 (Pa) 295 m	#249 (Pa); #254 (Pa) 467 m	#60 (Pl); #253 (Pa) 370 m	#397 (Gl) 249 m
16	**17**	**18**	**19**	**20**
#251 (Pa) 346 m	#252 (Pa) 295 m	#57 (Pl); #248 (Pa) 468 m	#55 (Pl); #267 (Pa) 381 m	#63 (Pl); #255(Pa) 473 m
21	**22**	**23**	**24**	**25**
#61 (Pl) 258 m	#54 (Pl); #250 (Pa) 461 m	#49 (Pl); #249 (Pa) 465 m	#254 (Pa) 185 m	#56 (Pl); #253 (Pa) 461 m
26	**27**	**28**	**29**	**30**
#81 (Pl) 48 m	#251 (Pa) 346 m	#252 (Pa) 295 m	#60(Pl);#248(Pa);#267(Pa) 453 m	#255 (Pa) 308 m

Fig. 14.7 Schedule for vehicle 7 in economical (**a**) and social (**b**) plans

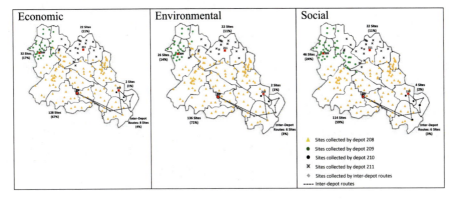

Fig. 14.8 Service areas for glass material for the economic, environmental, and social plans

environmental plans (Fig. 14.6). To balance the number of working hours in the social solution, more sites have to be assigned to this depot.

The environmental solution is the one with the highest number of sites assigned to depot 208 (also acts as the sorting station). The outbound transportation is performed by large vehicles that release more CO_2 than the collection vehicles. Therefore, since the objective is to minimize the CO_2 emissions, more sites are assigned to the sorting station to avoid the outbound transportation. Moreover, the environmental solution selects routes where vehicles travel shorter distances with heavy load given since it minimizes the CO_2 emissions.

Nine different solutions are obtained when applying the augmented ε-constraint method (S1 to S9 in Table 14.3). Such solutions can be visualized in Fig. 14.9 where

Table 14.3 Pareto optimal solutions

	S1	S2	S3	S4	S5	S6	S7	S8	S9
Economic (km)	30,118	28,445	27,676	27,489	27,412	27,345	27,287	27,337	27,261
Environmental (kg)	38,042	36,351	35,580	35,580	35,179	35,100	35,010	34,747	34,982
Social (h)	165	170	175	180	185	190	195	200	200

14.4 Results and Analysis

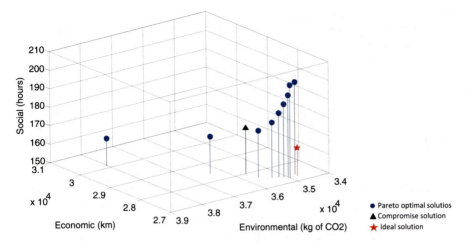

Fig. 14.9 Approximation to Pareto front considering the three objectives with the ideal point and the compromise solution highlighted

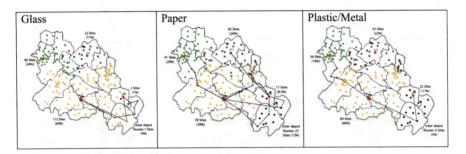

Fig. 14.10 Representation of the compromise solution for the three recyclable materials

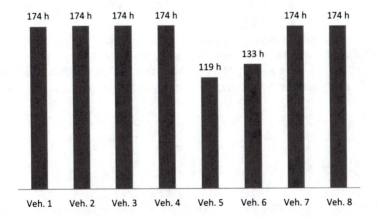

Fig. 14.11 Number of working hours by vehicle in the compromise solution

one can observe that to improve the social objective (reducing the number of maximum working hours), the economic and environmental objectives deteriorate. For instance, to increase the social objective in 17.5%, the economic and the environmental objectives deteriorate 10% and 9.5%, respectively (S1 versus S8). However, the economic objective only deteriorates 1.2% and the environmental 2.4% with an improvement of 12.5% in the social objective (S3 versus S8). Regarding the economic and environmental objectives, the trade-off only exists between S8 and S9. To improve 0.7% in the environmental objective, the economic objective deteriorates 0.3%. In the remaining solutions, these objectives are inversely proportional to the social objective.

Aiming to find a compromise solution between the three objectives to reach a sustainable plan for the logistics network, a compromise solution method is applied. The ideal point (z_I) is defined according to the individual minima of each objective. In this case, ideal point coordinates are $z_I = (27,261$ km, $34,747$ kg CO_2, 165 h). The nadir point (z_N) is defined according to the individual maxima of each objective, $z_N = (30,118$ km, $38,138$ kg CO_2, 200 h). Figure 14.9 also depicts the compromise solution and the ideal point. After normalizing the objective functions with the amplitude between the nadir and ideal points, the compromise solution (z_C) is obtained by minimizing the Tchebycheff distance to the ideal point. The compromise solution obtained is $z_C = (28,013$ km, $35,653$ kg CO_2, 174 h) – all details presented in Annex B.2.

In the compromise solution (depicted in Fig. 14.10), the economic objective deteriorates 2.7%, the environmental 2.6%, and the social 5.5% regarding each corresponding value when a single objective is optimized. For all materials, the number of sites assigned to the sorting station is smaller than the ones obtained for the economic and environmental solutions but higher than the one of social solution. For instance, in the compromise solution for paper, 39% of the sites are assigned to depot 208 (sorting station), while 45% are assigned when the economic and environmental objectives are minimized individually and 38% when considering the social objective. Also, more sites are collected in inter-depot routes. These differences increase the distance traveled and emitted CO_2 but balance the solution regarding workload among depots (Fig. 14.11).

The compromise solution represents a sustainable solution that has been presented to the company. Savings of about 10% in the distance and 9% in CO_2 emissions and a reduction of 21% in the maximum of driving hours are obtained with this sustainable solution, when comparing to the current company operation plan.

14.5 Conclusion

The planning a multi-depot logistics system has been taken into account considering the three dimensions of sustainability. Economic, environmental, and social objective functions have been modeled in a tactical routing and scheduling problem with

Annex A: Multi-objective Formulation for the MDPVRPI

multiple depots. In particular, this work addresses service areas and routes definition as well as routes scheduling, CO_2 emissions, and human resources working hours.

The solution approach has been applied to a real recyclable waste collection system where the trade-offs between the three objectives were highlighted and a compromise solution proposed. When economic and environmental objectives are minimized, unbalanced solutions are obtained regarding working hours by vehicle (and consequently be driver). On the contrary, when the social objective is minimized, a balanced solution is obtained where all drivers drive the same number of hours. However, this equity solution leads to a significant increase in distance and CO_2 emissions. Between environmental and economic objectives there are only minor trade-offs. An efficient solution taking into account the three objectives is obtained through the compromise solution method, where the distance to the ideal point is minimized.

Annex A: Multi-objective Formulation for the MDPVRPI

The multi-objective MDPVRPI is formulated as a set partitioning problem (Balas and Padberg 1976), where K represents the set of all feasible routes (closed and inter-depot routes) and τ_{ktg} is a binary variable that equals 1 if route k is performed on day t by vehicle g; and 0 otherwise. The mathematical formulation considers the following indices and sets.

Indices

k	Route indices
t	Time period (days) indices
g	Vehicle indices
i, j	Node indices
m	Recyclable material indices

Sets

K	Route set $K = \sum_{m \in M} K_m$, $K = K_{in} \cup K_{cl}$
K_m	Route subset to collect material m
K_{in}	Inter-depot route subset
K_{cl}	Closed route subset
T	Time period set
G	Vehicle set
V	Node set $V = V_c \cup V_d \cup V_s$
V_c	Collection site subset
V_d	Depot subset
V_s	Sorting station subset
M	Recyclable material set

Each route $k \in K$ is characterized by (1) a distance dis_k; (2) a duration dur_k including travel, service, and unloading times; (3) a load Lo_k; and (4) CO_2 emissions Co_k. The collection sites belonging to route k are given by a binary parameter μ_{ik} that equals 1 if collection site i belongs to route k and 0 otherwise. The starting and ending depots for route k are also given by binary parameters St_{ki} and En_{ki}, respectively; St_{ki} equals 1 if route k starts at depot i, and En_{ki} equals 1 if route k ends at depot I and 0 otherwise.

The vehicles are fixed at the depots. If vehicle g belongs to depot i, the binary parameter α_{gi} equals 1 and 0 otherwise.

The collection frequency of each collection site i with recyclable material m is given by fr_{im} representing the number of times a collection site needs to be visited within the planning horizon. The minimum and maximum interval between two consecutive collections for recyclable material m are given by $Imin_m$ and $Imax_m$, respectively.

Three objective functions are addressed in this work to tackle the three sustainability dimensions: the economic objective ($z^1(S)$), the environmental objective ($z^2(S)$), and the social objective ($z^3(S)$). Let S be the vector of decision variables; $z^1(S)$, $z^2(S)$, and $z^3(S)$ the three objective functions; and Ω the feasible region; the multi-objective problem can be written in the following form:

$$\min \quad \left\{ z^1(S), \ z^2(S), \ z^3(S) \right\}$$
$$\text{st} \quad S \in \Omega \tag{14.1}$$

The total distance traveled ($z^1(S)$) is given by Eq. (14.2).

$$z^1(S) = \sum_{k \in K} \sum_{t \in T} \sum_{g \in G} dis_k \tau_{ktg} + \tag{14.2a}$$

$$\sum_{j \in V_s} \sum_{i \in V_d} \sum_{m \in M} \sum_{k \in K_m} \sum_{t \in T} \sum_{g \in G} En_{ki} \tau_{ktg} Lo_k / QT_m 2 d_{ij} - \tag{14.2b}$$

$$\sum_{j \in V_s} \sum_{i \in V_d} \sum_{m \in M} \sum_{k \in K_m} \sum_{t \in T} \sum_{\substack{g \in G \\ \alpha_{gj} = 1}} St_{ki} En_{ki} \tau_{ktg} Lo_k / QT_m 2 d_{ij} + \tag{14.2c}$$

$$\sum_{\substack{g \in G \\ \alpha_{gi} = 1}} \sum_{\substack{k \in K \\ En_{ki} = 0 \\ St_{kj} = 1}} \sum_{t \in T} \sum_{i,j \in V_d} 2 \tau_{ktg} d_{ij} \tag{14.2d}$$

The total distance traveled involves, as mentioned, the inbound distance (14.2a), the outbound distance (14.2b and 14.2c), and also a possible extra distance since it is allowed to vehicles based at depot i to perform closed routes from and to depot j (14.2d). The distance (d_{ij}) of moving a vehicle between depots is then penalized. The outbound distance considers the ending depot of each route and the load collected, to compute the number of needed round-trips to the sorting station. Note that the number of round-trips is not round upward since it is being accounting for

Annex A: Multi-objective Formulation for the MDPVRPI

the number of round-trips that occur within a finite time period. These are to be repeated in the next period. When, for instance, 10.4 round trips are considered within the period, it means that 10 round trips occur within the period and the 11th occurs in the next period, but some of the load is related to the previous period. It is also considered that if a vehicle, belonging to the sorting station performs closed routes from depot i, the load collected will be unloaded at the sorting station and not at depot i. Therefore, no outbound distance will be accounted for. Term (14.2c) decreases the objective function of such value.

The environmental objective is related to the CO_2 emissions associated with the collection routes and the outbound transportation between depots and the sorting station. Its total value ($z^2(S)$) given by Eq. (14.3).

$$z^2(S) = \sum_{k \in K} \sum_{t \in T} \sum_{g \in G} Co_k \tau_{ktg} + \tag{14.3a}$$

$$\sum_{j \in V_s} \sum_{i \in V_d} \sum_{m \in M} \sum_{k \in K_m} \sum_{t \in T} \sum_{g \in G} En_{ki} \tau_{ktg} Lo_k / QT_m \left(CoF_{ijm} + CoE_{ji}\right) - \tag{14.3b}$$

$$\sum_{j \in V_s} \sum_{i \in V_d} \sum_{m \in M} \sum_{k \in K_m} \sum_{t \in T} \sum_{\substack{g \in G \\ \alpha_{gj} = 1}} St_{ki} En_{ki} \tau_{ktg} Lo_k / QT_m \left(CoF_{ijm} + CoE_{ji}\right) + \tag{14.3c}$$

$$\sum_{\substack{g \in G \\ \alpha_{gi} = 1}} \sum_{\substack{k \in K \\ En_{ki} = 0 \\ St_{kj} = 1}} \sum_{t \in T} \sum_{i,j \in V_d} 2\tau_{ktg} CoE_{ij} \tag{14.3d}$$

The CO_2 emissions for the inbound transportation (routes to collect all collection sites) are given by the first term (14.3a), where the emission value of each route k is given by parameter Co_k . The CO_2 emissions from the outbound transportation are also considered (terms 14.3b and 14.3c) where larger vehicles are used. Notice that round trips between the sorting station and the depots are performed, with vehicles traveling empty from the sorting station to the depot and in full truckload (FTL) back to the sorting station. The amount of CO_2 emissions for outbound transportation is given by parameter CoF_{ijm} when the vehicle travels in FTL from depot i to sorting station j with material m and CoE_{ij} when the vehicle travels empty in the opposite direction. The last term (14.3d) accounts for the CO_2 emissions of a vehicle, based at depot i, traveling empty to depot j to perform closed routes from and to depot j.

As mentioned above, the social objective minimizes the maximum working hours among drivers. The maximum value of vehicle's total working hours in the planning horizon is given by a positive decision variable $DMax$ when assuming a fixed driver-vehicle assignment (constraint 14.4).

$$DMax \geq \sum_{k\in K}\sum_{t\in T}\tau_{ktg}\mathrm{dur}_k + \sum_{\substack{k\in K \\ \mathrm{St}_{kj}=1 \\ \mathrm{En}_{ki}=0}} \sum_{\substack{i,j\in V_d \\ i\neq j}} \tau_{ktg}2b_{ij}, \quad \forall g \tag{14.4}$$

Then, the function for the social objective is given by Eq. (14.5).

$$z^3(S) = DMax \tag{14.5}$$

With the objective functions defined, the constraints for the multi-objective model of the MDPVRPI are expressed in constraints (14.6) to (14.13).

$$\sum_{k\in K_m}\sum_{t\in T}\sum_{g\in G}\tau_{ktg}\mu_{ik} = \mathrm{fr}_{im} \quad \forall i\in V_c, \forall m \tag{14.6}$$

$$\sum_{k\in K}\tau_{ktg}\mathrm{dur}_k + \sum_{\substack{k\in K \\ \mathrm{St}_{kj}=1 \\ \mathrm{En}_{ki}=0}} \sum_{\substack{j\in V_d \\ j\neq i}} \tau_{ktg}2b_{ij} \leq H \quad \forall t, \forall g, \forall i\in V_d : \alpha_{gi}=1 \tag{14.7}$$

$$\sum_{\substack{k\in K_{in} \\ \mathrm{St}_{ki}=1}} \tau_{ktg} = \sum_{\substack{k'\in K_{in} \\ \mathrm{En}_{k'i}=1}} \tau_{k'tg} \quad \forall g, \forall t, \forall i\in V_d \tag{14.8}$$

$$\sum_{g\in G}\tau_{ktg}\mu_{ik} + \sum_{g\in G}\tau_{kt'g}\mu_{ik} \leq 1 \quad \forall i\in V_c, \forall k\in K_m, \forall m, \forall t, t'\in T, t$$
$$> t', (t-t') \leq I\mathrm{min}_m \tag{14.9}$$

$$\sum_{g\in G}\tau_{ktg}\mu_{ik} + \sum_{g\in G}\tau_{k't'g}\mu_{ik'} \leq 1 \quad \forall i\in V_c, \forall k, k'\in K_m, \forall m, \forall t, t'\in T, t$$
$$> t', (t-t') \leq I\mathrm{min}_m \tag{14.10}$$

$$\sum_{g\in G}\tau_{ktg}\mu_{ik} + \sum_{g\in G}\tau_{kt'g}\mu_{ik} \leq 1 \quad \forall i\in V_c, \forall k\in K_m, \forall m, \forall t, t'\in T, t$$
$$> t', (t-t') > I\mathrm{max}_m, (t-t')$$
$$\leq I\mathrm{max}_m + I\mathrm{min}_m \tag{14.11}$$

$$\sum_{g \in G} \tau_{ktg}\mu_{ik} + \sum_{g \in G} \tau_{k't'g}\mu_{ik'} \leq 1 \quad \forall i \in V_c, \forall k, k' \in K_m, \forall m, \forall t, t' \in T, t$$

$$> t', (t - t') > Imax_m, (t - t')$$
$$\leq Imax_m + Imin_m \qquad (14.12)$$

$$\tau_{ktg} \in \{0, 1\} \quad \forall k \in K, \forall t \in T, \forall g \in G \qquad (14.13)$$

Constraint (14.6) ensures that a collection site i with material m has to be collected fr_{im} times over the time horizon. Constraint (14.7) states that the total route duration performed by vehicle g on day t will not exceed the maximum time allowed for a working day (H). If a vehicle g, belonging to depot i, performs a route starting at depot j, the travel time between i and j is considered.

Since all vehicles have to return to their origin depot, constraint (14.8) guarantees that an inter-depot route k, starting at depot i, is part of the solution only if another inter-depot route k' ends at depot i. Considering all depots $i \in V_d$, constraint (14.8) ensures continuity among inter-depot routes enabling a vehicle rotation.

Constraints (14.9) to (14.12) model the minimum and maximum intervals between consecutive collections which can be performed by the same route or by two different routes. Therefore, constraint (14.9) states that the same route for material m has to be performed with a minimum time interval of $Imin_m$, while constraint (14.10) considers the case of two different routes collecting the same site i at consecutive collections. Analogously, constraints (14.11) and (14.12) ensure the maximum interval $Imax_m$ between consecutive collections. Variable's domain is given in constraint (14.13).

Annex B: Solution Procedure

B.1 Step 1: Routes Generation Procedure

The set of recyclable materials M is involved, and given that each material has to be collected in separated routes, each procedure of step 1 is run independently for each material.

The models involved in each procedure are formulated through MILP formulations based on the two-commodity flow formulation (Baldacci et al. 2004). In such formulations, the network is defined by a direct graph $GR = (V, E)$ with $V = V_c \cup V_d \cup V_f \cup V_s$, being $V_c = \{1, \ldots, n\}$ a set of n customers, $V_d = \{n + 1, \ldots, n + w\}$ a set of w depots, $V_f = \{n + w + 1, \ldots, n + 2w\}$ a replica of the depots set, $V_s = \{n + 2w + 1, \ldots, n + 2w + s\}$ a set of s sorting stations, and $E = \{(i, j) : i, j \in V_c \cup V_d \cup V_f \cup V_s, i \neq j\}$ the edge set.

Each site $i \in V_c$ is characterized by a demand p_i and a service duration t_i. The service duration depends on the average time to collect a container (U), on the average distance between containers within a locality (B), on the average speed

278 14 A Sustainable Reverse Logistics System: A Retrofit Case

within localities (vw) and on the number of containers at each locality (c_i), being $t_i = c_i\left(U + \frac{B}{vw}\right)$. The inbound vehicles have a weight capacity of Q and the outbound vehicles QT. The maximum duration for a working day is given by H. Every edge (i, j) has an associated distance d_{ij} and a travel time b_{ij}, where $b_{ij} = \frac{d_{ij}}{vb}$ and vb is the average speed between localities. An unloading time L is also considered to account for the time to unload a vehicle at the end of each route.

The depot replica set (V_f) is needed since, in the two-commodity flow formulation, routes are defined by paths starting at the real depots and ending at the replica ones. To establish the routes, this formulation requires two flow variables defining two flow paths for any route. One path from the real depot to the replica one modeled by the flow variable representing the vehicle load (variable y_{ij}). In a collection problem, this load increases along the route. The other path from the replica depot to the real one is given by the second flow variable (y_{ji}) that models the vehicle empty space which decreases along the route.

These sets, parameters, and variables are the baseline to all route generating procedures which are briefly described in the next sections.

B.1.1 Procedure 1: MDVRP

In the MDVRP only closed routes are defined. A set of routes K is considered and partitioned by depot, $K = K_1 \cup \ldots \cup K_i$, where K_i is the subset of routes belonging to depot i. Decision variables are the binary variables x_{ijk} that equal 1 if site j is visited immediately after site i on route k ($x_{ijk} = 0$, otherwise) and the corresponding reverse variable x_{jik} when the reverse path is being defined and the flow variables y_{ijk} and y_{jik}; and a binary variable δ_{ik} is defined to assign site i to route k. The objective function also considers the distance to be traveled within each collection site (second term of Eq. (14.14)) and the outbound distance (third term of Eq. (14.14)).

$$\text{Min} \ \frac{1}{2} \sum_{i \in V} \sum_{j \in V} \sum_{k \in K} x_{ijk} d_{ij} + \sum_{i \in V_c} c_i S + 2 \sum_{i \in V_c} \sum_{j \in V_f} \sum_{h \in V_s} \sum_{k \in K} \frac{y_{ijk}}{QT} d_{hj} \qquad (14.14)$$

subject to

$$\sum_{\substack{j \in V \\ j \neq i}} \left(y_{ijk} - y_{jik}\right) = 2 p_i \delta_{ik}, \quad \forall i \in V_c, \forall k \qquad (14.15)$$

$$\sum_{i \in V_c} \sum_{j \in V_f} \sum_{k \in K} y_{ijk} = \sum_{i \in V_c} p_i \qquad (14.16)$$

Annex B: Solution Procedure

$$\sum_{i \in V_c} \sum_{j \in V_f} \sum_{k \in K} y_{jik} \leq |K| Q - \sum_{i \in V_c} p_i \tag{14.17}$$

$$\sum_{i \in V_c} y_{ijk} \leq Q \quad \forall j \in V_f, \forall k \in K_j \tag{14.18}$$

$$\sum_{\substack{i \in V \\ i \neq j}} x_{ijk} = 2\delta_{jk}, \quad \forall j \in V_c, \forall k \tag{14.19}$$

$$y_{ijk} + y_{jik} = Q x_{ijk} \quad \forall i, j \in V, i \neq j, \forall k \tag{14.20}$$

$$\sum_{k \in K} \delta_{ik} = 1 \quad \forall i \in V_c : p_i > 0 \tag{14.21}$$

$$\delta_{ik} = \delta_{(i+w)k} \quad \forall i \in V_d, \forall k \in K_i \tag{14.22}$$

$$\sum_{i \in V_c} \sum_{j \in V} t_i x_{ijk} + \sum_{i \in V} \sum_{j \in V} b_{ij} x_{ijk} \leq 2(H - L) \quad \forall k \in K \tag{14.23}$$

$$\sum_{j \in V_c} x_{ijk} \leq 1 \quad \forall i \in V_d, \forall k \in K_i \tag{14.24}$$

$$\sum_{i \in V_c} x_{ijk} = 0 \quad \forall j \in V_f, \forall k \notin K_j \tag{14.25}$$

$$\sum_{j \in V_c} x_{ijk} = 0 \quad \forall i \in V_d, \forall k \notin K_i \tag{14.26}$$

$$y_{ijk} \geq 0 \quad \forall i, j \in V, k \in K \tag{14.27}$$

$$x_{ijk} \in \{0, 1\} \quad \forall i, j \in V, k \in K \tag{14.28}$$

$$\delta_{ik} \in \{0, 1\} \quad \forall i \in V_c, k \in K \tag{14.29}$$

The above formulation is an extension for the MDVRP of the formulation proposed by Baldacci et al. (2004) for the CVRP. Constraints (14.15) to (14.20) are rewritten since it is considered index k and the binary variable δ_{ik}. Constraints (14.21) to (14.26) are new constraints that deal with multiple depots and duration constraints. Equation (14.21) guarantees that each locality with positive demand has to be visited by a single route. Constraint (14.22) matches the real depots with their replica, ensuring that a route will start at the real depot and will end at the corresponding replica. Constraint (14.23) guarantees that the duration of each

Fig. 14.12 Solution method for the MDVRP

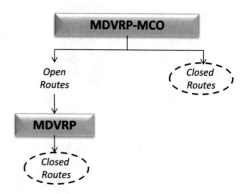

route does not exceed the maximum allowed routing time. Constraint (14.24) ensures that each route will leave its home depot at most once. Finally, constraints (14.25) and (14.26) jointly ensure that a vehicle route cannot leave and return to a depot other than its home depot (real and replica depot). The new variable definition is given in Eq. (14.29).

The proposed formulation, when applied to large instances, is computationally difficult to solve. Therefore, a solution method is proposed to solve the MDVRP (see Fig. 14.12). First, a problem where both closed and open routes are allowed, is solves, the MDVRP with mixed closed and open routes (MDVRP-MCO). The MDVRP-MCO formulation is proposed in the work of Ramos et al. (2013) and is capable of dealing with large instances. Moreover, the majority of the routes in the solution for the MDVRP-MCO are feasible for the MDVRP – the closed routes. For "(the open routes)", the MDVRP formulation is applied having, as input data, only the sites belonging to each open route.

B.1.2 Procedure 2: MDVRPI

The MDVRPI allows inter-depot routes, where vehicles have to return to the home depot on the same working day. Therefore, a vehicle rotation is limited by the maximum duration of a working day (H). To solve the MDVRPI, the solution methodology proposed by Ramos (2012) was used, considering an unlimited vehicle fleet. A MDVRPI Relaxation is solved where inter-depot and closed routes are obtained (see Fig. 14.13). This formulation corresponds to the MDVRP-MCO formulation to which adds constraint (14.30).

$$\sum_{j \in V} x_{ij} + \sum_{j \in V} x_{ji} = \sum_{j \in V} x_{(i+w)j} + \sum_{j \in V} x_{j(i+w)} \quad \forall i \in V_d \quad (14.30)$$

Constraint (14.30) guarantees that the number of routes departing from one depot is equal to the number of routes arriving at that depot. This ensures connectivity between the inter-depot routes and the rotation concept, i.e., a vehicle returns to its

Annex B: Solution Procedure

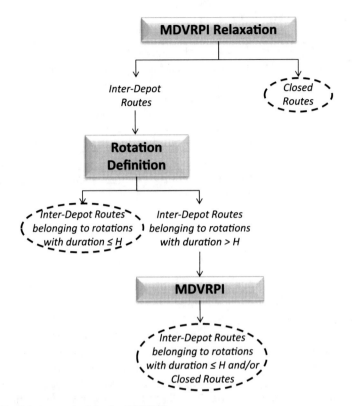

Fig. 14.13 Solution method for the MDVRPI

home depot. However, it is not guaranteed that the vehicle returns within a working day since no duration constraints for rotation are considered in the MDVRPI Relaxation. Notice that in the two-commodity formulation, to any real depot $i \in V_d$, a corresponding copy depot assumed $i + w \in V_f$ (w is the number of depots), and the x_{ij} and x_{ji} modeled the opposite paths.

For the inter-depot routes obtained from the solution of the MDVRPI Relaxation, rotations are defined by linking the inter-depot routes until one reaches the starting depot. The duration of each rotation is then assessed. For rotations that do not respect the working-day time limit, the MDVRPI formulation is solved and rotations redefined to comply with the imposed limit. As a solution, one can have inter-depot routes belonging to rotations that satisfy the maximum duration for a working day and/or closed routes. More details can be founded in Ramos (2012).

B.1.3 Procedure 3: MDVRPI Extension

The MDVRPI Extension solves the problem by visiting all sites only by inter-depot routes. For that, the MDVRPI Relaxation is used, but instead of considering all depots and all collection sites at the same time, only two depots are considering in each run, and only the closest sites to those depots are made available to be collected. Moreover, a constraint is added to enforce routes to start and end at different depots. As a result, only inter-depot routes are defined.

A pair of depots $[dp, dp'] \in V_d$ is considered at a time, and constraints (14.31) and (14.32) are added to the MDVRPI Relaxation formulation, imposing that all routes have to start at depot dp and end at depot dp' to obtain a solution with only inter-depot routes between each pair of depots.

$$x_{ij} = 0, \quad \forall i \in V_c, j = dp + w \tag{14.31}$$

$$x_{ij} = 0, \quad \forall j \in V_c, i = dp' \tag{14.32}$$

Regarding the maximum duration for each inter-depot route in this procedure, it is considered the value $\left(H - L - b_{dp,dp'}\right)$ to guarantee that the vehicle can return to the origin depot within a working day.

After running the three procedures, the set K is built. Each route $k \in K$ is characterized by mileage (dis_k), duration (dur_k), load (Lo_k), and CO_2 emissions (Co_k). The first three parameters are provided by the solutions of the problems solved. The last one, the CO_2 emissions, has to be assessed a posteriori. For that, the emission model proposed by Barth et al. (2004) was used. When a vehicle travels over an arc (i,j), it is assumed that it emits a certain amount of CO_2, which depends on the fuel consumption that, in turn, is a function of many factors (such as, distance traveled, vehicle load – curb weight plus load – speed, road angle, engine features, vehicle frontal surface area, coefficients of rolling resistance and drag, and air density, among others (see Barth et al. 2004)). The conversion factor of 1 l of diesel fuel containing 2.6676 kg of CO_2 was assumed (as proposed in Defra). Note that CO_2 emissions were considered on arcs and nodes since nodes represent collection sites aggregating one or more containers and a certain mileage is traveled within each node.

The computation of the CO_2 emissions for all routes $k \in K$ concludes step 1.

B.2 Step 2: Solution Method for the Multi-objective Problem

In step 2 the multi-objective problem defined in Sect. 14.3 is solved (Fig. 14.4). In such problems it is rarely the case a single point optimizes simultaneously all objective functions (Coello and Romero 2003); therefore trade-offs between the objectives have to be analyzed in line with the notion of Pareto optimality. A solution is Pareto optimal if there exists no feasible solution, which improves

Annex B: Solution Procedure

one objective without causing a deterioration in at least one other objective. This concept generally does not apply to a single solution, but rather a set of solutions called the Pareto optimal set. The image of the Pareto optimal set under the objective functions is called Pareto front.

The improved version of the traditional ε-constraint method is applied to the problem so that the Pareto front is generated. Mavrotas (2009) proposes that the objective function constraints are transformed into equations (instead of inequalities as in the conventional method) by incorporating slack or surplus nonnegative variables, which are then used as penalization factors in the objective function. This augmented ε-constraint method produces only efficient solutions. In this work three objective functions exists; therefore a total of $(q_2 + 1) \times (q_3 + 1)$ runs are performed to obtain the Pareto front, when q_2 and q_3 are the equal amplitude intervals partitioning the range of each objective function. When the problem becomes infeasible, it means that there is no need to further constrain the corresponding objective function as it will from then on lead to infeasibility (more details in Mavrotas 2009).

When solving the problem under analysis in this work, where three objectives are being tackled, an approximation to the Pareto front is designed by using the augmented ε-constraint method, where the economic objective is optimized and the social and environmental constrained (see Table 14.4).

Finally, to propose a sustainable solution, that is, a compromise solution between the three objectives, a compromise solution method (Yu 1985) is applied, where the Pareto optimal solution closest to the ideal point is obtained. The ideal point (z_I) is defined according to the individual minima of each objective $z_I = \left(z_{min}^1, z_{min}^2, z_{min}^3 \right)$, while the nadir point (z_N) is defined according to the worst values obtained for each objective ($z_N = \left(z_{max}^1, z_{max}^2, z_{max}^3 \right)$. To apply this method, the objective functions are normalized by the differences between the nadir and ideal points, measuring the variability of the objective function within the Pareto set. Afterward, the compromise solution is obtained by minimizing the distance from the Pareto front to the ideal point, where the Tchebycheff metric is used as distance measure:

$$\min \left\{ \max_{j=1}^{\phi} \left\{ \lambda_j \left| z^j(S) - z_I^j \right| \right\} : S \in \Omega \right\} \qquad (14.33)$$

where ϕ is the number of objective functions in study and λ_j the normalized factor for each objective function:

$$\lambda_j = \frac{1}{r_j} \left[\sum_{i=1}^{\phi} \frac{1}{r_i} \right]^{-1} \qquad (14.34)$$

$$r_j = z_{max}^j - z_{min}^j \qquad (14.35)$$

Table 14.4 Pseudo-code of the augment ε-constraint method

1. *Lexicographic optimization to create the payoff table*
    1.1. min $z^1(S)$

       st

       eqs. (14.4), (14.6)–(14.13)
       Output: solution $s_1 = (z^{1*}, z^2, z^3)$
    1.2. min $z^2(S)$

       st

       eqs. (14.4), (14.6)–(14.13), $z^1(S) = z^{1*}$
       Output: solution $s_2 = (z^{1*}, z^{2*}, z^3)$
    1.3. min $z^3(S)$

       st

       eqs. (14.4), (14.6)–(14.13), $z^1(S) = z^{1*}$, $z^2(S) = z^{2*}$
       Output: solution $s_2 = (z^{1*}, z^{2*}, z^{3*})$
    1.4. Repeat 1.1 to 1.3 for $z^2(s)$ and $z^3(S)$
    1.5. Write the payoff table for the three objectives

2. *Set ε values*
    2.1. Set ranges of the objective functions:

    $$r_2 = z^2_{max} - z^2_{min}$$
    $$r_3 = z^3_{max} - z^3_{min}$$
    2.2. Set number of grid points q_2 and q_3
    2.3. Set the variation of ε_2 and ε_3:
    $$\Delta\varepsilon_2 = \frac{r_2}{q_2}$$
    $$\Delta\varepsilon_3 = \frac{r_3}{q_3}$$

3. *Solve Problem* (where v_2, v_3 are the surplus variables and *eps* is a small number, usually between 10^{-6} and 10^{-3})

    $n_2 = 0$, $n_3 = 0$
    while $n_2 \leq q_2$ and $n_3 \leq q_3$
       do

          min $\left(z^1(S) - eps\left(\frac{v_2}{r_2} + \frac{v_3}{r_3}\right)\right)$

          st

          eqs. (14.4), (14.6)–(14.13)
          $z^2(S) + v_2 = z^2_{max} - n_2\Delta\varepsilon_2$
          $z^3(S) + v_3 = z^3_{max} - n_3\Delta\varepsilon_3$
       end do
       $n_2 = n_2 + 1$
       $n_3 = n_3 + 1$
    end while

References

Azi N, Gendreau M, Potvin JY (2010) An exact algorithm for a vehicle routing problem with time windows and multiple use of vehicles. Eur J Oper Res 202:756–763

Balas E, Padberg M (1976) Set partitioning: a survey. SIAM Rev 18:710–760

Baldacci R, Hadjiconstantinou E, Mingozzi A (2004) An exact algorithm for the capacitated vehicle routing problem based on a two-commodity network flow formulation. Oper Res 52:723–738

Barth M, Scora G, Younglove T (2004) Modal emissions model for heavy-duty diesel vehicles. Trans Res Rec 1880:10–20

References

Coello CAC, Romero CEM. Evolutionary algorithms and multiple objective optimization. In: M. Ehrgott, X. Gandibleux, editors, Multiple criteria optimization: state of the art annotated bibliographic surveys; 2003. p. 277–331

Department for Environment, Food and Rural Affairs (Defra). Guidelines to Defra/DECC's GHG conversion factors for company reporting. http://www.defra.gov.uk/environment/economy/business-efficiency/reporting. Accessed 17 Mar 2018

Laporte G. What you should know about the vehicle routing problem. Naval Research Logistics 2007;54:811–819

Mavrotas G (2009) Effective implementation of the epsilon-constraint method in multi-objective mathematical programming problems. Appl Math Comput 213:455–465

Oliveira A, Vieira O (2007) Adaptive memory programming for the vehicle routing problem with multiple trips. Comput Oper Res 34:28–47

Petch RJ, Salhi S (2003) A multi-phase constructive heuristic for the vehicle routing problem with multiple trips. Discret Appl Math 133:69–92

Ramos TRP (2012) Tactical and operational planning in reverse logistics systems with multiple depots. Dissertation, Universidade Técnica de Lisboa

Ramos TRP, Gomes MI, Barbosa-Povoa AP (2013) Planning waste cooking oil collection systems. Waste Manag 33:1691–1703

Rieck J, Zimmermann J (2010) A new mixed integer linear model for a rich vehicle routing problem with docking constraints. Ann Oper Res 181:337–358

Yu PL (1985) Multiple criteria decision making: concepts, techniques and extensions. Plenum Press, New York

Chapter 15
Collection of Used or Unrecoverable Products: The Case of Used Cooking Oil

Abstract The planning of collection routes in a waste cooking oil system is studied in this chapter. The problem is modeled as a multi-depot vehicle routing problem with mixed closed and open routes. This is to mean that all routes start at one depot but can end at the same or at a different one, depending on what minimizes the decision-maker goal (cost, distance travel, among others). The developed approach allowed the company to redesign its collection system, which resulted in a decrease of 13% on mileage and 11% on fleet hiring cost when compared to the current company operation mode. The impact of expanding the collection network is also studied allowing for a reduction of the collection cost per client.

Keywords Waste cooking oil · Routing · Multiple depots · Open routes · Time duration constraints · Waste collection

15.1 Introduction

The collection of used or unrecoverable products for recycling has been growing in recent years mainly due to the increase of society awareness toward environmental aspects. One of these examples is the case of the cooking oil waste collection that has been debated due to the negative environmental impacts caused by the uncontrolled disposal of such products. Diverting waste cooking oil from landfills extends the product life cycle and prevents the contamination of groundwater supplies with this harmful liquid waste. Besides, waste cooking oil can be used to produce some other products such as biodiesel (Zhang et al. 2003; Felizardo et al. 2006; Phan and Phan 2008; Sabudak and Yildiz 2010), which represents an alternative to nonrenewable and limited fossil fuels, while being less polluting. According to Demirbas (2009), there is a net reduction in CO_2 emissions when using biodiesel fuel as an energy source in a diesel engine, since it reduces the consumption of diesel fuels. Additionally, biodiesel reduces particular matter (PM), hydrocarbons (HC), and carbon monoxide (CO) emissions but increases nitrogen oxide (NOx) emissions when compared with diesel fuel (EPA 2002). Despite the environmental benefits of using biodiesel (renewability, biodegradability, nontoxicity, and low emissions),

---

© Springer International Publishing AG, part of Springer Nature 2019
A. Pires et al., *Sustainable Solid Waste Collection and Management*,
https://doi.org/10.1007/978-3-319-93200-2_15

its cost is the main obstacle to commercialization. Since the feedstock represents up to 75% of the total manufacturing cost of biodiesel, using cooking oils as raw materials can reduce this cost as it is 2–3 times cheaper than virgin vegetables oils (Zhang et al. 2003). Yaakob et al. (2013) supported by an extensive review concluded that waste cooking oil could be a promising feedstock in biodiesel production. Van Kasteren and Nisworo (2007) estimate the required selling price of biodiesel produced from waste cooking oil at 0.17 US$/l to 0.52 US$/l, depending on the plant capacity. Moreover, with the fossil fuel increasing prices, the biodiesel from waste cooking oil tends to become also economically attractive.

Some companies have been investing in the collection and treatment of used cooking oil from restaurants, schools, and canteens, seeking for the business opportunity behind this waste stream. After being collected, the oil can be used in biofuel production units or in the chemical industry to produce soap, detergents, lubricants, paint, and grease, among others. Moreover, collecting and recycling this type of waste contribute to solve simultaneously three environmental problems: waste reduction by product reuse/recovery, reduction of the fossil fuels energy dependence, and reduction of pollutant emissions.

The present work is developed following a study of one company currently operating in Portugal. This company is responsible for the collection, sorting, recovery, and treatment of a large diversity of solid waste produced by 11 municipalities of mainland Portugal. Among them are non-recyclable domestic waste, recyclable waste packaging, electrical and electronic equipment waste, used tires, construction and demolition waste, and waste cooking oil. The company operates three depots where the collection vehicles start and end their routes. The specialized vehicles for the collection of non-recyclable domestic waste and recyclable waste packaging are owned by the company. The non-specialized collection vehicles are outsourced implying a fixed cost per vehicle route and a variable cost per kilometer traveled. These latter vehicles are the ones used to collect used tires or cooking oil waste.

There is a total of 303 sites (188 restaurants, 80 schools, and 35 canteens) where oil containers have to be collected. When picking up full containers, empty ones are delivered. The company decided they should be collected once a week, and, therefore, the number of containers delivered to each site is the one required to accomplish a weekly collection target, i.e., according to the oil volume consumed per week. For example, if a site consumes 50 liters per week, and since each container has a capacity of 30 liters, two containers are delivered to this site. These are collected and exchanged for empty and cleaned containers every week. In a collection day, vehicles leave depots with empty containers and return to a depot to unload and clean the full oil containers. Since the oil volume consumed in each site varies over the year, two time periods are considered by this company: *Normal Period*, September to June, and *Seasonal Period*, July and August. In the *Normal Period*, every site is visited for collection once a week. In the *Seasonal Period*, schools are not visited; the number of containers increases in restaurants located in vacation areas and decreases in business canteens while maintaining the weekly collection. Therefore, there are two routing plans. The *Normal Period* which involves 432 oil containers over 303 clients (restaurants, schools, and canteens), and the collections are to be repeated every week from September to June (43 weeks). The *Seasonal Period* which involves

292 containers over 223 clients (restaurants and canteens), and collections are to be repeated every week from July to August (9 weeks).

Strategically, the company has decided to contract independent vehicles to collect this waste stream. As mentioned before, such vehicles involve the payment of a fixed cost per route and a variable cost per kilometer traveled. The company aims at minimizing the total cost, the number of vehicle routes performed, and the total distance traveled when visiting all the collection sites. Furthermore, the company is striving to expand the collection network in terms of the number of visited clients and wants to assess the impact of this potential growth on the defined objectives.

A mixed-integer linear programming (MILP) model has been developed to address this case study (all modeling details are presented in Annex A). All routes start at one depot but can end at the same or at a different one, depending on what minimizes the considered objective. Such problem is referred as a Multi-Depot Vehicle Routing Problem with Mixed Closed and Open Inter-Depot Routes where capacity and duration constraints are taken into account.

This chapter unfolds as follows. In the next section, the company current operation mode is described. Section 15.3 presents all the analysis that can be performed with the model presented in Annex A. Namely, the current operation mode is optimized, and an expansion analysis is made so as access where and when to add new customers. Lastly, some conclusions are drawn.

15.2 Company Current Operation Mode

The waste cooking oil collection system in study has 3 depots and 303 clients (restaurants, schools, and canteens) and operates under 2 collection periods along the year caused by the pattern observed on the waste oil collecting demand: *Normal Period*, September to June, and *Seasonal Period*, July to August. The 303 clients are clustered into collection sites, corresponding to urban areas or isolated locations. A collection site is characterized by the number of clients and the oil containers (e.g., collection site "24" is an area with 28 clients and a total of 30 oil containers in the *Normal Period*). The traveled distance and the time spent at a collection site takes into account the average distance between clients (0.5 km), the road velocity in urban areas (30 km/h), and the average time spent at each client (6 min). Taking again collection site "24" as an example, one sees the total distance traveled is 14 km and the total time spent adds up to 196 min.[1] Since collection periods occur on non-rush hours with low traffic, an average of 30 km/h is observed based on historical data combined with the daily experience of the company. Moreover, the same average speed is assumed for all collection sites since all visited areas have similar characteristics regarding road and traffic network.

---

[1] 198 min = 28 clients × 6 min/client + 14 km × 2 min/km

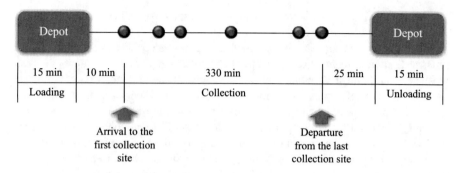

Fig. 15.1 Example of vehicle route timings

As mentioned, the company works with a different number of clients and containers along the year. For the *Normal Period*, 432 containers are spread over the 303 current clients that, in turn, are clustered into 100 collection sites. Forty-four percent of the collection sites have only one client with a single oil container, and 11% have more than five clients with more than five oil containers to be collected. Regarding the *Seasonal Period*, 292 containers are spread over 223 clients (schools are not considered) and clustered in 86 collection sites.

Each vehicle is hired for a 7.5-h period, including half an hour of lunch break and has a capacity to collect up to 45 oil containers. Since collection sites are restaurants, schools, and canteens, the visits must occur within the kitchens' working period, a 6-h period from 10 a.m. to 4 p.m. At depots, when the outsourced vehicles arrive, they are loaded with empty oil containers, and upon their return, the collected oil containers are unloaded. The loading/unloading operation has an average duration of 15 min each. Figure 15.1 depicts possible vehicle route timings.

In the *Normal Period*, the company operates 11 vehicles covering a total of 1220 km per week. It represents weekly hiring costs of 2540€ corresponding to 120€ as a fixed cost per hired vehicle and 1€ per km traveled per vehicle. The collection is performed with six closed routes and five open ones (Fig. 15.2a). In the *Seasonal Period*, eight vehicles are weekly hired, traveling 1039 km and representing a total hiring cost of 1999€. Five closed routes and three open ones are used during this period (Fig. 15.2b).

Table 15.1 shows the main key performance indicators of the current operation. On average 28 clients are visited per week, where at the *Normal Period* an average of 39 containers are collected, representing a capacity utilization rate of 87%, and at the *Seasonal Period* 37 containers are collected, corresponding to a capacity utilization rate of 81%. Route length ranges between 68 and 196 km. On average, collection takes a total of 272 min in the *Normal Period* and 291 min in the *Seasonal Period* (82% and 88% of usage rate, respectively). No route reaches the maximum time available (420 min). The longest one takes 97% of this time (408 min).

15.3 The New Collection Network

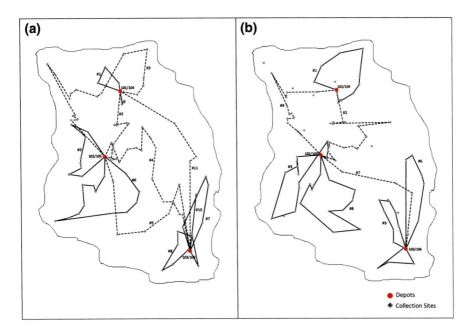

Fig. 15.2 Routes operated during (**a**) *Normal Period* and (**b**) *Seasonal Period*

Table 15.1 Current routes characteristics

Key performance indicators	Normal Period				Seasonal Period			
	Total per week	Per route			Total per week	Per route		
		Min	Avg	Max		Min	Avg	Max
No. clients visited	303	16	28	37	223	13	28	40
No. containers collected	432	31	39	45	292	23	37	45
Distance travelled (km)	1220	68	111	196	1039	93	130	161
Collection duration[a] (m)	2989	165	272	328	2324	182	291	330
Total duration (m)	3733	255	339	399	2876	250	360	408
Used capacity (%)	–	69	87	100	–	51	81	100
Used collection duration (%)	–	50	82	99	–	55	88	100
Used total duration (%)	–	61	81	95	–	60	86	97

[a]Collection Duration involves the elapse time between arriving to the first collection site and leaving from the last collection site

15.3 The New Collection Network

15.3.1 If the Current Network Is Optimized

The best found collection network at the *Normal Period* comprises ten routes, one route less than the current company operation mode. Six are closed routes and four

Table 15.2 Solution proposed for the *Normal Period*

Route No.	Route sequence	Type
#1	101 - 11 - 9 - 10 - 12 - 13 - 1 - 36 - 35 - 85 - 84 - 37 - 38 - 2 - 90 - 87 - 89 - 86 - 7 - 88 - 65 - 76 - 105	Open
#2	101 - 14 - 22 - 25 - 33 - 29 - 30 - 31 - 32 - 74 - 73 - 72 - 71 - 70 - 105	Open
#3	101 - 34 - 17 - 18 - 16 - 26 - 23 - 24 - 104	Closed
#4	102 - 69 - 68 - 67 - 64 - 66 - 20 - 15 - 21 - 28 - 27 - 19 - 104	Open
#5	102 - 82 - 99 - 100 - 98 - 4 - 94 - 95 - 96 - 8 - 3 - 78 - 6 - 79 - 77 - 80 - 81 - 5 - 105	Closed
#6	103 - 39 - 42 - 46 - 47 - 40 - 106	Closed
#7	103 - 45 - 54 - 61 - 57 - 75 - 62 - 63 - 92 - 97 - 91 - 83 - 105	Open
#8	103 - 48 - 93 - 44 - 43 - 41 - 106	Closed
#9	103 - 52 - 53 - 51 - 50 - 56 - 58 - 59 - 49 - 60 - 106	Closed
#10	103 - 55 - 106	Closed

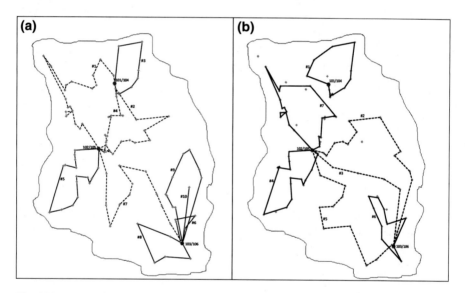

Fig. 15.3 Solution proposed for the (**a**) *Normal Period* and (**b**) *Seasonal Period*

are open (Table 15.2 and Fig. 15.3a). At the *Seasonal Period*, seven routes are defined where five are closed and two are open (Fig. 15.3b).

As mentioned, a weekly collection frequency is considered, and, consequently, routes are to be repeated every week. For the open inter-depot routes, the starting depot will alternate each week so that the number of empty and full containers matches every week. For instance, considering route #2 for the *Normal Period*, when the vehicle starts the route at depot 101 and ends at depot 102, depot 101 needs to have at least 45 empty oil containers that will be exchanged for 45 full containers and

15.3 The New Collection Network

293

unloaded at depot 102. Therefore, in the next week, this route will start at depot 102, since the 45 empty containers are now available at this location.

The total distance traveled to collect the 432 oil containers scattered over the 303 clients in the *Normal Period* is of 1067 km, and only 10 vehicles are required against the 11 currently hired by the company. At the *Seasonal Period*, the distance traveled to collect the 292 containers is 895 km and again, one less vehicle is needed. It represents a decrease of 13% on the annual mileage. The proposed solution entails a weekly hiring fleet cost of 2267€ at the *Normal Period* and 1735€ at the *Seasonal Period*, a reduction of 11% regarding the current annual hiring cost.

The utilization capacity and collection duration rates have increased from 87% to 96% and from 82% to 90%, respectively, in the *Normal Period* (Table 15.3). The average number of clients visited per route is now higher than in the current operation mode as well as the number of collected containers. Route maximum length and duration have reduced to 163 km and to 374 min, respectively. Since vehicles are currently hired for a 420-minute period, negotiations can be performed to reduce the hiring period to 375 min, potentially diminishing the fixed cost per vehicle.

15.3.2 Network Expansion

The company plans to increase the number of clients and, consequently, the number of oil containers spread over the 11 municipalities. This marketing effort is to be executed in two phases. First, increase the number of clients within the existent collection sites, i.e., attract more restaurants, schools, and canteens in the areas already being visited. Second, increase the number of collection sites, i.e., add new areas to the current network. The goal has been set to enlarge the number of clients in 25% (corresponding to 76 clients) in a medium-term horizon. At phase one, the company aims to attract 50% of the target clients (38 clients) in the current visited areas, and, in the second phase, the company aims to attract the remaining 50% of clients in new areas.

Phase 1

One assesses the areas (or collection sites) where the marketing effort should be invested by the company. In fact, the idea is to take advantage of the time availability in the new proposed routes. The available capacity and duration were analyzed and showed the space to add a total of 12 new clients[2] by only increasing the usage rates (no change needs to be done on the new routes).

---

[2]Assuming each client needs only one container, it will correspond to 12 containers.

Table 15.3 Routes' characteristics for the *Normal Period*

Route no.	No. clients visited	No. containers collected	Distance travelled (km)	Collection duration[a] (min)	Total duration (min)	Used capacity (%)	Used collection duration (%)	Used total duration (%)
#1	23	42	163	324	372	93	98	89
#2	28	45	117	311	350	100	94	83
#3	35	45	97	317	370	100	96	88
#4	29	44	64	251	292	98	76	70
#5	28	44	133	326	369	98	99	88
#6	37	41	56	293	334	91	89	79
#7	22	44	141	282	340	98	85	81
#8	35	43	100	327	374	96	99	89
#9	29	43	129	282	370	96	85	88
#10	37	41	68	259	349	91	78	83
Total	303	432	1067	2971	3520	–	–	–
Avg	30	43	107	297	352	96	90	84
Min	22	41	56	251	292	91	76	70
Max	37	45	163	327	374	100	99	89

[a]Collection Duration involves the elapse time between arriving to the first collection site and leaving from the last collection site

Table 15.4 Potential clients to be added considering capacity and duration availability of the proposed routes for the *Normal Period* and the impact on the usage rates

Route no.	Capacity available (no. clients)[a]	Collection duration available (min)	Collection duration available (no. clients)	Potential clients to be added	Used capacity (%)	Used collection duration (%)	Used total duration (%)
#1	3	6	0	0	93	98	89
#2	0	19	2	0	100	94	83
#3	0	13	1	0	100	96	88
#4	1	79	11	1	100	78	71
#5	1	4	0	0	98	99	88
#6	4	37	5	4	100	97	86
#7	1	48	6	1	100	87	83
#8	2	3	0	0	96	99	89
#9	2	48	6	2	100	90	91
#10	4	71	10	4	100	87	90
Total	18	329	41	12	–	–	–
Avg	–	–	–	–	99	93	86
Min	–	–	–	–	93	78	71
Max	–	–	–	–	100	99	91

[a]It is considered that the number of clients is equal to the number of containers

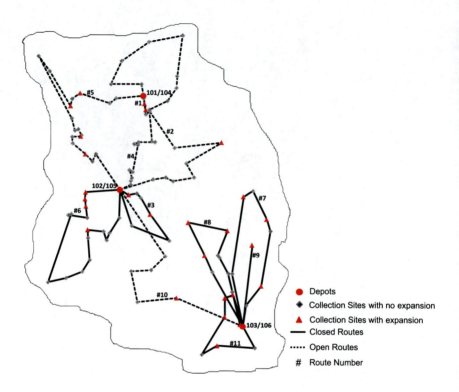

Fig. 15.4 New routes for a network with 341 clients located at 100 collection sites

Table 15.4 reports, for each proposed route, the available capacity and time (*number of clients = number of collected containers* and *collection duration available in minutes*) to perform additional collection and how this availability can be converted into additional clients to visit[3] (*collection duration available - no. clients*), as well as into the number of potential clients to be added.[4] In short, the proposed routes have, regarding time, available capacity to accommodate a total of 41 new customers. However, regarding capacity for new containers, only a total of 18 new containers can be loaded to the vehicles.

An oriented marketing effort on the areas covered by routes 4, 6, 7, 9, and 10 could lead to an increase in the capacity usage rates from 96% to 99% and an

---

[3]Collection duration available (no. of clients) = collection duration available in minutes/average time to spent at a client

[4]No. of potential clients to add = Min {Capacity available (no. of clients), collection duration available (no. of clients)}

The average time spent at a client is 7 min: average time spent on each client (6 min) + average time to travel between two clients (0.5 km × 2 min/km = 1 min).

15.3 The New Collection Network

increase in the used collection duration rates from 90% to 93%. However, due to the vehicles capacity to new containers, only 12 clients can be added. To reach the company's goal for phase 1, there are still 38 new clients to consider.

After the identification by the company where to locate these potential 38 new clients, a new route plan was computed. The new operational plan has 11 routes, where 4 are open and the remaining 7 closed (Fig. 15.4). The total hiring fleet cost in this scenario reaches 2442€ per week and covers 1122 km.

Phase 2

In the second phase, the company identified 10 new areas of interest to be added to the current network, where 38 new potential clients are located. In this case, 12 routes are needed to visit 379 clients located at 110 collection sites. Four are open and eight are closed (Fig. 15.5). The total distance traveled is now 1205 km, and the total hiring fleet cost is 2645€ per week. The impact of expanding the current network is summarized in Table 15.5. The expansion of clients reached without modifying the

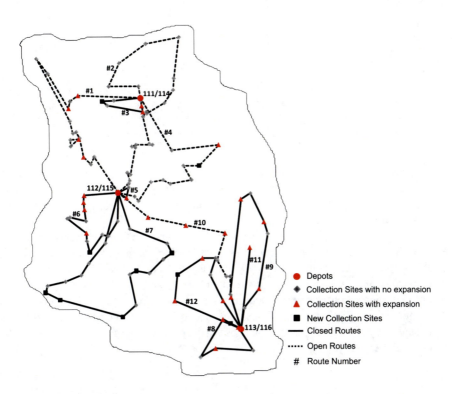

Fig. 15.5 Routes proposed for a network with 379 clients located at 110 collection sites

Table 15.5 Results comparison between the three phases to the current network expansion

	Current network (solution proposed)	Phase 0 - Add 12 clients, maintaining collection sites	Phase 1 - Add 38 clients, maintaining collection sites	Phase 2 - Add 76 clients, increasing collection sites
Number of collection sites	100	100	100	110
Number of clients	303	315	341	379
Number of containers	432	444	470	508
Number of routes	10	10	11	12
Number of closed routes	6	6	7	8
Number of open routes	4	4	4	4
Distance travelled (km)	1067	1073	1122	1205
Total cost (€)	2267	2273	2442	2645
Average used capacity rate (%)	96	99	95	94
Average used collection duration rate (%)	90	93	89	88
Average total duration rate (%)	84	86	84	84

proposed solution with the 12-customer increase (phase 0) was considered, which can be seen as a short-term locality-oriented marketing action. Phases 1 and 2 imply a more significant effort from the marketing and operations departments and are seen as medium-term actions.

The cost per client decreased with the growth of the collection network (Fig. 15.6), meaning the company should pursue these economies of density. Whenever the company intends to increase the number of clients on existing collection sites (areas), the analysis of the capacity and duration availability on the optimal routes should be performed to guide the marketing effort from an operations point of view, gaining on efficiency. For instance, phase 0 where this analysis was considered represents the largest decrease in the cost per client when compared to the other phases of expansion (-3.5%).

Annex A

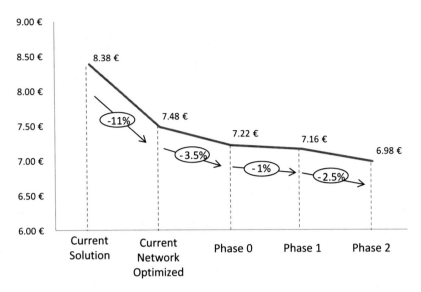

Fig. 15.6 Hiring fleet cost per client for all scenarios presented

15.4 Conclusions

This chapter addressed a real problem of waste collection using outsourced vehicles where the company wanted to increase its market share. In this case, all routes have to start at a depot to load clean containers and end at the same or different depot to unload the full containers. The new vehicle routes allow a decrease of 13% on mileage and 11% on fleet hiring cost when compared to the current company operation mode. The impact of expanding the collection network has also been studied leading to a reduction of the collection cost per client. Such results clearly show the impact optimization models can have in supporting decision-making, allowing for a straightforward study of different management scenarios.

Annex A

The MDVRP-MCO is formulated on a graph $G = (V, A)$, where $V = \{1, \ldots, N + W\}$ is the vertex set and $A = \{(i, j) : i, j \in V, i \neq j\}$ is the edge set. The vertex set V is partitioned into two subsets $V_c = \{1, \ldots, N\}$ and $V_d = \{N + 1, \ldots, N + W\}$, representing the set of N collection sites and the set of W depots, respectively. Each vertex $i \in V_c$ has a number of containers to be collected (p_i) and a non-negative service duration (s_i). A symmetric distance matrix $D = (d_{ij})$ is associated to the edge set A.

The MDVRP-MCO builds k vehicle routes such that (1) each route starts and ends at a depot (not necessarily the same); (2) each collection site is visited exactly

Fig. 15.7 Routes illustration for the MDVRP-OR

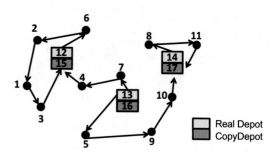

once by a vehicle; (3) the total demand of each route does not exceed the vehicle capacity Q; (4) the total duration of each route, including travel and service times, does not exceed a preset time limit T; so (5) the total routing cost is minimized.

To apply the two-commodity flow formulation, the graph G is extended. The graph $\bar{G} = (\bar{V}, \bar{A})$ is now obtained by adding the vertex set $V_f = \{N + W + 1, \ldots, N + 2W\}$ which is a copy of the depots set. Thus, $\bar{V} = V \cup V_f$, $\bar{A} = A \cup \{(i,j) : i \in V_c, j \in V_f\}$, and $d_{ij} = d_{ih}$, $i \in V_c$, $j \in V_f$, $h \in V_d$.

To formulate the MDVRP-MCO based on the two-commodity flow formulation, the same decision variables as Baldacci et al. (2004) were used – x_{ij} and y_{ij} – and add two decision variables to carry out the duration constraints, e_{ij} and a_{ij}. A third variable k is introduced allowing for the minimization of the number of vehicles or vehicle routes (in this context vehicles and vehicle routes are equivalent). Therefore, the decision variables in this formulation are:

- x_{ij}, a binary variable representing the routing solution:
  $= 1$, if site j is visited immediately after site i; 0, otherwise.
- y_{ij}, a flow variable representing the load in the vehicle route when edge (i,j) is crossed. The flow y_{ji} represents the empty space on vehicle route when edge (i,j) is crossed; therefore, $y_{ij} + y_{ji} = Q$, at any edge (i,j).
- e_{ij}, a continuous variable representing the exit time from site i to site j.
- a_{ij}, a continuous variable representing the arrival time to site j from site i.
- k, an integer variable representing the number of vehicles needed.

All routes start at a real depot (set V_d) and end at a copy one (set V_f) as shown in Fig. 15.7.

As mentioned, each route is defined by two flow paths: one path, from a real depot to a copy depot, defined by variables y_{ij}, modeling the vehicle load; the other path, the reverse path starts at a copy depot and ends at a real depot, defined by variables y_{ji}, modeling the empty space on the vehicle. An illustration of such paths can be seen in Fig. 15.8.

The MILP model is formulated as follows:

Annex A

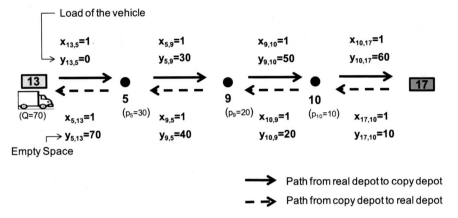

Fig. 15.8 Illustration of the two paths defining a route solution for the MDVRP-MCO

$$\text{Minimize} \quad \frac{1}{2} \sum_{i \in \bar{V}} \sum_{j \in \bar{V}} x_{ij} c_{ij} + f.k \tag{15.0}$$

Subject to

$$\sum_{\substack{j \in \bar{V} \\ j \neq i}} (y_{ij} - y_{ji}) = 2p_i, \quad \forall i \in V_c \tag{15.1}$$

$$\sum_{i \in V_c} \sum_{j \in V_f} y_{ij} = \sum_{i \in V_c} p_i \tag{15.2}$$

$$\sum_{j \in V_f} \sum_{i \in V_c} y_{ji} = k.Q - \sum_{i \in V_c} p_i \tag{15.3}$$

$$\sum_{i \in V_c} \sum_{j \in V_d} y_{ij} = k.Q \tag{15.4}$$

$$\sum_{\substack{i \in \bar{V} \\ i \neq j}} x_{ij} = 2, \quad \forall j \in V_c \tag{15.5}$$

$$y_{ij} + y_{ji} = Q x_{ij} \quad \forall i, j \in \bar{V}, \ i \neq j \tag{15.6}$$

$$e_{ij} + t v_{ij} x_{ij} = a_{ij} \quad \forall i, j \in \bar{V}, \ i \neq j \tag{15.7}$$

$$\sum_{\substack{i \in \bar{V} \\ i \neq j}} \left(e_{ji} - a_{ij} \right) = 2s_j \quad \forall j \in V_c \tag{15.8}$$

$$e_{ij} - \sum_{h \in \bar{V} \setminus \{i,j\}} a_{hi} \leq s_i x_{ij} \quad \forall i \in V_c, \ \forall j \in \bar{V}, \ i \neq j \tag{15.9}$$

$$e_{ij} \leq BigMx_{ij} \quad \forall i,j \in \bar{V} \tag{15.10}$$

$$a_{ij} \leq BigMx_{ij} \quad \forall i,j \in \bar{V} \tag{15.11}$$

$$e_{ij} - a_{j'i'} \leq E \quad \forall i,i' \in V_c, \ i \neq i', \ \forall j \in V_f, \ \forall j' \in V_d \tag{15.12}$$

$$a_{ji} + L \leq T \quad \forall i \in V_d \cup V_f, \forall j \in V_c \tag{15.13}$$

$$e_{ij} = 0 \quad \forall i \in V_d \cup V_f, \forall j \in V_c \tag{15.14}$$

$$x_{ij} \in \{0,1\} \quad \forall i,j \in \bar{V} \tag{15.15}$$

$$y_{ij}, y_{ji} \geq 0 \quad \forall i,j \in \bar{V} \tag{15.16}$$

$$e_{ij}, a_{ij} \geq 0 \quad \forall i,j \in \bar{V} \tag{15.17}$$

$$k \quad \text{integer} \tag{15.18}$$

The objective function (15.0) models the minimization of the total cost involving variable and fixed costs. Here, c_{ij} is the cost of crossing edge (i,j) which is a function of the distance as $c_{ij} = \beta \cdot d_{ij}$, with β the unit cost per kilometer. Since the collection routes are defined by two paths (see Fig. 15.8), each solution edge is counted twice, doubling the distance traveled and, consequently, doubling the cost. For example, at Fig. 15.8, the cost of edge [13,5] ($c_{13,\,5}$) is counted twice, since $x_{13,\,5} = 1$ and $x_{5,\,13} = 1$. Therefore, to identify the real variable cost value, the first term has to be divided by 2 to eliminate the value of the second path. In the second term, f is the vehicle fixed hiring cost.

The first six equations are related to the flows rules along the routes definition. Constraint (15.1) states that the outflow minus the inflow at each collection site is equal to twice the amount to be collected at each collection site. Constraint (15.2) ensures that the total inflow of the copy depots is equal to the total amount to be collected. The total outflow of copy depots corresponds to the residual capacity of the used vehicles. Constraint (15.3) ensures that the total outflow of the copy depots is equal to the residual capacity of the vehicle fleet, while constraint (15.4) states that the total inflow of the real depots is equal to the total capacity. Constraint (15.5) guarantees that each collection site has two incident edges, and constraint (15.6) ensures that the inflow plus the outflow of any node is equal to the vehicle capacity.

On the other hand, constraints (15.7) to (15.14) are related to the timings associated with routes definition. Constraint (15.7) guarantees that the exit time from node i plus traveling time from i to j (denoted by tv_{ij}) is equal to the arrival time at node j. Constraint (15.8) states that the exit time minus the arrival time at each collection site is equal to twice the amount of time spent at each collection site, while constraint (15.9) ensures time continuity. Constraints (15.10) and (15.11) ensure that if the arc (i, j) is not crossed, then the exit and arrival times from i to j are equal to zero. Constraint (15.12) ensures that the time elapsed between arriving to the first site and leaving from the last site does not exceed the available time to visit the collection sites, denoted by E. Constraint (15.13) guarantees that the route duration plus the time to load and unload the vehicle at depots (denoted by L) does not exceed the time available for a working day (denoted by T). Constraint (15.14) states that all routes start at time zero.

Finally, the variables domains are given at Eqs. (15.15) to (15.18). To tightening the lower bound, the flow inequalities (15.19) and (15.20) proposed by Baldacci et al. (2004) tailored here to a collection problem were also considered.

$$(Q - p_j)y_{ji} - p_jy_{ij} \geq 0 \quad \forall i, j \in \bar{V} \tag{15.19}$$

$$(Q - p_i)y_{ij} - p_iy_{ji} \geq 0 \quad \forall i, j \in \bar{V} \tag{15.20}$$

All the results presented were obtained by the model presented above, implemented in GAMS 23.6 and solved by the CPLEX Optimizer 12.1.0, on an Intel Xeon CPU X5680 @ 3.33GHz.

References

Baldacci R, Hadjiconstantinou E, Mingozzi A et al (2004) An exact algorithm for the capacitated vehicle routing problem based on a two-commodity network flow formulation. Oper Res 52:723–738

Demirbas A (2009) Political, economic and environmental impacts of biofuels: a review. Appl Energy 86:S108–S117

EPA - Environmental Protection Agency U.S. (2002) A comprehensive analysis of biodiesel impacts on exhaust emissions. Draft technical report EPA420-P-02-001

Felizardo P, Correia MJN, Raposo I et al (2006) Production of biodiesel from waste frying oil. Waste Manag 26:487–494

Phan AN, Phan TM (2008) Biodiesel production from waste cooking oils. Fuel 87:3490–3496

Sabudak T, Yildiz M (2010) Biodiesel production from waste frying oils and its quality control. Waste Manag 30:799–803

Van Kasteren JMN, Nisworo AP (2007) A process model to estimate the cost of industrial scale biodiesel production from waste cooking oil by supercritical transesterification. Resour Conserv Recycl 50:442–458

Yaakob Z, Mohammad M, Alherbawi M et al (2013) Overview of the production of biodiesel from waste cooking oil. Rene Sust Energ Rev 18:184–193

Zhang Y, Dubé MA, McLean DD et al (2003) Biodiesel production from waste cooking oil: 2. Economic assessment and sensitivity analysis. Bioresour Technol 90:229–240

Part IV
Challenges and Perspectives for Sustainable Waste Management Through Waste Collection

Chapter 16
The Evolution of the Waste Collection

Abstract This chapter discusses the concept of the integrated waste collection that could bring the collection operational unit of the integrated solid waste management into the center of it. The integrated waste collection concept is based on the operational units existing in the waste management system, having in mind the waste hierarchy principle and the contribution to the circular economy. The integrated waste collection has impacts on the economy, society, and environment, and its conception needs to look at the way how it can be sustainable and contribute to the sustainability of the waste management where it belongs.

Keywords Biological treatment · Energy recovery · ISWM · Recycling · Source separate collection · Sustainability · Waste hierarchy · Waste prevention

16.1 Definition of Integrated Waste Collection Concept

Until now, waste management systems have been focusing on the options of treatment, recovery, and disposal of waste, looking to waste collection as an operational unit only needed to make waste management system work. A more prominent role has been given to waste collection nowadays, making this operational unit each day more relevant to help to increase sustainability in the use of resources, to economically optimize the waste management system, and to reduce environmental impacts resulting from the collection of waste. The concept of integrated waste collection intends to put the focus of waste management system in the collection operational unit, where waste hierarchy principle, under the umbrella of the circular economy, is implemented in the waste management system by the waste collection. The critical items under the concept of the integrated waste collection are:

- A holistic view, where the city and region (urban and industrial) consider all symbiosis roles.
- Use of system analysis methodologies to find solutions to the problems.
- Waste hierarchy under the umbrella of the circular economy.

© Springer International Publishing AG, part of Springer Nature 2019 307
A. Pires et al., *Sustainable Solid Waste Collection and Management*,
https://doi.org/10.1007/978-3-319-93200-2_16

A holistic view is needed to embrace all the aspects related to the waste collection and effects on the waste management system. Industrial ecology is the basis to embrace the aspects, where the vision of ecosystem includes the human system (urban and industrial). This approach is needed to understand the waste management system with an organismic approach of wholeness, connectedness, and ordered complexity (Naveh 2000). Industrial ecology is capable of helping to implement the holistic approach, being strategies as industrial symbiosis used to maximize waste collection as the vector for a holistic and sustainable waste collection system, because holistic requires seeing the system as a whole, with the pillars of sustainability. According to the Mulvaney and Robbins (2011), industrial ecology is a:

> Systems-oriented study of the physical, chemical, and biological interactions and interrelationships both within industrial systems and between industrial and natural ecological systems.

Industrial symbiosis is a strategy to implement industrial ecology, where enterprises, employees, and the community collaborates to all enjoy benefits, being synergistic possibilities offered by geographic proximity (Mirata 2004; Glavič and Lukman 2007). An integrated waste collection needs to be tough, designed, planned, and implemented having as basis the industrial ecology and symbiosis in the geographic area where it is located, at different levels. The connections between all elements of the system must be driven by the type of waste/secondary resource to be collected, regarding amounts and waste features and destination features, and having in mind the geographic scales which could exist to potentiate a sustainable collection.

Systems analysis is capable of dealing with the complexity of waste management systems, by making simplifications of the real waste management systems during modeling without losing their essential features that characterize waste management (Chang et al. 2011). Systems analysis in waste management field has started since the 1960s, firstly to solve single objective optimization schemes like cost minimization of waste transport from transfer stations to landfills (Anderson 1968; Anderson and Nigam 1967). Significant evolution of the systems analysis techniques applied has been identified by Chang et al. (2011) and Pires et al. (2011). The evolution is mostly due to the natural evolution of systems analysis techniques but also to the increased complexity of waste management nowadays, where all different kinds of mandatory regulation, together with sustainability requirements, are included in the day-to-day management of waste. The types of systems analysis methods are (1) systems engineering models including cost-benefit analysis, forecasting model, simulation model, optimization model, and integrated modeling system; (2) systems analysis platforms, including management information system/decision support system/expert system, and finally system assessment tools such as scenario development, material flow analysis, life cycle assessment or life cycle inventory, risk assessment, environmental impact assessment, strategic environmental assessment, socioeconomic assessment, and sustainable assessment. The use of those systems analysis techniques is useful and crucial to implement and develop integrated waste collection concept.

16.1 Definition of Integrated Waste Collection Concept

According to the EU Waste Framework Directive, waste hierarchy principle is divided into prevention, reuse, preparing for reuse, recycling, recovery, and landfill on a preferential scale (EC 2008). Besides its implementation and diverse variations, waste hierarchy principle alone has issues, such as lack of guidance for choosing among the levels of the hierarchy and the absence of a distinction between open-loop and closed-loop recycling, and does not support decisions that could affect other sectors as well as waste management (Ewijk and Stegemann 2016). The waste hierarchy could be promoted under the umbrella of the circular economy, i.e., to dematerialize the economy by reintroducing waste as resources (if the waste is not avoidable), through the waste collection. How? The interactions between waste collection and each of the waste hierarchy steps need to be decomposed and analyzed to visualize its contribution and how waste managers can, in fact, manipulate waste collection to address better management of waste, in the light of waste hierarchy and circular economy.

Looking at waste hierarchy principle, the same structure is not being reflected nowadays in an integrated solid waste management (ISWM) system. Prevention is not part of ISWM because waste managers see it as a drawback – they need waste for their business, so any attempt to reduce it is a threat to themselves. However, prevention needs to be included because it is present in both waste hierarchy and circular economy. Reuse intends that the products have not yet reached waste stage not be within the scope of waste management. According to Waste Framework Directive definition, reuse means (European Parliament and Council 2008):

> Any operation by which products or components that are not waste are used again for the same purpose for which they were conceived.

Preparation for reuse can be influenced by the waste management system; however, the preparation is not part of the ISWM like what happens to prevention. Waste management systems are mostly devoted to sorting for recycling and treatment of waste. Waste treatment can be mechanical and biological (composting, anaerobic digestion), and waste collection can influence those processes. Energy recovery (also known as waste-to-energy), the less exciting option regarding the circular economy, can also be influenced by waste collection. In the end, mass burning and disposal are the less attractive options on waste hierarchy principle, and the interactions with the waste collection are the oldest, because waste management systems were, in the beginning, made by waste collection and transportation to the final destination – incineration or landfill.

The integrated waste collection needs to fulfill the purpose of a sustainable solid waste management system (McDougall et al. 2001): environmentally effective, economically affordable, and socially acceptable. Figure 16.1 shows the integrated waste collection hub which highlights the interactions between the collection and the other waste hierarchy-based processes and sustainability aspects considered in the development and implementation of the integrated waste collection.

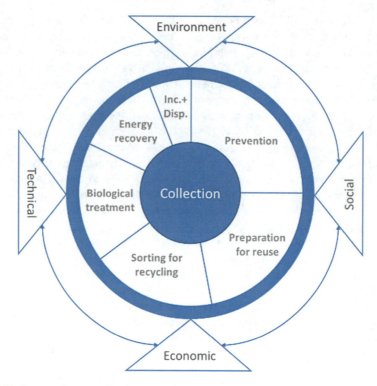

Fig. 16.1 Integrated waste collection

16.2 The Functioning of the Integrated Waste Collection (IWC)

The waste collection can influence each of the processes occurring in a waste management system. For each process, an interaction occurs, and the way how the collection performs influences positively and negatively the process. The description of those interactions is in the next sections.

16.2.1 How Collection Interacts with Waste Prevention

Waste prevention intends to avoid the generation of waste, to reduce its hazardousness, and to reduce its impacts during its existence as waste (Directive 2008/98/EC). To do so, there is a need to increase the lifetime of products and its durability and to define new methods of production to contain less amount of hazardous substances. When reaching waste phase, the entire management chain is conceived to not have an adverse impact on the environment and human health (from collection to disposal). According to Ewijk and Stegemann (2016), waste collection is the only life cycle phase concerned with prevention that waste managers may influence. Waste managers can influence the amount of waste generated by applying economic instruments like pay-as-you-throw. Reichenbach (2008) highlighted the role of PAYT in increasing recycling activity and reducing overall waste generation, mostly residual waste stream. Dahlén and Lagerkvist (2010) verified that the implementation of weight-based billing in household waste collection contributed to less household waste per capita generation around 20% in Sweden, although the higher recycling rate does not explain such difference. In a meta-analysis made by Bel and Gradus (2016), the efficiency of unit-based pricing on household waste collection demand is more efficient (i.e., reduces waste generation) if a weight-based pricing system exists together with priced compostable waste.

A different way to promote waste prevention and reuse is proposed by Ewijk and Stegemann (2016), where it is defended that integrating collection of materials for reuse into collection schemes can be a solution to reach preparation for reuse, to avoid that materials and products become controlled as waste.

Another contribution of waste collection to waste prevention is in the optimization toward reducing socioeconomic and environmental impacts. Optimization algorithms applied to waste collection leads to its optimization, where aspects such as travel distance, waste collected, fuel consumption and cost, tightness, efficiency, and emissions (namely, CO_2) can be considered in the algorithm, proposing different scheduling models to be implemented in the field.

16.2.2 How Collection Interacts with Preparation for Reuse

The preparation for reuse is devoted to products which have become waste, i.e., those that have been disposed in the temporary container, and includes (European Parliament and Council 2008):

> Checking, cleaning or repairing for recovery operations, by which products or components of products that have become waste are prepared so that they can be re-used without any other pre-processing.

The waste collection has to be made in such a way that the waste keeps its features to be repaired to be reused again, which can be complicated, not only because the product is now under the waste legislation but also because the collection was not made to preserve functioning features to allow its preparation for reuse. A source separate collection system can be implemented to specific waste streams, like waste of electric and electronic equipment (WEEE), in such a way that they are collected to be prepared for reuse and not for recycling, which is the current practice. In the light of the circular economy and waste hierarchy, the preparation for reuse gets more relevance because it can create more value for end-of-life products. However, existing convention systems (like containers for source separation at street and recycling centers and take-back systems) are focusing on reducing the cost of collection and disposal (Stahel 2016). Waste collection companies need to rethink waste collection to make end-of-life products reusable. For example, in the case of WEEE, Parajuly and Wenzel (2017) assessed WEEE collection system and verified that there is the need to improve them to exploit the possibilities of reuse of discarded products and the subsequent recovery of the material, defending that a change of paradigm is necessary: to no collect WEEE as *waste* but as *products*. Sing and Ordoñez (2016) also reinforce it for other products/wastes, pointing out that existing collection is for "material collection system" and not for "manufacturing-centered take-back system." Integrating collection of materials for reuse into collection schemes proposed by Ewijk and Stegemann (2016) also can be used for preparation for reuse to prevent environmental harm during materials handling and storage.

16.2.3 How Collection Interacts with Sorting for Recycling

To sort waste for recycling, there is a need to collect high-quality materials to ensure recycling without losing material or properties. The quality of recyclables not only supports high-quality recycling but can also contribute to reducing, reusing, and keeping products out of landfills (Parker 2017).

Until now, the key to reaching good-quality recyclables depended on citizens' participation in recycling schemes. Public awareness was needed to teach citizens to put recyclable fractions in the specific recyclable containers, for collection and transport for sorting plants (or material recovery facilities), through curbside or door-to-door collection schemes or drop-off stations or collection points. The collection systems have different performances regarding getting recyclables: drop-off systems are more accessible but suffer from low and unpredictable throughput and higher contamination when the curbside collection has recyclables with lower contamination (Martinho et al. 2017; Parker 2017).

Besides the type of collection, there are other variants in the separate collection, namely, dual-stream, single-stream, and mixed collection. Dual-stream collection is usual in this situation, where one bin is to collect paper/cardboard and the other bin is to collect commingled plastic, metal, and glass, being collected in separate trucks or separate compartments in the same vehicle (Fitzgerald et al. 2012). Also, refill/deposit systems recover mono-materials, usually beverage bottles (like single-use PET bottles), where they were sold with the surcharge of a refundable deposit, which is given back to the user upon return of the empty bottle after use (Villanueva and Eder 2014). To be successful, waste producers need to place correctly waste in the recycling bins, to avoid contamination, and the transportation should not compact thoroughly the waste to allow subsequent separation in sorting plant. According to Parker (2017), depending on the types of waste materials collected mixed, extra effort is needed to sort waste into separated materials, which can significantly reduce the quality of recycled product.

Due to the technological evolution of sorting technology, reflected at sorting plants, single-stream collection is nowadays a new practice in obtaining recyclables.

In this type of collection, all recyclables are collected together but kept separated from residual waste. Both collection systems have pros and cons. According to Parker (2017), for a dual system, more public education is needed to avoid contamination, but it presents a lower sorting post collection, obtaining the purest recyclables, although the operation cost per each separate material is higher. For single system, a reduction for post-collection cleaning occurs, but some public awareness is needed, and a higher amount of recyclables is reached (Parker 2017) (according to Fitzgerald et al. (2012), a change from dual to single stream in the USA could result in a 50% increase in recyclables and in avoiding 710 kg CO_2 eq. per metric ton of collection). In the cases of mixed collection, where there is no separate collection of recyclables, there are advantages in the number of recyclables collected and in the operation costs (there is no need to pay for a separate collection system), and no awareness campaigns are needed, although the quality of the recyclables is lower at mechanical treatment units (Parker 2017).

After collection, transportation and sorting can also affect sorting for recycling of materials. According to Parker (2017), transportation and the compaction of materials can make it more challenging to separate material by material for recycling, and sorting is not 100% efficient, and reprocessors still need to remove contaminants from recyclables separated by waste managers.

16.2.4 How Collection Interacts with Biological Treatment

The European Union legislation on waste (Landfill Directive, Waste Framework Directive) is committed to promoting the Organic matter application into soil works as a sink of CO2 helping to reduce global warming, and at the same time, reduce the consumption of fertilizer and peat (Boldrin et al. 2009).

Studies have reflected the difference in composition of organic matter collected commingled with MSW and source-separated collection. Huerta-Pujol et al. (2011) compared both types of biowaste and verified differences: in general nutrient contents are higher in separate collection than in mixed waste collection, and heavy metal contents are significantly lower in separate collection, supporting the idea that the heavy metal migration from non-compostable materials to decomposable matrix occurs when the material is in contact (Table 16.1). Those results have also been reached by Malamis et al. (2015, 2017) and Schüch et al. (2016), showing that a separate collection of biowaste is needed to reach high-quality compost. López et al. (2010) have verified that organic matter from the separate collection has a higher moisture content, total organic matter, and organic nitrogen than organic matter from mixed municipal waste and lower pH, electric conductivity, and C/N ratio in the separate collected organic matter. Other parameters have presented no differences, which was the case of soluble inorganic nitrogen, organic resistant nitrogen, stability degree, and the relation between organic and inorganic nitrogen (López et al. 2010). Malamis et al. (2017) have analyzed compost produced at a mechanical-biological treatment (MBT) resulting from the separate collection of

Table 16.1 Nutrient and heavy metal contents of MSW fraction for composting from source separated (SC) and commingled collection (MS), after mechanical processing, expressed in dry matter basis (dmb)

Parameters	SC				MS				
	n	Mean	SE	Median	n	Mean	SE	Median	p
P (%)	21	0.58	0.07	0.45	13	0.44	0.04	0.40	0.1027
K (%)	21	**1.14**	0.08	1.01	13	**0.56**	0.05	0.55	<0.0001
Na (%)	21	0.69	0.06	0.64	13	**0.56**	0.05	0.61	0.1055
Ca (%)	20	3.11	0.42	2.55	13	3.52	0.41	3.84	0.4870
Mg (%)	21	**0.19**	0.02	0.16	13	**0.50**	0.10	0.41	0.0122
Fe (%)	21	**0.10**	0.10	0.03	0.05	**0.33**	0.05	0.35	0.0007
Mn (mg kg^{-1})	21	**32**	5.1	25	13	**73**	9.2	74	0.0010
Zn (mg kg^{-1})	18	**34**	4.4	29	12	**82**	10.2	75	0.0007
Cu (mg kg^{-1})	21	**15**	2.8	11	12	**33**	4.0	34	0.0015
Ni (mg kg^{-1})	21	**2**	0.2	2	13	**10**	3.2	6	0.0224
Cr (mg kg^{-1})	20	**2**	0.4	1	13	**9**	1.9	8	0.0024
Pb (mg kg^{-1})	16	**4**	0.8	3	12	**33**	8.6	22	0.0070
Cd (mg kg^{-1})	20	0.3	0.05	0	13	0.3	0.06	0.30	0.7017

Source: Huerta-Pujol et al. (2011)

n number of samples, SE standard error, p significance of the *t-test*. SC and MS values in bold are significantly different ($p < 0.05$)

biowaste and biowaste from mixed MSW, where the MBT was adjusted to receive and treat separate biowaste with green waste in a dedicated composting tunnel. The results have shown that adequate source separation is a prerequisite for the sound quality production of compost, and units can be adjusted to process source separated biowaste (Malamis et al. 2017).

The type of containing system used can also influence the quality of organic matter. Studies from Huerta-Pujol et al. (2010) have shown that some polyethylene bags used in campaigns for separate collection of organic waste presented a high amount of heavy metals, while compostable bags presented low quantities (Table 16.2). Also, contaminants like glass or plastic occur in the mixed collection of biowaste, and levels of impurity are highly variable depending on the containing device like individual residential receptacles, street bins, and other containers (Cerda et al. 2017).

The separate collection of biowaste with other waste types showed results that could enable their co-collection. Colón et al. (2013) have tested the presence at 3% of compostable diapers in a full-scale composting door-to-door collected biowaste. The results reached show that the presence of compostable diapers does not alter the composting and compost obtained regarding pathogenic content, stability, and elemental composition (including nutrients and heavy metals). Espinosa-Valdemar et al. (2014) have also reached similar results: the quality of compost waste is not affected by the presence of diapers.

Table 16.2 Heavy metal content (dry basis) by each group of bags considered

Bag type	n	Fe (mg kg^{-1})		Zn (mg kg^{-1})		Cu (mg kg^{-1})		Ni (mg kg^{-1})		Cr (mg kg^{-1})		Pb (mg kg^{-1})	
		Mean	SE	Mean	SE	Mean	SE	Mean	SE	Mean	SE	Mean	SE
C	8	13	4	16	14	20	7	1.2	0.1	2	0.5	8	3
OF	8	443	410	34	9	43	27	1.1	0.2	1138	402	4779	1458
R	13	106	57	97	23	87	42	3.7	1.2	87	43	458	254
S	4	44	8	131	45	112	42	8.3	2.4	1.5	0.5	34	11

Source: Huerta-Pujol et al. (2010)
C compostable bags, OF commercially available bags for biowaste source-sorted collection, R rubbish bags, S supermarket bags, SE standard error

There are also studies which tried to relate the type of building and the compost quality. Plahl et al. (2002) have verified that compost resulting from high building density biowaste presented significantly higher contamination than that of compost of the whole city of Vienna. Such finding may dictate a new factor when establishing waste collection routes.

16.2.5 How Collection Interacts with Energy Recovery

The need for a separate collection is vital also for energy recovery, similar to what happens to material recycling and biological recycling. Myrin et al. (2014) showed that the existence of a significant level of biowaste collection (around 65–70%) changes the characteristics of refuse-derived fuel (RDF) from residual waste, by presenting lower values of chemical substances (including heavy metals) compared to RDF made of mixed MSW (without biowaste source separation) and RDF made of wood. Also, Schüch et al. (2016) support the same evidence that biowaste source separation is benefitting energy recovery. The emissions of dioxins are also lower (Myrin et al. 2014). Di Leonardo et al. (2016) also highlighted that the characteristics of RDF are site-specific, depending upon the source separate collection Fig. 16.2. For instance, an effective collection of PVC and thermometers could reduce the amounts of chlorine and mercury at residual waste, improving SRF class related to those substances (Rada and Ragazzi 2014). Table 16.3 presents the composition of RDF with residual waste from a separate collection of biowaste compared with RDF from residual waste without source separation of biowaste and waste wood fuel. The reduced amount of ashes and in most of heavy metals shows the importance of source separation of biowaste to improve quality of refuse-derived fuels. Besides biowaste, other relevant wastes need to be removed to increase the quality of residual waste, namely, PVC materials and thermometers, to reduce Cl and Hg contents (Rada and Ragazzi 2014). Regarding low heating value, the separate collection can be a supportive measure to enhance the generation of solid recovered fuel with high calorific value, especially if the separate collection is highly efficient, making direct classification of residual waste into solid recovered fuel possible (Rada and Ragazzi 2014).

The existence of high levels of separate collection for recycling (material and biological) has shown to be difficult to be compatible with energy recovery actions, leading to overcapacity of energy recovery installations (Cimpan et al. 2015; Rada 2015). Countries like Austria, Germany, Netherlands, Belgium, Sweden, and Denmark have an overcapacity of WtE facilities issues due to the increase of recycling efforts together with the stagnation of waste generation from 2008 (Friege and Fendel 2011; Sora 2013; Vos 2012). Cimpan et al. (2015) have analyzed the overcapacity of WtE from Denmark regarding resources, global warming, and cumulative energy demand (CED) Fig. 16.3. The results reached verified that substantial global warming potential and CED savings could be credited in waste management systems with overcapacity of WtE since the freed WtE capacity was

16.2 The Functioning of the Integrated Waste Collection (IWC)

Table 16.3 Elemental composition of the refuse-derived fuel materials

Parameters	RDF1	RDF2	RW
Ash content (% of sample)	5.5	14.0	5.6
Moisture content (%)	16.8	17.4	26.7
Heating value (MJ/kg)	20.4	21.3	19.5
Cl (%)	0.21	0.64	0.06
S (%)	0.10	0.30	0.09
Cd (mg kg^{-1})	0.79	5.5	0.30
Co (mg kg^{-1})	1.8	53	1.2
Ni (mg kg^{-1})	3.7	10	1.8
Pb (mg kg^{-1})	36	160	48
Cr (mg kg^{-1})	27	100	68
Cu (mg kg^{-1})	120	76	68
Zn (mg kg^{-1})	530	540	360
Na (mg kg^{-1})	1500	1600	430
Fe (mg kg^{-1})	1300	3800	750
Al (mg kg^{-1})	2000	2700	990
Ca (mg kg^{-1})	8800	20,000	3400
Si (mg kg^{-1})	15,700	6500	8700

Source: Myrin et al. (2014)
Note: *RDF1* biowaste waste source separation, *RDF2* biowaste from commingled waste collection, *RW* wood waste

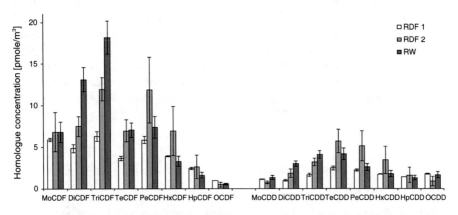

Fig. 16.2 PCDF and PCDD homologue profiles for the combustion of the tested fuels. The bars show duplicate sample averages and the error bars indicate the two measured concentration in each case. (Source: Myrin et al. (2014))

used to treat imported waste, which would end at the landfill although the savings are not verified for WtE plans with efficiencies near the threshold defined in Waste Framework Directive (0.6 for installations in operation and permitted before first January 2009 and 0.65 for installations permitted after 31 December 2008) (Figs. 16.2 and 16.3).

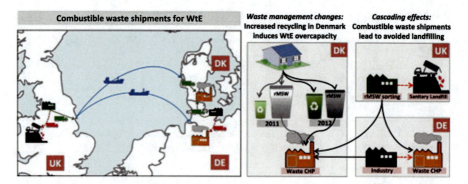

Fig. 16.3 Overcapacity of WtE installations and consecutive shipment of waste to fulfill freed capacity. (Source: Cimpan et al. (2015))

16.2.6 How Collection Interacts with Disposal

Similar to what happens with the separate collection and WtE plants, also a great separate collection for recycling will promote a reduction of waste going to landfill, regarding amount but also regarding composition, leading to different impacts. A reduced amount of waste going to landfill will avoid the use of land for this purpose, increasing the lifetime and reducing the need for more landfill infrastructures in the future (regarding WtE, this is an overcapacity of the landfill). Regarding composition, reducing the amount of biodegradable waste and moisture will reduce the release of landfill gas (reach in methane, a GHG) and the generation of leachate, which has several environmental risks due to its pollution potential. Leachate generated at a landfill requires collection and treatment with considerable technical expenditure over many decades, not excluding the damages on the landfills due to waste settling, leading to damage on surface lines, gas extraction, and leachate collection systems (Stegmann 2005).

16.3 Sustainability in Integrated Waste Collection

Integrated waste collection sustainability has to consider the different goals of each pillar of sustainability, also having in mind not to jeopardize future generations but also the interactions with the other waste management operations/processes described. The three pillars of sustainability – environment, economy, and society – are in constant conflict. The goal for reaching a low-cost collection may not be in accordance with the needs of the population and regulation that requires an adequate collection of waste to avoid public health issues; and a low-cost collection may not allow a source separation collection to collect valuable materials, and efficient collection vehicles in terms of air emissions (by hypothesis). The goal to reach a low environmental impact of waste collection requires the less collection possible with high-efficiency vehicles but may require a considerable capital and operation costs that citizens may not be able to afford them. Also, the multiplicity of collection streams needed to collect high-value materials for recycling (if recycling includes it in the analysis) can also be costly for citizens. Society may have the intention to recycle and separate waste at source, but may not be available to participate in a way that is cost-effective for waste collection managers. However, society may not be keen on having to separate waste at homes without any visible benefit, which may be a problem when the intention is to promote a circular economy. The complexity is huge, and system analysis is needed to help reach IWC sustainably.

16.4 Final Remarks

In the light of sustainability, waste hierarchy, and circular economy, the several goals of an IWC can be conflicting. The conflict starts right on waste generation issue: if the intention is to reduce waste by waste prevention, which is in accordance to a circular economy and waste hierarchy principle, the same is not in the mind of waste collection company, which needs waste for financial support of its activity. Even that at a long term, the benefit to the waste manager is not to have to invest in waste infrastructure; in the short term, the break in waste generation will increase collection and treatment costs (Zacho and Mosgaard 2016). Integrating collection in the rest of the waste management system is one way to support the diversion of the collection of waste financially, for example, by taxing the disposal of waste at the landfill, and part of the tax is used to finance the loss of money due to prevention. Another way to compensate a waste prevention revenue break can be reached by changing their core business and including in their system the preparation for reuse, which is not the case nowadays. In any case, the responsibility for waste needs to be improved by a better sharing of responsibilities by all stakeholders, including producers who need to develop only products that could be reused or recycled, users and consumers that do know what to do with their products and where is the final destination, and waste managers, who need to change to promote waste prevention, increase reuse, and preparation for reuse and recycle.

There is no single and sustainable way to collect waste in an integrated perspective. Collecting by material type instead of providing the collection service by the type of producer (domestic, industrial, commercial), having in mind the technology available in the waste management system to prepare for recycling and in contamination risk during collection, is the new paradigm to have in mind when defining IWC. Also, IWC needs to focus on waste prevention and reuse options, which can also bring new business sectors for waste managers. System analysis tools and engineered solutions are needed to help reach IWC solutions. The complexity is considerable, and environmental impacts (direct and indirect) from waste collection need to be taken into account when deciding the IWC to implement. According to Reinhart et al. (2016), the falling sales revenue due to China's declining economy, depressed oil prices, and the firm US dollar make recycling cost recovery a challenge.

Although the proposed change in the collection paradigm, there are also underlying issues that waste collection intended to solve but which are increasing environmental impacts, like in the case of littering and marine litter issue. Reinhart et al. (2016) highlighted that uncollected waste remains a significant issue due to the quantities of litter ending up in marine environments and gathering in remote gyres, polluting the oceans and seas.

References

Anderson L (1968) A mathematical model for the optimization of a waste management system. SERL Report. Sanitary Engineering Research Laboratory, University of California, Berkeley, CA

Anderson LE, Nigam AK (1967) A mathematical model for the optimization of a waste management system, ORC 67–25, Operations Research Center, University of California, Berkeley, CA

Bel G, Gradus R (2016) Effects of unit-based pricing on household waste collection demand: a meta-regression analysis. Resour Energy Econ 44:169–182

Boldrin A, Andersen JK, Moller J, Christensen TH, Favoino E (2009) Composting and compost utilization: accounting of greenhouse gases and global warming contributions. Waste Manag Resour 27:800–812

Cerda A, Artola A, Font X, Barrena R, Gea T, Sánchez A (2017) Composting of food wastes: status and challenges. Bioresour Technol 248:57–67

Chang NB, Pires A, Martinho G (2011) Empowering systems analysis for solid waste management: challenges, trends, and perspectives. Critical Reviews in Environmental Science and Technology 41:1449–1530

Cimpan C, Rothmann M, Hamelin L, Wenzel H (2015) Towards increased recycling of household waste: documenting cascading effects and material efficiency of commingled recyclables and biowaste collection. J Environ Manag 157:69–83

Colón J, Mestre-Monserrate M, Puig-Ventosa I, Sánchez A (2013) Performance of compostable baby used diapers in the composting process with the organic fraction of municipal solid waste. Waste Manag 33:1097–1103

Dahlén L, Lagerkvist A (2010) Pay as you throw: strengths and weaknesses of weight-based billing in household waste collection systems in Sweden. Waste Manag 30:23–31

Di Leonardo MC, Franzese M, Costa G, Gavasci R, Lombardi F (2016) The application of SRF vs RDF classification and specifications to the material flows of two mechanical-biological treatment plants of Rome: comparison and implications. Waste Manag 47:195–205

Espinosa-Valdemar RM, Sotelo-Navarro PX, Quecholac-Piña X, García-Rivera MA, Beltrán-Villavicencio M, Ojeda-Benítez S, Vázquez-Morillas A (2014) Biological recycling of used baby diapers in a small-scale composting system. Resour Conserv Recycl 87:153–157

European Parliament, Council (2008) Directive 2008/98/EC of the European Parliament and of the Council of 19 November 2008 on waste and repealing certain directives. Off J Eur Union L312:3–30

Fitzgerald GC, Krones JS, Themelis NJ (2012) Greenhouse gas impact of dual stream and single stream collection and separation of recyclables. Resour Conserv Recycl 69:50–56

Friege H, Fendel A (2011) Competition of different methods for recovering energy from waste. Waste Manag Res 29:S30–S38

Glavič P, Lukman R (2007) Review of sustainability terms and their definitions. J Clean Prod 15:1875–1885

Huerta-Pujol O, Soliva M, Giró F, López M (2010) Heavy metal content in rubbish bags used for separate collection of biowaste. Waste Manag 30:1450–1456

References

Huerta-Pujol O, Gallart M, Soliva M, Martínez-Farré FX, López M (2011) Effect of collection system on mineral content of biowaste. Resour Conserv Recycl 55:1095–1099

López M, Soliva M, Martínez-Farré FX, Fernández M, HuertaPujol O (2010) Evaluation of MSW organic fraction for composting: separate collection or mechanical sorting. Resour Conserv Recycl 54:222–228

Malamis D, Moustakas K, Bourka A, Valta K, Papadaskalopoulou C, Panaretou V, Skiadi O, Sotiropoulos A (2015) Compositional analysis of biowaste from study sites in Greek municipalities. Waste Biomass Valoriz 6:637–646

Malamis D, Bourka A, Stamatopoulou E, Moustakas K, Skiadi O, Loizidou M (2017) Study and assessment of segregated biowaste composting: the case study of Attica municipalities. J Environ Manag 203:664–669

Martinho G, Gomes A, Santos P, Ramos M, Cardoso J, Silveira A, Pires A (2017) A case study of packaging waste collection systems in Portugal – Part I: Performance and operation analysis. Waste Management, 61, 96–107

McDougall F, White P, Franke M, Hindle P (2001) Integrated solid waste management: A life cycle inventory. Blackwell Science Ltd, Oxford

Mirata M (2004) Experiences from early stages of a national industrial symbiosis programme in the UK: determinants and coordination challenges. J Clean Prod 12:967–983

Mulvaney D, Robbins P (eds) (2011) Green politics – an A-to-Z guide. SAGE Publications, Los Angeles

Myrin ES, Persson PE, Jansson S (2014) The influence of food waste on dioxin formation during incineration of refuse-derived fuels. Fuel 132:165–169

Naveh Z (2000) What is holistic landscape ecology? A conceptual introduction. Landsc Urban Plan 50:7–26

Parajuly K, Wenzel H (2017) Potential for circular economy in household WEEE management. J Clean Prod 151:272–285

Parker R (2017) Essentials of environmental science, 2nd edn. National Agriculture Institute, Inc, Rupert

Pires A, Martinho G, Chang NB (2011) Solid waste management in European countries: A review of systems analysis techniques. Journal of Environmental Management, 92:1033–1050

Plahl F, Rogalski W, Gilnreiner G, Erhart E (2002) Vienna's biowaste compost – quality development and effects of input materials. Waste Manag Res 20(2):127–133

Rada EC (2015) Selective collection and waste–to–energy strategies: which interaction? In: Syngellakis S (ed) Waste to energy. WIT Press, Southampton, pp 241–250

Rada EC, Ragazzi M (2014) Selective collection as a pretreatment for indirect solid recovered fuel generation. Waste Manag 34:291–297

Reichenbach J (2008) Status and prospects of pay–as–you–throw in Europe – a review of pilot research and implementation studies. Waste Manag 28:2809–2814

Reinhart D, Bolyard SC, Berge N (2016) Grand challenges – management of municipal solid waste. Waste Manag 49:1–2

Schüch A, Morscheck G, Lemke A, Nelles M (2016) Bio–waste recycling in Germany – further challenges. Procedia Environ Sci 35:308–318

Sing J, Ordoñez I (2016) Resource recovery from post–consumer waste: important lessons for the upcoming circular economy. J Clean Prod 134:342–353

Sora MJ (2013) Incineration overcapacity and waste shipping in Europe: the end of the proximity principle. http://www.no-burn.org/wp-content/uploads/Overcapacity_report_2013.pdf. Accessed 15 Jan 2018

Stahel WR (2016) The circular economy. Nature 531:435–438

Stegmann R (2005) Mechanical biological pretreatment of municipal solid waste. In: Abstracts of the Sardinia 2005, tenth international waste management and landfill symposium, S. Margherita di Pula, Cagliari, 3–7 Oct 2005 van Ewijk S, Stegemann JA (2016) Limitations of the waste hierarchy for achieving absolute reductions in material throughput. J Clean Prod 132:122–128

Villanueva A, Eder P (2014) End–of–waste criteria for waste plastic for conversion –technical report. Publications Office of the European Union, Luxembourg

Vos M (2012) Recycling benefits from combustible waste imports. Wasteforum November: 1–2

Zacho KO, Mosgaard MA (2016) Understanding the role of waste prevention in local waste management: a literature review. Waste Manag Res 34:980–994

Chapter 17
Trend Analysis on Sustainable Waste Collection

Abstract To make waste collection desirable, waste collection paradigm needs to change in such way that could promote waste hierarchy and circular economy. Reverse logistics, crowd logistics, physical Internet, and Freight on Transit are innovative solutions applied to waste, sub-products, and end-of-life products capable of bringing the change to the waste collection sector and promote sustainable waste collection behind the frontiers of the system itself.

Keywords Crowd logistics · Freight on transit · IoT · Physical Internet · Reverse logistics · Waste collection · Waste transportation

17.1 Reverse Logistics

According to Thürer et al. (2016), logistics means the management of the flow of things between the point of origin and the point of consumption – customer or consumer; reverse logistics can be the management of things from the point of consumption back to the origin. The logistics is also identified as the forward supply chain, where it combines the processes to fulfill costumer's request and includes all entities involved such as suppliers, manufacturers, transporters, warehouses, retailers, and customers themselves (Chopra and Meindl 2010). In this respect, the reverse logistics divides into two parts: the collection part and the recycling/remanufacturing part (Thürer et al. 2016).

Reverse logistics, together with the formal supply chain, form a closed-loop supply chain, like the one presented in Fig. 17.1. Solid lines show the forward chain; dashes show the reverse logistics. In this respect, a closed-loop supply chain (CLSP) is a management philosophy that design, controls, and operates a system to maximize value creation over the entire life cycle of a product with the dynamic recovery of value from different types and volumes of returns over time (Guide and van Wassenhove 2009).

In both RL and CLSC, end-of-life products play a vital role, being collected from customers, being repaired, disassembled, remanufactured, recycled, and disposed of

---

© Springer International Publishing AG, part of Springer Nature 2019
A. Pires et al., *Sustainable Solid Waste Collection and Management*,
https://doi.org/10.1007/978-3-319-93200-2_17

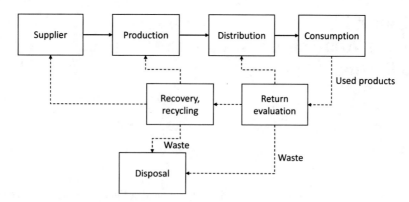

Fig. 17.1 A generic form of forward/reverse logistics. (Source: Adapted from Tonanont 2009 and Pumpinyo and Nitivattananon, 2014)

(Govindan and Soleimani 2017; Soleimani and Kannan 2015). The key to their work is the customer, and many papers have studied the importance of consumer's return practices (Choi et al. 2013).

Reverse logistics is divided into the following activities (Ayvaz and Görener 2015): collecting or collection, inspection and sorting, and reprocessing or direct recovery. Collecting or collection is the first and very important stage, referent to the activities involved with the collection of the used products or materials from the consumers/users for processing and transporting to the place for processing (e.g., rework and remanufacturing) (Fleischmann 2000; Srivastava and Srivastava 2006). Inspection and sorting include the activities to assess the proper reuse, being such activities also named as local eliminating, conducted at the collecting point (Ayvaz and Görener 2015). It is in this step that inspection, sorting, testing, disassembly, separating into small parts, and storage operations are performed (Fleischmann 2000). Direct recovery intends to send the product again for selling locals without any treatment, and reprocessing includes the conversion of used product into a usable product again.

Reverse logistics and circular economy have several similar characteristics, in particular, the repair, refurbishing, remanufacturing, recycling, and disposal cycles (EMF 2016; Thierry 1995). Although some of the CE principles (e.g., leakage minimization) are considerably broader than the reverse logistics activities (Ripanti et al. 2015), several reverse logistics models have been proposed by EMF (2016) to promote circular economy, presented in Figs. 17.2, 17.3, and 17.4. In all reverse logistics schemes proposed by EMF (2016), they are divided into *front-end*, *engine*, and *back-end*. Front-end includes reverse logistics itself, their processes, and network; engine refers to the recovery of returned products, like methodologies for evaluation, and inventory control; back-end is the remarketing of recovered products in secondary markets.

In recovery service provider collection scheme (Fig. 17.2), the type of products/ waste which fits in this reverse logistic model are mass production products, with

17.1 Reverse Logistics

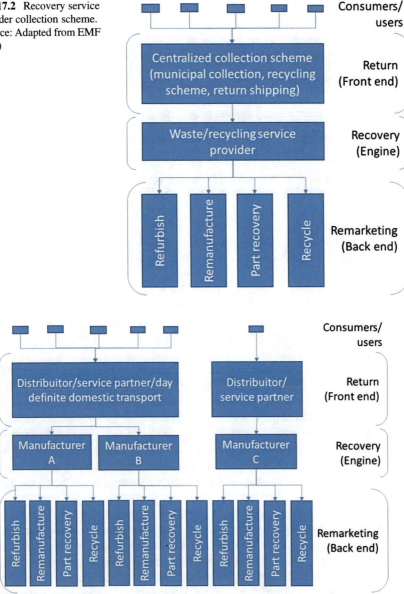

Fig. 17.2 Recovery service provider collection scheme. (Source: Adapted from EMF 2016)

Fig. 17.3 Service parts logistics. (Source: Adapted from EMF 2016)

distribution via retail networks, with comparably low residual value at the of product life cycle, like tires, shipping pallets, and consumer electronics (EMF 2016). The type of products/waste should be under extended producer responsibility registration. This type of reverse logistics could be a centralized collection scheme handled

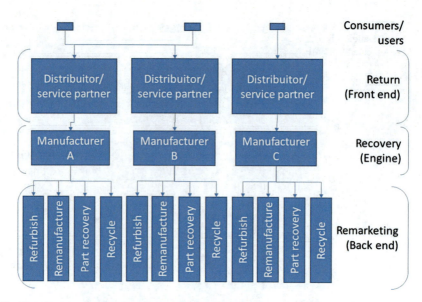

Fig. 17.4 Reverse logistics for advanced industrial products. (Source: Adapted from EMF 2016)

by a recovery service provider, for a cost-effective collection, and could use underused forward logistics network capacities to enable recovery of returned goods and waste.

In Fig. 17.3 the service parts logistics is presented, which is devoted to products with comparably higher residual value with moderate expected return rates, like machinery and automotive parts (EMF 2016). The reverse logistics has to combine the return of disposed of parts with the supply of new or refurbished parts to ensure an adequate replacement of service parts (EMF 2016). The distributor/service partner should collect parts from different customers or collect through customer-dedicated transports and deliver them to manufacturers, which would have the market of the reuse, refurbish, remanufacture, or recycling of collected parts (EMF 2016).

In Fig. 17.4 is the reverse logistics model for advanced industrial products. The service provider has to implement a direct or trusted collection since the products in this reverse logistics model are complex and have high residual value with relatively low return volumes, such as medical equipment and information and communication technology (ICT) (EMF 2016). The collection has to be made to preserve and maximize the product return value, and collection should combine with the replacement of the asset by a new or refurbished product (EMF 2016).

More than defining waste models, to make reverse logistics/CLSC possible, it has to be legally enforced. A country where reverse logistics is regulated is in Brazil. With the Law n.12.305/2010, the National Policy on Solid Waste formulated the requirements for industries to incorporate reverse logistics of post-consumption for the industrial sectors: tires, lubricants, batteries, bulbs, electronics pesticides, and

17.1 Reverse Logistics

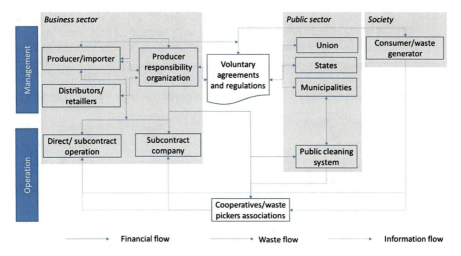

Fig. 17.5 Reverse logistics system proposed by Brazilian law. (Source: Adapted from Couto and Lange 2017)

Table 17.1 Reverse logistics implementation barriers

Management	Financial	Policy	Infrastructure
Understanding significance of RL	Lack of initial capital	Lack of enforceable laws and directives on take-back of end-of-life products	Lack of sufficient in-house facilities (storage equipment and vehicles)
Lack of trained personnel	Lack of funds for training	Lack of government-supported economic policies	Lack of a system to monitor returns
Management commitment	Lack of funds for storage and handling	RL not considered as a critical aspect of competitive performance	Lack of coordination with third-party logistics (3PL) providers
Lack of expert at management level	Lack of fund for return monitoring systems	Customers not informed of taking back	
Lack of shared understanding of best practices		Lack of public awareness of environmental protection	
Lack of waste management practices		Don't have green design implementation for end-of-life products	
		No practice in place for recycling	

Source: Abdulrahman et al. (2014)

packaging made primarily of plastic, metal, and glass (Silva et al. 2013; Jabbour et al. 2014). The idealized system of reverse logistics in national law is in Fig. 17.5.

Despite the evolution in reverse logistics, there are barriers to its implementation. According to the review on reverse logistics from Abdulrahman et al. (2014), there are four types of implementation barriers in the Chinese context, but they could also be valid in other countries: management, financial, policy, and infrastructure (Table 17.1).

17.2 Crowd Logistics

Due to the current empty passenger vehicles, an opportunity to transport anything else exists. Crowd logistics is the delivery operation that is carried out by using passenger's excess capacity on journeys that are occurring, with economic, social, and environmental benefits (Rai et al. 2017). Crowd logistics is also known as crowdshipping, crowdsourced delivery, cargo-hitching, or collaborative logistics, having the term originated from crowdsourcing, which covers both the words "crowd" or a mass of unorganized individuals and "outsourcing" or shift of processes, functions, and duties to third parties (Howe 2006; Mehmann et al. 2015; Sampaio et al. 2017; Verdonck et al. 2013). Rai et al. (2017) revised the definition of crowd logistics as:

> An information connectivity enabled marketplace concept that matches supply and demand for logistics services with an undefined and external crowd that has the free capacity with regards to time and/or space, participates on a voluntary basis and is compensated accordingly.

Continuing from Rai et al. (2017), the crowd logistics is divided into several stakeholders: the receivers which receive the goods, commissioners that send the goods, logistics service providers which execute the transportation service in a traditional way (to ensure timely completing of tasks in the case that crowd may not ensure it), platform providers which match all parties (being the marketplace where the transportation service is requested and paid), and the crowd which will implement the logistics in several different ways – subcontractors, professional drivers, and casual drivers.

Crowd logistics is a developing concept, being tested in different fields. In the case of a waste collection, few cases exist. The H2020 CityLab project includes a task conducted in Rome, where an innovative system for integrating crowd logistics in the urban area is implemented with the aim of improving clean waste collection (CityLab 2017a). The postal operator makes the crowd logistics, already delivering mail/parcels all around Rome. At the same time, the postal operator also collects recyclable materials (plastic caps) during the same transportation route, and using electric vehicles, avoiding dedicated collection trips. The results reached for the collection of plastic caps have shown to be technically feasible and environmentally sustainable but not financially profitable (CityLab 2017b).

Chen et al. (2017) proposed the delivery of returned goods from e-commerce (clothes, e-waste) from final consumption points (shops) back to retailers by taxis, which are in constant mobility. The strategy has several advantages regarding sustainability (Chen et al. 2017):

- Generates extra environmental footprints by avoiding collection trucks.
- The crowdsourcing itself avoids the dedicated collection costs.
- The solution also helps taxi drivers to earn a little extra money and to consumers an alternative and facilitated way to return products as well.

17.3 Physical Internet

The concept of Physical Internet urges from the need to try to solve and organize the supply chain network using modular containers and a systemic open infrastructure (Montreil 2011). The last step of the supply chain is in the urban logistics, which involves several stakeholders: the carriers, the citizens, the public administration, public transport operators, and retailers, at the least (Biggi and Tretola 2015). According to Ballot et al. (2014) and Mervis (2014), Physical Internet intends to transport containers encapsulating freights via interconnected transport services – transportation relay – instead of transporting freights by a single long-haul truck from origin to destination. The intention is to innovate in the way how logistics operators provide logistics; changing from a hub and spoke approach, with few distribution centers and many connections between them to a network system based on multiple nodes, interconnected with multimodal links (Biggi and Tretola 2015). The modular tracked containers (π-containers), needed in Physical Internet to decrease the number of vehicles, are standardized containers easy to handle, store, transport, interlock, load, and unload, allowing optimal filling and increasing automation of goods movement (Biggi and Tretola 2015).

How can waste collection implement Physical Internet? There are two ways how Physical Internet can be applied. The first is related with underground tube network (or vacuum system) applied for waste collection, where the network would act in a similar way to a packet switched telecommunication network, transporting one kind of waste at a time (Biggi and Tretola 2015):

Waste collection points are placed outdoors or indoors and are accessible 24 h a day. Waste inlets store refuse temporarily until the next emptying cycle. The pipelines transport the refuse into containers at the waste station. When full, the containers are sent away for further processing using, for example, the city's existing underground railway network.

The implementation of this view of Physical Internet can develop the concept of pneumatic tube in the light of the conceptual high-speed transportation system envisioned by entrepreneur Elon Musk, incorporating reduced-pressure tubes in which pressurized capsules ride on a cushion of air, driven by a combination of linear induction motors and air compressors, the Hyperloop system (Biggi and

Tretola 2015). A second way that waste collection can use Physical Internet concept is in light of the circular economy. The case study presented by Chen et al. (2017) on crowdsourcing is based on a taxi, where the Physical Internet concept inspired routing strategies. In this respect, a specific waste (probably clean waste like e-waste) is forwarded to a peering network, which can hand it over to another peer to reach the destination; then, a second parcel for the same destination may (or may not) use a different route, depending on the options and load at the particular time (den Boer et al. 2017). A simple example from den Boer et al. (2017) helps to clarify it:

A simple example is buying a secondhand washing machine online from someone living on the other side of the country. A local specialist collects the machine. It gets picked up by one of the "peer" networks of the specialist who does regular backhaul runs between the area of origin and destination. Another specialist delivers the machine locally. The difference from current practice is that the route and the (intermediate) carriers are not predetermined but selected as the parcel goes. The parcel selects the route.

Both ways can be implemented and change the way how waste and end-of-life products can reenter again in the economy, although both are quite different regarding concept, costs, and possible public acceptance.

17.4   Freight on Transit

According to Cochrane et al. (2017), Freight on Transit (FOT) refers to:

An operational strategy where public transit vehicles and/or infrastructure are used to move freight.

FOT can also be designed as urban rail transit-based city logistics system (Liu et al. 2008), light rail freight (Arvidsson 2010), cargo tram operations (Regué and Bristow 2013), shared track operations (Resor 2003), and mixed goods service (Sivakumaran et al. 2010).

There are few experiences of FOT applied to waste collection. The one most known is the Zurich cargo tram. In this collection system, the Cargo-Tram service offers a car-free alternative to disposing of bulky waste since 2003 and since 2006 collect electrical and electronic waste, being called E-Tram, being free of charge and makes about 18 round trips every month (Eltis 2015). The creation of this service with a 165 km tram network was related to needing to offer an attractive and inexpensive collection system and, at the same time, avoiding the private car trips to disposal yards and illegal dumping of those waste making 43% of household waste to remain in the recycling process (Eltis 2015). Neuhold (2005) verified that collection waste by road transport requires 5020 km covered by lorries and three times longer to move in the city at peak hours, corresponding to 960 running-time hours, hence 37,500 l of diesel per year. With the E-Tram, the release of harmful substances (including greenhouse gases emissions) can be avoided and the illegal dumping can be reduced (from 3000 l at year 1997 to 1200 l at year 2004) Neuhold (2005). Other evidence of success reflects on the amount of bulky waste collected in 2003, 380 t (twice the tonnage of uncontrolled waste left on pavements before the implementation of the Cargo-Tram) and 785 t in 2004, with a cost of €3200 per ride (including operative and back-office costs) (INTERREG IVC 2011).

17.5 Final Remarks

This chapter intended to bring a bold paradigm breaking vision for the future of collection and transport of waste. Pioneer studies are being tested to assess the novelties and new concepts for waste collection that could be more sustainable. Reverse logistics, crowd logistics, Physical Internet, and Freight on Transit are possible solutions probably more applicable to clean waste, from source separated collection, and not for biodegradable waste or packaging waste.

Through this chapter, a small step has been given. A lot more is needed to shape the vision of waste collection and management of the future and, much more importantly, to give it flesh through real initiatives and projects. Multidisciplinary collaboration between academia, waste, and waste equipment industry and government also would allow the regulation of new paradigms to occur. In China, the lack of enforceable legislation on take-back end-of-life products and lack of market instruments are the most prominent policy barriers to RL implementation (Abdulrahman et al. 2014). Probably the other technologies would suffer from the same need of legislation to push them forward.

Implementing the mentioned techniques/methodologies requires financial resources, namely, initial capital and funds to implement and monitor the collection systems, and requires experts from different fields at business management levels to implement the break technologies.

References

Abdulrahman MD, Gunasekaran A, Subramanian N (2014) Critical barriers in implementing reverse logistics in the Chinese manufacturing sectors. Int J Prod Econ 147:460–471

Arvidsson N (2010) New perspectives on sustainable urban freight distribution: a potential zero emissions concept using electric cars on trams. In: Abstracts of the 12th world conference on transport research, Lisbon, 11–15 June 2010

Ayvaz B, Görener A (2015) Reverse logistics in the electronics waste industry. In: Akkucuk U (ed) Handbook of research on waste management techniques for sustainability. IGI Global, Hershey, pp 155–171

Ballot E, Montreuil B, Meller R (2014) The physical internet: the network of logistics networks. La Documentation, Paris

Biggi D, Tretola G (2015) Addressing urban congestion with modular logistics and collaborative networks. In: Abstracts of the URban freight and BEhavior change (URBE) conference, Roma Tre University, Rome, 1–2 Oct 2015 den Boer E, Kok R, van Amstel WP, Quak H, Wagter H (2017) Outlook City Logistics 2017. Report for Topsector Logistiek

Chen C, Pan S, Wang Z, Zhong RY (2017) Using taxis to collect citywide E-commerce reverse flows: a crowdsourcing solution. Int J Prod Res 55:1833–1844

Choi TM, Li Y, Xu L (2013) Channel leadership, performance and coordination in closed loop supply chains. Int J Prod Econ 146:371–380

Chopra S, Meindl P (2010) Supply chain management: strategy, planning and operation, 4th edn. Pearson Prentice Hall, Englewood Cliffs

CityLab (2017a) ICT platform for direct and reverse logistics integration. CityLab http://www.citylab-project.eu/presentations/171020_Rome/Meware.pdf. Accessed 28 Dec 2017

CityLab (2017b) Living lab updates. CityLab http://www.citylab-project.eu/documents/Living_labs_300917.pdf. Accessed 28 Dec 2017

Cochrane K, Saxe S, Roorda MJ, Shalaby A (2017) Moving freight on public transit: best practices, challenges, and opportunities. Int J Sustain Transp 11:120–132

Couto MCL, Lange LC (2017) Análise dos sistemas de logística reversa no Brasil. Eng Sanit Ambiental. https://doi.org/10.1590/s1413-41522017149403

Eltis (2015) Cargo-Tram and E-Tram bulky and electric waste collection by tram in Zurich (Switzerland). Eltis. http://www.eltis.org/discover/case-studies/cargo-tram-and-e-tram-bulky-and-electric-waste-collection-tram-zurich. Accessed 28 Dec 2017

Ellen MacArthur Foundation (EMF) (2016) Waste not, Want not – Capturing the value of the circular economy through reverse logistics: An introduction to the reverser logistics maturity model

Fleischmann M (2000) Quantitative models for reverse logistics. Dissertation, Erasmus University

Govindan K, Soleimani H (2017) A review of reverse logistics and closed-loop supply chains: a journal of cleaner production focus. J Clean Prod 142:371–384

Guide VDR, van Wassenhove LN (2009) OR Forum – the evolution of closed-loop supply chain research. Oper Res 57:10–18

Howe J (2006) The rise of crowdsourcing. Wired. https://www.wired.com/2006/06/crowds/. Accessed 28 Dec 2017

INTERREG IVC (2011) Cargotram, Zurich. INTERREG. http://www.interreg4c.eu/good-practices/practice-details/index-practice=115-cargotram-zurich&.html. Accessed 28 Dec 2017

Jabbour ABLS, Jabbour CJC, Sarkis J, Govindan K (2014) Brazil's new national policy on solid waste: challenges and opportunities. Clean Techn Environ Policy 16:7–9

Liu Y, Ketai H, Liu J, Xu Y (2008) Analysis of the concept of urban rail transit based city logistics system. In: Abstracts of the international conference on smart manufacturing application. IEEE, Gyeonggi-do, South Korea. pp 288–292

Mehmann J, Frehe V, Teuteberg F (2015) Crowd logistics – a literature review and maturity model. In: Abstracts of the Hamburg international conference of logistics, Hamburg, Germany, Sept 2015

Mervis J (2014) The information highway gets physical: the Physical Internet would move goods the way its name sake moves data. Science 344:1057–1196

Montreil B (2011) Towards a Physical Internet: meeting the global logistics sustainability grand challenge. Interuniversity Research Centre on Enterprise Networks, Logistics and Transportation

Neuhold G (2005) Cargo Tram Zurich - The environmental savings of using other modes. http://www.bestufs.net/conferences/2005-06-23_amsterdam.html. Accessed 30 Apr 2018

Pumpinyo S, Nitivattananon V (2014) Investigation of Barriers and Factors Affecting the Reverse Logistics of Waste Management Practice: A Case Study in Thailand. Sustainability 6:7048–7062

Rai HB, Verlinde S, Merckx J, Macharis C (2017) Crowd logistics: an opportunity for more sustainable urban freight transport? Eur Transp Res Rev 9. https://doi.org/10.1007/s12544-017-0256-6

References

Regué R, Bristow AL (2013) Appraising freight tram schemes: a case study of Barcelona. Eur J Transp Infrastruct Res 13:56–78

Resor RR (2003) Catalog of 'common use' rail corridors. Publication DOT/FRA/ORD-03-16 US Department of Transportation, Washington, DC

Ripanti EF, Tjahojono B, Fan I (2015) Circular economy in reverse logistics: Relationships and potential applications in product remanufacturing. In: Abstracts of the 20th logistics research network (LRN) conference, Derby, UK, Sep 2015

Sampaio A, Savelsbergh M, Veelenturf L, van Woensel T (2017) Crowd-based city logistics. Optimization online. Available via http://www.optimization-online.org/DB_FILE/2017/11/6346.pdf Accessed 28 Dec 2017

Silva DAL, Renó GWS, Sevegnani G, Sevegnani TB, Truzzi OMS (2013) Comparison of disposable and returnable packaging a case study of reverse logistics in Brazil. J Clean Prod 47:377–387

Sivakumaran K, Lu XY, Hanson M (2010) Use of passenger rail infrastructure for goods movement: Economic feasibility study from the California's San Francisco Bay area. Transp Res Rec 2162:44–52

Soleimani H, Kannan G (2015) A hybrid particle swarm optimization and genetic algorithm for closed-loop supply chain network design in large-scale networks. Appl Math Model 39:3990–4012

Srivastava SK, Srivastava RK (2006) Managing product returns for reverse logistics. Int J Phys Distrib Logist Manag 36:524–546

Thierry M, Salomon MJN, van Nunen J, van Wassenhove L (1995) Strategic issues in product recovery management. California Management Review 27:114–136

Thürer M, Pan YH, Qu T Luo H, Li CD, Huang GQ (2016) Internet of Things (IoT) driven kanban system for reverse logistics: solid waste collection. J Intell Manuf. https://doi.org/10.1007/s10845-016-1278-y

Tonanont A (2009) Performance evaluation in reverse logistics with data envelopment analysis. Dissertation, University of Texas at Arlington

Verdonck L, Caris A, Ramaekers K, Janssens GK (2013) Collaborative logistics from the perspective of road transportation companies. Transp Rev 33:700–719

Chapter 18
Technical Barriers and Socioeconomic Challenges

Abstract Solid waste collection and its management are an increasing issue in cities in the future, where the urban population is still growing and consumption patterns are changing all the time. Strategies implemented at developed countries to solve waste collection and management issues are different from developing countries, and those differences need to be detailed to better understand the barriers to the implementation of sustainable waste collection and management in both socioeconomic backgrounds. In this chapter analysis of challenges to sustainable waste collection and management in developing and developed countries will be conducted to better improve the application of this new paradigm in the next decades.

Keywords Developing countries · Developed countries · Environmental impacts · Industrial countries · Informal sector · Separate collection · Social impacts · Waste management system

18.1 Developed Countries

The challenges and barriers of developed countries are mostly related to the new paradigms on waste management but also related to the basis of waste management. Here are addressed, in particular, each of the considered barriers and challenges.

18.1.1 Advancements in Environmental Informatics

Lu et al. (2013) have realized that, although environmental informatics have been capable of helping on planning and operation of the waste management system, synergies were missing concerning public involvement. Public participation in decision-making processes on waste management is needed and few cases of environmental informatics application in the field exist. The project Urban Wins, financed by H2020, was devoted to involving stakeholders on live meeting (agoras)

© Springer International Publishing AG, part of Springer Nature 2019

A. Pires et al., *Sustainable Solid Waste Collection and Management*,

https://doi.org/10.1007/978-3-319-93200-2_18

335

and online agoras to decide a strategic plan for prevention and management of solid waste. An online platform exists for stakeholders from each European city involved in a project to participate in the planning stage. Besides this platform, other emerging technologies such as automatic knowledge acquisition, machine learning, auto-reasoning, and semantic web could be employed to promote environmental informatics in waste management (Lu et al. 2013). Looking at the waste collection, there is a constraint in the use of environmental informatics to manage waste collection systems with different types of constraints for sustainability in both developed and developing countries (Lu et al. 2017).

18.1.2 Advancements in Information and Communication Technology

The potentialities that information and communication technology (ICT) can bring to the way how waste is managed can open new worlds concerning the existent paradigm. ICTs automated several elements of waste management, namely data gathering, identification, communication, storage, and analysis, comparatively to regular computing (Hannan et al. 2015). Hannan et al. (2015) classify in four categories the application of ICT in waste management, spatial technologies, identification technologies, data acquisition technologies, and data communication technologies, which are detailed in Table 18.1.

Looking at Table 18.1, there are around 20 different applications of ICT in waste management. Although they can be condensed in two drivers: to control, monitor, and optimize waste processes related to sorting, treatment, recycling, and waste recovery and to monitor and track waste collection and transportation, including import and export of waste.

ICT can be used to control waste treatment processes by allowing the connection of equipment and devices into the Internet. Besides data collection and processing and helping in the optimization of equipment, another perspective can be brought by the machine learning implementation to devices developing the same functions. Machine learning allows computers to learn from experiences based on data, information, and training from experts (Kurniawan, 2018). Machine learning algorithms can be Naive Bayes, support vector machines, and nearest neighbor has been used in automated sorting equipment (Gundupalli et al. 2017).

In the case of monitoring and tracking waste transportation, including import and export of waste, ICT through RFID and GPS with GSM are techniques applied to monitor waste transportation. In the study of Lee et al. (2018), the wireless GPS location trackers were used to track computer monitors and printers from the USA, being located in Asia. RFID has been used to ensure that waste of electric and electronic equipment is correctly processed and the reuse of WEEE components is controlled and quantified in the European project WEEE TRACE (European Commission, 2013). Common to the applications mentioned so far is the use of data

18.1 Developed Countries

Table 18.1 Examples of ICT application in solid waste management

ICT classification	ICT subclasses	Applications
Spatial technologies	GIS	Site selection, planning, management, estimation, optimization
	GPS	Route and collection optimization, vehicle tracking, planning, scheduling, billing
	Remote sensing	Site selection, environmental impacts assessment, features monitoring
Identification technologies	Barcode	Intelligent recycling, waste disposal, reduce landfill space, risk management
	RFID	Bin and driver tracking, optimization, sorting, and recycling
Data acquisition technologies	Sensors	Sorting: optimization, moisture, energy, and odor measurement, scheduling
	Imaging	Waste sorting, route and collection optimization, monitoring
Data communication technologies	GSM/ GPRS	Long-range communication
	Zigbee	Short-range communication
	Wi-Fi	Short-range communication
	Bluetooth	Short-range communication
	VHFR	Long-range communication

Note: GIS, geographic information system; GPS, global positioning systems; RFID, radio frequency identification; GSM, Global System for Mobile Communications, GPRS, General Packet Radio Service; VHFR, very high-frequency radio
Source: Hannan et al. (2015)

communication technologies. GSM, Wi-Fi, and the other techniques are being used in the cases of machine learning and in tracking WEEE.

18.1.3 Waste Infrastructure Synergies

The need to save natural habitats in nonurban areas and make more sustainable use of valuables at urban areas poses the need to promote waste infrastructure sharing and infrastructure network, allowing a cut on infrastructure costs and sustainable development of the city and the waste sector (Neuman, 2011). Sharing waste infrastructures can be a type of industrial symbiosis, where interfirm cooperation occurs by the exchange of by-products (waste) to be used as raw materials (Posch et al. 2011). By changing their wastes (material or energy types), the impact on the natural environment is minimized, and financial savings with the acquisition of raw materials will be reached but also creates knowledge and innovation capacity at the region (Posch et al. 2011; Mirata and Emtairah, 2005).

Sharing waste infrastructure is usually associated with the most high-risk waste treatment technology and expensive – waste treatment facilities for nuclear waste (IAEA 2011) and incineration plants (Lee and Hur 2017). In the case of nuclear waste facilities, the sharing of those infrastructures is justified by the difficulty in countries with low generation of nuclear waste which are not capable of supporting such high-cost infrastructure, being safer to send their nuclear waste and spent nuclear fuel to other countries (IAEA 2011). In the case of incineration, specifically the case of Seoul, the drivers for the sharing of the 4 incineration plants were the direct landfill ban of food waste and extended producer responsibility implemented that have reduced significantly waste generation, reducing the number of incinerators from 11 to 4, processing waste for 25 districts (from initial 4 districts) (Lee and Hur 2017).

18.1.4 Reaching All-in-One: Citizens Satisfied and Participative, Cost Affordable, and Low Environmental Impact of the Waste Management System

The trigger to promote integrated solid waste management in developed countries have been public health, environment, resource scarcity, climate change public awareness, and participation (Marshall and Farahbakhsh 2013). In addition to those goals, the regulations that impose constraints to the activities (motivated by the goals mentioned), the limits in terms of costs and financing of the integrated waste management, and limitations of the public participation and interest on the subject make the task of reaching an integrated and sustainable management of waste almost unreachable.

European regulation is quite demanding regarding what has to be done to manage waste sustainably. The legislation has a framework concerning the management of waste, legal requirements concerning waste management operations (treatment, recovery, and disposal), legislation specific for waste streams, how reporting on waste management should be made to allow benchmarking between countries, and evaluate each country concerning the implementation of the regulation. At a national level, countries need to drive such requirements to national law and national practice, instructing municipalities to manage waste according to the European Union's terms by elaborating waste management plans and putting them into practice. When waste management is put into practice, other concerns reach waste managers, complicating, even more, the challenge. Aesthetic values in the urban environment, historical areas, dense construction, and at the same time, budget constraints pose challenges to the type of waste collection system to be implemented cost-effectively, providing the service at a low cost (Schott et al. 2013).

The diversity of goals and interests makes it difficult for a city to adapt their strategy to comply with legal requirements concerning waste management. The waste management itself creates competition regarding the final destination of waste streams. In Malmö, Sweden, the increased competition for organic waste with high biogas potential in open market makes difficult the planning of organic waste management, particularly plans related with technology investment or the case of recyclables with high energy content where the investment in incineration plants and district heating systems might decrease the interest in their recycling (Schott et al. 2013).

No formula could help waste managers in how they could integrate all the factors mentioned so far to help them in managing waste. The application of systems analysis can help to bring the systematic and holistic view needed to look at the entire picture, but involving the citizens and stakeholders related to waste management and product consumption is to be included in any planning exercise.

18.2 Developing Countries

The barriers to integrate solid waste management at developing countries are urbanization, inequality, economic growth, cultural and socioeconomic aspects, policy, governance and institutional issues, and international influences, limiting not only the applicability of approaches that were successful at developed countries (Marshall and Farahbakhsh 2013), but also due to the increasing generation of waste, the burden posed on the municipal budget increased the difficulty in making an integrated solid waste management a reality. In fact, for developing countries, solid waste management represents 20–50% of their available budget on waste management, and less than 30% of urban waste is collected and disposed of appropriately in African developing countries (World Bank 2017; Ziraba et al. 2016). This economic challenge is a broader view of the problems of implemented integrated solid waste management in developing countries. Other difficulties are addressed in this section.

18.2.1 Basics on Waste Collection System Are Still in Development

For the generality of developing countries, to implement financially sustainable solid waste management services is a significant challenge, which is the case of Bahir Dar, Ethiopia (Lohri et al. 2014). In this case, a public-private partnership (PPP) was created, to provide waste collection and management. With the PPP, a significant increase in waste management costs related to waste transportation occurred, where only one revenue exists – the waste collection fee - highlighting the need to enhance cost efficiency and balance the cost-revenues towards cost recovery (Lohri et al. 2014). To improve financial sustainability from the waste services, Lohri et al. (2014) proposed four options: improve fee collection efficiency by linking fees to

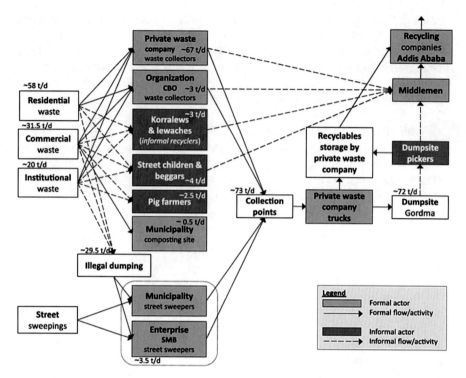

Fig. 18.1 Process flow diagram of municipal solid waste in Bahir Dar, Ethiopia. (Source: Adapted from Lohri et al. 2014)

water consumption, increase revenues by selling organic waste recycling products, diversifying revenues, and cost reduction. To a better understanding of the challenge, Fig. 18.1 shows the waste flow diagram in Bahir Dar (Lohri et al. 2014).

18.2.2 The Conversion of Informal Sector into Formal Waste Management Sector

In developing countries, the informal sector, represented by the denominations of "scavengers," "waste pickers," or "rag pickers," is responsible for the existence of recycling of household recyclable waste, due to the fact that formal waste management organization strategies for source separation of waste is inexistent or is still in its infancy (Fei et al. 2016; Medina 2000). The informal sector is composed by small-scale, low-technology, low-paid, mostly unregulated and unregistered, without tax payment (Harriss-White 2010; Wilson et al. 2006). Such sector is also notorious by the serious social problems, like poor work and living conditions, older people without any other work opportunities, child labor, school absences,

18.2 Developing Countries

incomplete education of adults, and willing workers incapable to find another job (Brix-Asala et al. 2016; Medina 2000; Wilson et al. 2006). Waste workers from informal sector expose themselves to a significant health risk than from formal sector, and the society (citizens, authorities) are often hostile to informal waste workers (Medina 2000).

In countries like China and Brazil, both formal and informal sectors coexist, and there are cases where the domain of informal sector is notable, for example, the case of e-waste in China (Gu et al. 2016). The recognition of the work made by informal sector has made authorities to include them in the implementation of modernization of waste management. The most common approaches to formalize informal waste sector are (Aparcana 2017):

- Associations or cooperatives composed of informal sector workers.
- Community-based organization (CBO) or micro- and small enterprises (MSE) formed by the informal sector workers.
- Contract the individual workers by the formal waste sector.

In the case of associations of cooperatives, Brazil is a compelling case. In this model, the associations or cooperatives, composed of municipalities and waste workers, establish contracts or cooperation agreements to conduct collection services and source separation of recyclables (Aparcana 2017). Those cooperatives or associations receive waste and separate and resell the materials, and the municipality may cover the costs of the machinery and depots, like what happens in Belo Horizonte City (Colombjin and Morbidini 2017). Although the effort in organizing the cooperatives and associations, many Brazilian waste pickers choose not to join a cooperative, with the justification of being free to make their market choices (Colombjin and Morbidini 2017).

CBO or MSE establishes when the individuals get organized to provide waste services, representing an opportunity to define a source of income and create a clean environment in their community (Aparcana 2017; Muller and Hoffman 2001). MSW involves groups up to 10 (micro) or 20 (small) persons, operating with low investment to provide primarily collection and processing of recyclables to intermediate or final products, depending on market demands (Ahmed and Ali 2004; Wilson et al. 2006). Although not being the most common conversion of informal to formal sector, the case study of Lima, Peru, reported by Baud et al. (2001) on the creation of 140 MSE by the nongovernmental organization IPES, is mainly run by women from poor communities.

The contract of individual workers by the formal sector to conduct collection or at recycling facilities can be seen as a way of formal sector in helping "poor people" (Aparcana 2017). One example of this practice is in Ghana, where a plastic waste recycling company employs 400 plastic waste pickers because they have difficulty in getting plastic waste for its activity (Oteng-Ababio 2012).

According to Aparcana (2017), these strategies to convert informal to formal waste activity requires, most of all, the inclusion of measures country specific at economic, regulatory, and institutional levels to increase the levels of success. However, another type of formalization of informal sector needs to be developed and tested, in the same time that their acknowledge on the subject is brought into the waste management system, giving a considerable contribution in the implementation of integrated solid waste management in developing countries.

18.2.3 The Importation of Hazardous Waste and Trade of Hazardous Waste

Hazardous waste is present in all activities, including in domestic activities. Hazardous waste in the domestic sector at developing countries includes not only remnants of chemical fertilizers, pesticides, herbicides, and cleaning products but also other wastes which are corrosive, toxic, and reactive (Al-Khatib et al. 2015; Hennebert et al. 2013). Measures to control and environmentally manage such waste have been made by developed countries, but for developing countries that is not the case (Otoniel et al. 2008). Most of the time, citizens are not aware of the hazardousness of MSW, and not separate collection of those wastes occur, what can be a problem if scavengers exist in open dumps, where people may cut themselves in sharp objects, burns from toxics, and diseases from genotoxics (Al-Khatib et al. 2015).

The most known municipal hazardous waste is e-waste (or waste of electrical and electronic equipment (WEEE)). Due to the presence of materials with economic value, they are managed to obtain such resources. In developing countries such as China, India, Pakistan, and Nigeria, standard practice is to smolder plastic off cables to recover copper; precious metals such as gold, platinum, palladium, and silver from printed circuit boards are leached to acid baths, and the used acid is released to the ground or nearby stream, intoxicating population around those areas (Leung et al. 2006; Robinson 2009; Sthiannopkao and Wong 2013).

The travel of e-waste from developed countries into developing countries is still a subject of matter. Although the existence of import bans in developing countries, the low costs of recycling, and disposal of e-waste at developing countries comparatively to developed countries seem to ensure the flow of e-waste from developed to developing countries by labeling them as used goods, which are not covered by the Basel Convention or to declare it as for reexport (Sthiannopkao and Wong 2013). However, e-waste is not only resulting from importation but it is also resulting from national consumption of electronics (Nnorom and Osibanjo 2008). Heeks et al. (2015) resumed the significant issues that e-waste management has on developing countries: the risk of mismanagement of e-waste is higher than in developed countries, there is no formal recycling, and recycling legislation is weak or absent.

The problem of e-waste management present in MSW is realized by waste management official, expert, practitioners, and academics at developing countries, concerning the lack of waste separation of e-waste, the safe disposal of e-waste, the lack of proper recycling, the need of technical support, and the increase of population' awareness on the toxic nature of e-waste (Ikhlayel 2018). The integrated management of e-waste for developing countries proposed by Ikhlayel (2018) is based on the MSW management system, where a source separation of e-waste is promoted by drop-off containers being sent to material recovery facilities from the MSW system to be dismantled by waste pickers (allowing the inclusion of informal sector) and the flows of the waste products are the same as the MSW management, allowing the regulation of e-waste management at first and, second, to use the existing technology for MSW to treat e-waste, due to the limitations of technology present in those countries.

18.2.4 Public Health Related to Mismanagement of Waste and Its Dependents

Dumpsites in developing countries are in reality, being considered the most polluted places on Earth by the International Solid Waste Association (ISWA), receiving roughly 40% of the world's waste, serving 3–4 billion people, counting with 8–10% of the global anthropogenic greenhouse gas emissions by 2025 (Mavropoulos et al. 2016). Generically, the global problems of significant dumpsites are the marine litter, the contribution to climate change, the way how waste trafficking occurs, and the amount of population that is served by those dumpsites (Fig. 18.2). Although the global problems exist, direct issues are resulting from dumpsites. One of the most visible ones is the fatalities occurring due to mechanical instability (Laner et al. 2009). From December 2015 to June 2016, 750 deaths of scavengers and their relatives related to poor management in dumpsites were recorded (Mavropoulos et al. 2016).

Besides the fatalities related to scavengers, including children, in dumpsites, there are also other health effects resulting from this waste deposits. All people living at dumpsites are exposed to environmental pollutants via accidental ingestion and inhalation of contaminated dust, and dermal absorption and citizens living away from the dumpsite are exposed to pollutants by air, water, and food (Tongesayi et al. 2018). In Dhapa, India, near the closed dump, diseases such as hepatitis, diarrhea, vomiting, abdominal pain, dysentery, and others are occurring in residents due to groundwater contamination (Maiti et al. 2016). In Okhla landfill site open dumpsite in Delhi, the high prevalence of respiratory symptoms, inflammation of the airways, decrement of the lung function, increased susceptibility to tissue damage, and cardiovascular diseases, and many other health problems were detected at disposal workers by Ray et al. (2005, 2009).

Although noncontrolled dumpsites are addressed mainly by developing countries, situations of illegal dumping of waste also occur at developed countries, with implications for human health, at least. The most known is the "Triangle of Death" in Italy, related to the three vertex cities of Naples, Marigliano, and Nola, one of the most polluted areas in Europe due to the illegal management of hazardous waste by criminal organizations (Basile et al. 2017). Diffused and severe contamination by

Fig. 18.2 Dumpsite as a global challenge. (Source: Adapted from Mavropoulos et al. 2016)

hazardous substances has been linked to brain, liver, lung, stomach, and intestine cancer (Senior and Mazza 2004).

18.2.5 Social Apathy for Participation

The way of living in developing countries does not allow that environmental problems resulting from waste management deserve particular attention. Social apathy concerning waste problems and recycling is a reality exposed by Guerrero et al. (2013), Buenrostro and Bocco (2003), and Moghadam et al. (2009) in developing countries. In particular, do Carmo and de Oliveira (2010), the recyclers at Rio de Janeiro, Brazil, look at their work as dirty and nasty, making them feel apathetic and hoping that it is a temporary work. The social apathy on waste can also be seen when there is no waste collection promoted by authorities, leading to aesthetic and environmental deterioration (Shekdar 2009). Vij (2012) highlighted the social apathy concerning waste management by Indian citizens, where they throw away their waste outside bins, alongside the road, even with effort in keeping streets clean, citizens do not respect the effort of authorities.

Waste management requires the collaboration of all stakeholders to increase awareness near waste producers including citizens, industries, commerce and services, business and public entities, where non-governmental agencies can have a role in exposing the importance of a correct waste management. The training of municipalities' staff can increase the importance of waste management employment (Singh 2010), reducing the apathy vies of waste and all related to waste.

18.3 Final Remarks

The shift of the paradigm in developing countries is needed to reduce and eliminate impacts from mismanagement of waste. Waste governance, evocated by Rodic (2015) and reinforced by Mavropoulos et al. (2016), is the key solution to involve all responsible stakeholders in the cleanliness of the open spaces of the cities as well as in the protection of natural resources, pushing forward good waste governance onto the realms of production and consumption. Waste governance requires the identification of waste stakeholders and their involvement, where the national authorities define the goals or driving forces: public health, environment, resource recovery, and waste prevention. Planning of waste strategies through the design and implementation in time of policy instruments to respond to the goals such as direct regulation, economic instruments, and social instruments (to raise awareness on waste near stakeholders) is needed to ensure an integrated solid waste management. The integrated solid waste management should start by closing dumpsites and construct sanitary landfills, preceded by the regulations to define the requirements for construction of such infrastructures. All the other waste management options, like incineration, recycling, reuse, and prevention, require time to be defined and, possibility, the financial support of the private sector.

References

Ahmed AS, Ali M (2004) Partnerships for solid waste management in developing countries: linking theories to realities. Habitat Int 28:467–479

Al-Khatib IA, Kontogianni S, Nabaa HA, Alshami N, Al-Sari' MI (2015) Public perception of hazardousness caused by current trends of municipal solid waste management. Waste Manag 36:323–330

Aparcana S (2017) Approaches to formalization of the informal waste sector into municipal solid waste management systems in low- and middle-income countries: review of barriers and success factors. Waste Manag 61:593–607

Basile A, Loppi S, Piscopo M, Paoli L, Vannini A, Monaci F, Sorbo S, Lentini M, Esposito S (2017) The biological response chain to pollution: a case study from the "Italian Triangle of Death" assessed with the liverwort *Lunularia cruciate*. Environ Sci Pollut Res 24:26185–26193

Baud I, Grafakos S, Hordijk M, Post J (2001) Quality of life and alliances in solid waste management: contributions to urban sustainable development. Cities 18:3–12

Brix-Asala C, Hahn R, Seuring S (2016) Reverse logistics and informal valorisation at the Base of the Pyramid: a case study on sustainability synergies and trade-offs. Eur Manag J 34:414–423

Buenrostro O, Bocco G (2003) Solid waste management in municipalities in Mexico: goals and perspectives. Resour Conserv Recycl 39:251–263

do Carmo MS, de Oliveira JAP (2010) The Semantics of Garbage and the organization of the recyclers: implementation challenges for establishing recycling cooperatives in the city of Rio de Janeiro, Brazil. Resour Conserv Recycl 54:1261–1268

Colombjin F, Morbidini M (2017) Pros and cons of the formation of waste-pickers' cooperatives: a comparison between Brazil and Indonesia. Decision 44:91–101

European Commission (2013) Keeping track of WEEE Eco-Innovation at the heart of European policies https://www.ec.europa.eu/environment/ecoap/about-eco-innovation/good-practices/eu/20130326-keeping-track-of-weee_en. Accessed 10 Mar 2018

Fei F, Qu L, Wen Z, Xue Y, Zhang H (2016) How to integrate the informal recycling system into municipal solid waste management in developing countries: based on a China's case in Suzhou urban area. Resour Conserv Recycl 110:74–86

Gu Y, Wu Y, Xu M, Wang H, Zuo T (2016) The stability and profitability of the informal WEEE collector in developing countries: a case study of China. Resour Conserv Recycl 107:18–26

Guerrero LA, Maas G, Hogland W (2013) Solid waste management challenges for cities in developing countries. Waste Manag 33:220–232

Gundupalli S, Hait S, Thakur A (2017) A review on automated sorting of source-separated municipal solid waste for recycling. Waste Manag 60:56–74

Hannan MA, Al Mamun MA, Hussain A, Basri H (2015) A review on technologies and their usage in solid waste monitoring and management systems: issues and challenges. Waste Manag 43:509–523

Harriss-White B (2010) Work and wellbeing in informal economies: the regulative roles of institutions of identity and the state. World Dev 38:170–183

Heeks R, Subramanian L, Jones C (2015) Understanding e-waste management in developing countries: strategies, determinants, and policy implications in the Indian ICT sector. Inf Technol Dev 21:653–667

Hennebert P, Papin A, Padox JM, Hasebrouck B (2013) The evaluation of an analytical protocol for the determination of substances in waste for hazard classification. Waste Manag 33:1577–1588

Ikhlayel M (2018) An integrated approach to establish e-waste management systems for developing countries. J Clean Prod 170:119–130

International Atomic Energy Agency (IAEA) (2011) Viability of sharing facilities for the disposition of spent fuel and nuclear waste: an assessment of recent proposals. IAEA, Vienna

Kurniawan A (2018) Learning AWS IoT. Packt Publishing, Birmingham

Laner D, Fellner J, Brunner PH (2009) Flooding of municipal solid waste landfills — an environmental hazard? Sci Total Environ 407:3674–3680

Lee D, Offenhuber D, Duarte F, Biderman A, Ratti C (2018) Monitour: tracking global routes of electronic waste. Waste Manag 72:362–370

Lee S, Hur YG (2017) Waste heat recovery project Seoul Solution http://susa.or.kr/sites/default/files/resources/%ED%99%98%EA%B2%BD_10_Waste%20Heat%20Recovery%20Project.pdf. Accessed 10 Mar 2018

Leung A, Cai ZW, Wong MH (2006) Environmental contamination from electronic waste recycling at Guiyu, southeast China. J Mater Cycles Waste Manag 8:21–33

Lohri CR, Camenzind EJ, Zurbrügg C (2014) Financial sustainability in municipal solid waste management – costs and revenues in Bahir Dar, Ethiopia. Waste Manag 34:542–552

Lu J-W, Chang NB, Liao L (2013) Environmental informatics for solid and hazardous waste management: advances, challenges, and perspectives. Crit Rev Environ Sci Technol 43 (15):1557–1656

Lu JW, Chang NB, Liao L, Liao MY (2017) Smart and green urban solid waste collection systems: advances, challenges, and perspectives. IEEE Syst J 11:2804–2817

References

Maiti SK, De S, Hazra T, Debsarkar A, Dutta A (2016) Characterization of leachate and its impact on surface and groundwater quality of a closed dumpsite - a case study at Dhapa, Kolkata, India. Procedia Environ Sci 35:391–399

Marshall RE, Farahbakhsh K (2013) Systems approaches to integrated solid waste management in developing countries. Waste Manag 33:988–1003

Mavropoulos A, Cohen P, Greedy D, Plimakis S, Marinheiro L, Law J, Loureiro A (2016) A roadmap for closing waste dumpsites: the world's most polluted places ISWA report http://www.iswa.org/fileadmin/galleries/About%20ISWA/ISWA_Roadmap_Report.pdf. Accessed 12 Mar 2018

Medina M (2000) Scavenger cooperatives in Asia and Latin America. Resour Conserv Recycl 31:51–69

Mirata M, Emtairah T (2005) Industrial symbiosis networks and the contribution to environmental innovation: the case of the Landskrona industrial symbiosis programme. J Clean Prod 13:993–1002

Moghadam MRA, Mokhtarani N, Mokhtarani B (2009) Municipal solid waste management in Rasht City Iran. J Waste Manag 29:485–489

Muller M, Hoffman L (2001) Community partnerships in integrated sustainable waste management – tools for decision-makers experiences from the urban waste expertise programme (1995-2001). WASTE, Gouda

Neuman M (2011) Infrastructure planning for sustainable cities. Geogr Helv 66:100–107

Nnorom IC, Osibanjo O (2008) Overview of electronic waste (e-waste) management practices and legislations, and their poor applications in the developing countries. Resour Conserv Recycl 52:843–858

Oteng-Ababio M (2012) The role of the informal sector in solid waste management in the GAMA; Ghana: challenges and opportunities. J Econ Social Geogr 103:412–425

Otoniel BD, Márquez-Benavides L, Pinette GF (2008) Consumption patterns and household hazardous solid waste generation in an urban settlement in México. Waste Manag 28:S2–S6

Posch A, Agarwal A, Strachan P (2011) Editorial: managing industrial symbiosis (IS) networks. Bus Strateg Environ 20:421–427

Ray MR, Roychoudhury S, Mukherjee S, Siddique S, Banerjee M, Akolkar AB, Sengupta B, Lahiri T (2009) Airway inflammation and upregulation of beta2 Mac-1 integrin expression on circulating leukocytes of female ragpickers in India. J Occup Health 51:232–238

Ray MR, Roychoudhury S, Mukherjee G, Roy S, Lahiri T (2005) Respiratory and general health impairments of workers employed in a municipal solid waste disposal at an open landfill site in Delhi. Int J Hyg Environ Health 208:255–262

Robinson BH (2009) E-waste: an assessment of global production and environmental impacts. Sci Total Environ 408:183–191

Rodic L (2015) Waste governance. In: Wilson DC (ed) Global waste management outlook. United National Environment Programme, Osaka, pp 125–202

Schott ABS, Aspegren H, Bissmont M, JLC J (2013) Modern solid waste management in practice - the city of Malmö experience, Briefs in applied sciences and technology. Springer, Greven, Deutschland

Senior K, Mazza A (2004) Italian "triangle of death" linked to waste crisis. Lancet Oncol 5:525–527

Shekdar A (2009) Sustainable solid waste management: an integrated approach for Asian countries. Waste Manag 29:1438–1448

Singh J (2010) Different methods in solid waste management. In: Singh J, Ramanathan A (eds) Solid waste management: present and future challenges. IK International Publishing House, New Delhi, pp 84–89

Sthiannopkao S, Wong M (2013) Handling e-waste in developed and developing countries: initiatives, practices, and consequences. Sci Total Environ 463-464:1147–1153

Tongesayi T, Kugara J, Tongesayi S (2018) Waste dumpsites and public health: a case for lead exposure in Zimbabwe and potential global implications. Environ Geochem Health 40:375–381

Vij D (2012) Urbanization and solid waste management in India: present practices and future challenges. Procedia - Soc Behav Sci 37:437–447

Wilson DC, Velis C, Cheeseman C (2006) Role of informal sector recycling in waste management in developing countries. Habitat Int 30:797–808

World Bank (2017) Solid waste management. Urban development. http://www.worldbank.org/en/topic/urbandevelopment/brief/solid-waste-management. Accessed 10 Mar 2018

Ziraba AK, Haregu TN, Mberu B (2016) A review and framework for understanding the potential impact of poor solid waste management on health in developing countries. Arch Public Health 74:55

Chapter 19
Future Perspectives

Abstract The ultimate challenge that solid waste management needs to comply in the future is the 2030 Agenda for Sustainable Development. The adoption of the agenda by the United Nations in 2016 intends to transform the way how the world has evolved. Seventeen goals decomposed in 169 targets are listed to ensure a better world for all, with poverty alleviation, human well-being, and environmental protection in new and more integrated ways (Dermatas, Waste Manag Res 35: 453–455, 2017; UN, Sustainable development goals – 17 goals to transform our world. United Nations. Available via http://www.un.org/sustainabledevelopment/sustainable-development-goals/, 2016). Discussing the way how solid waste management can contribute to improving sustainable development and how solid waste management can be shaped to answer to this call was the intention of this section.

Keywords Consumption · Economic growth · Environmental impacts · Production · Recycling · Resource use · SDGs · SWM

19.1 The Goals of the 2030 Agenda for Sustainable Development

Solid waste management (SWM) activity is direct and indirectly related to the sustainable development. The need to ensure proper waste collection to ensure public health is vital to any society to grow in sustainable development; indirectly, all the economic activity related to waste transactions can also create a business that will improve the life of the ones living directly from waste but also the society in question. Direct and indirect influence of SWM in sustainable development reaches ecology, economy, and society as a whole. From the 17 Sustainable Development Goals (SDG) of the 2030 Agenda for Sustainable Development adopted by the 193 UN countries (UN 2016), at least 13 SDG and their pertinent targets are related to SWM. More influences of SWM in SDG exist and will be addressed in the next sections.

© Springer International Publishing AG, part of Springer Nature 2019 349
A. Pires et al., *Sustainable Solid Waste Collection and Management*,
https://doi.org/10.1007/978-3-319-93200-2_19

19.1.1 SGD 1 "No Poverty" and SWM

The "no poverty" goal intends to eradicate extreme poverty for everyone, ensuring that men and women have equal rights to economic resources. In developing countries, 20% of people still live on less than \$1.90 a day and millions a little more than that (UN 2016). The sub-goals on this SDG potentially related with SWM are (UN 2016) as follows:

- By 2030, eradicate extreme poverty for all people everywhere, currently measured as people living on less than \$1.25 a day.
- By 2030, reduce at least by half the proportion of men, women, and children of all ages living in poverty in all its dimensions according to national definitions.
- By 2030, ensure that all men and women, in particular, the poor and the vulnerable, have equal rights to economic resources, as well as access to basic services, ownership, and control over land and other forms of property, inheritance, natural resources, appropriate new technology, and financial services, including microfinance.
- Ensure significant mobilization of resources from a variety of sources, including through enhanced development cooperation, to provide adequate and predictable means for developing countries, in particular, least developed countries, to implement programs and policies to end poverty in all its dimensions.

Solid waste management can contribute to the "no poverty" goal by the creation of green jobs relayed with reuse and recycling activities. European waste legislation is devoted to promote recycling, and if all recycling targets were reached as well as landfill diversion rates, around 180,000 direct jobs could be created by 2030, in addition to the estimated 400,000 jobs (European Commission 2014a, b).

In another perspective, the waste sector can contribute to jobs, although people are not keen on working in the sector because jobs are not well paid. In collection and transport, manual sorting and other manual occupations occur in least-quality jobs, when higher-quality jobs are associated with more specialized activities and high-technology treatment, like incineration plant managers (European Commission 2001).

19.1.2 SDG 2: Zero Hunger

This SDG intends to reform how agriculture, forestry, and fishery are made to ensure an adequate nutritious food for all and generate income while not damaging the environment. Combat degradation derived from food and agriculture system is needed to ensure nourishing of the future two billion people expected by 2050 (UN 2016). The sub-goals related to SWM are (UN 2016) as follows:

- By 2030, ensure sustainable food production systems and implement resilient agricultural practices that increase productivity and production; that help maintain ecosystems; that strengthen capacity for adaptation to climate change, extreme weather, drought, flooding, and other disasters; and that progressively improve land and soil quality.

In a sustainable food production, agriculture waste is a resource to explore, instead of just being material without interest to be disposed. Depending on the technique/technology, waste from food production can be used as a source of energy, as a natural fertilizer, or as a feed for animals. SWM can help to reach this goal by treating agriculture waste through technology which is used for biological waste: anaerobic digestion and composting. Although incineration could be also used, it is not capable to provide the land and soil quality as is capable anaerobic digestion and composting. The production of compost/digestate has to be performed having in mind the nutrients to be provided to the soil and plants, to generate the best and safe food products. However, today's agriculture has chemicals that can be a risk to public health and to the environment, and contamination with heavy metals also occurs during waste collection. New ways to produce food products and to collect waste are needed to comply with this sub-goal.

19.1.3 SDG 3: Good Health and Well-being

The good health and well-being goal intends to ensure healthy lives; increase life expectancy, preventing deaths of children under 5 years and reducing some of the common killers not only related with child and maternal mortality but also malaria and water-borne diseases; and reduce illness from hazardous chemicals (UN 2016). Sub-goals for SDG 3 related to SWM are the following:

- By 2030, substantially reduce the number of deaths and illnesses from hazardous chemicals and air, water, and soil pollution and contamination.

Solid waste management can bring a significant help to reach such SDG. The implementation of waste collection is a measure to ensure public health. When waste is not collected, is often dumped in waterways, or is burned in open air, it causes pollution and contamination; if waste is not collected, it can also clog the drains causing floods, contributing to water-borne diseases and malaria (Rodić and Wilson 2017). In developing countries, waste collection is determinant to reach this goal, but also for developed countries, the need to reduce the impacts from waste is also notorious, mainly due to the climate change and the hot waves that are expected to occur. High temperature will make biowaste to degrade faster, increasing the need to collect more often or to find new ways to deal with such waste.

19.1.4 SDG 6: Clean Water and Sanitation

The goal intends to ensure the availability of fresh, unpolluted water to everyone. The sub-goals are as follows:

- By 2030, improve water quality by reducing pollution, eliminating dumping and minimizing release of hazardous chemicals and materials, halving the proportion of untreated wastewater, and substantially increasing recycling and safe reuse globally.
- By 2030, substantially increase water-use efficiency across all sectors and ensure sustainable withdrawals and supply of freshwater to address water scarcity and substantially reduce the number of people suffering from water scarcity.

To do so, the control and minimization of sources of water pollution, like uncontrolled dumping, open burning, littering, as well promote the environmentally sound management of all wastes, particularly hazardous wastes (Rodić and Wilson 2017). Waste technologies should also be rethought to be efficient regarding water consumption and use. Just, For example, one mechanical-biological treatment unit in Portugal was conceived to use rainwater in the process, avoiding the consumption of clean water for less noble uses. In the case of plastic recycling, there is much consumption of water to wash waste plastics, which also needs to be reconsidered to improve such recycling technology.

19.1.5 SDG 7: Affordable and Clean Energy

The need to encounter sustainable energy and make energy accessible to all is the goal. The sub-goals are (UN 2016) as follows:

- By 2030, increase substantially the share of renewable energy in the global energy mix.
- By 2030, enhance international cooperation to facilitate access to clean energy research and technology, including renewable energy, energy efficiency, and advanced and cleaner fossil-fuel technology, and promote investment in energy infrastructure and clean energy technology.
- By 2030, expand infrastructure and upgrade technology for supplying modern and sustainable energy services for all in developing countries, in particular, least developed countries, Small Island Developing States, and landlocked developing countries, by their respective programs of support.

In this respect, the biodegradable fraction of waste can be a source of renewable, easy access, energy. The biodegradable waste fraction can be used to produce methane and consequently as fuel. Solutions of energy made of renewable waste sources are the production of combustible gas (like methane), or direct energy recovery in waste-to-energy plants or other processes that could recover the energy content of waste in their processes, replacing other fuels. The energy recovery of waste is also dependent on the technical efficiency, to ensure that energy is recovered and not just burned.

19.1.6 SDG 8: Decent Work and Economic Growth

This goal intends to ensure decent work opportunities and sustainable economic growth without harming the environment. The sub-goals are (UN 2016) as follows:

19.1 The Goals of the 2030 Agenda for Sustainable Development

- Sustain per capita economic growth by national circumstances and, in particular, at least 7 percent gross domestic product growth per annum in the least developed countries.
- Achieve higher levels of economic productivity through diversification, technological upgrading, and innovation, including through a focus on high value-added and labor-intensive sectors.
- Promote development-oriented policies that support productive activities, decent job creation, entrepreneurship, creativity, and innovation and encourage the formalization and growth of micro-, small-, and medium-sized enterprises, including thorough access to financial services.
- Improve progressively, through 2030, global resource efficiency in consumption and production and endeavor to decouple economic growth from environmental degradation, by the 10-year framework of programs on sustainable consumption and production, with developed countries taking the lead.
- By 2030, achieve full and productive employment and decent work for all women and men, including for young people and persons with disabilities, and equal pay for work of equal value.
- Protect labor rights and promote safe and secure working environments for all workers, including migrant workers, in particular, women migrants, and those in precarious employment.
- By 2030, devise and implement policies to promote sustainable tourism that creates jobs and promotes local culture and products.

SWM services, in developing countries, are provided by individuals and small/microenterprises who need support to improve livelihoods and contribute to reaching both SDG 8 and 1 (Rodić and Wilson 2017). Like already mentioned in SDG 1 "no poverty," the green economy recycling based can be a possible direction to promote economic growth. However, Gregson et al. (2016) have highlighted the risks of the European Green economy not to bring decent jobs. In Europe, recycling is being fostered to stop the waste of being sent to developing countries, to reuse them to decouple economic growth from demands of global resources and to create recycling jobs (Gregson et al. 2016). However, waste management works are not as clean and green as authorities and politicians intend to veneer them. According to Gregson et al. (2016):

> waste work globally has long been seen as a means to marking ethnic and racial, as well as gendered, differences.

Such marking has led waste and recycling green job types connected as "dirty work," performed by migrant labor, itinerant, or no EU-nationals to low cost (Gregson et al. 2016). Bonatti (2018) have verified the precarious immigration and employment statuses push migrant workers to take the problematic recycling work for their employers. The sub-goals on SDG 8 need to be brought into the waste management sector to improve the conditions of workers regarding health, work conditions, remuneration, and rights, which were questioned by Gregson et al. (2016).

Regarding sustainable tourism, the way how touristic areas and events are managing waste needs to be reviewed in the vision of sustainability. Also, events need to be though to promote waste prevention and zero-waste measures, which will reduce littering and environmental impacts related to the misleading management of waste in those events (Cierjacks et al. 2012; Martinho et al. 2018).

19.1.7 SGD 9: Industry, Innovation, and Infrastructure

The investment in infrastructure is needed to empower countries and reach a more sustainable world by technology/industry investment in an environmentally sound way. SGD sub-goals related to SWM are the following:

- Increase the access of small-scale industrial and other enterprises, in particular in developing countries, to financial services, including affordable credit, and their integration into value chains and markets.
- By 2030, upgrade infrastructure and retrofit industries to make them sustainable, with increased resource-use efficiency and greater adoption of clean and environmentally sound technologies and industrial processes, with all countries taking action by their respective capabilities.
- By 2030, upgrade infrastructure and retrofit industries to make them sustainable, with increased resource-use efficiency and greater adoption of clean and environmentally sound technologies and industrial processes, with all countries taking action by their respective capabilities.

SDG 9 can incorporate SWM sector in developing countries. Converting the informal recycling sector in developing countries to become formal will provide new access to funding and improvements in their operation. Improving the waste management infrastructures can also improve environmental quality. Passing from dumpsites, where scavengers work to recover recyclables and where leachates are generated, contaminating soil and water resources, to sanitary landfills, where leachates are adequately managed and treated, and where scavengers can constitute the formal recycling sector, improving their life quality, is one example of the benefits of infrastructure and innovation investment at SWM sector.

19.1.8 SDG 11: Sustainable Cities

UN (2016) intends that, in 2030, cities could grow in such way that resources used would be optimized, pollution and poverty minimized, facing the challenges of congestion lack of funds for essential services, shortage of housing, and declining infrastructure. In this SDG, the sub-goals are (UN 2016) as follows:

- By 2030, ensure access for all to adequate, safe, and affordable housing and basic services and upgrade slums.
- By 2030, enhance inclusive and sustainable urbanization and capacity for participatory, integrated, and sustainable human settlement planning and management in all countries.
- Strengthen efforts to protect and safeguard the world's cultural and natural heritage.
- By 2030, reduce the adverse per capita environmental impact of cities, including paying special attention to air quality and municipal and other waste management.
- By 2020, substantially increase the number of cities and human settlements adopting and implementing integrated policies and plans toward inclusion, resource efficiency, mitigation, and adaptation to climate change and resilience to disasters and develop and implement, in line with the Sendai Framework for Disaster Risk Reduction 2015–2030, holistic disaster risk management at all levels.

19.1 The Goals of the 2030 Agenda for Sustainable Development 355

- Support least developed countries, including through financial and technical assistance, in building sustainable and resilient buildings utilizing local materials.

Integrated SWM should be implemented all over the world with a sustainable view. An appropriate collection of waste that removes waste from streets (in a public health vision), eliminates open dumpsites and open burning, and implements waste hierarchy principle (if it is better for the environment) to nonhazardous and hazardous waste can be a vision of integrated SWM. Sub-goals of SDG 11 to be reached require that the waste collection system should be provided equally to all, with the same options to conduct the separate collection, although the needed adaptations are case by case (due to technical constraints). The need to implement participatory processes to define how SWM should be managed, giving the guidelines on how the SWM should be conceived to answer to waste hierarchy principle. Concerning the collection system, there might be the need to develop or adopt the waste collection system to avoid damaging heritage. Also, the waste treatment technologies are not needed if there is already technology used for other purposes that could treat/recycle waste, and this way, natural heritage could be preserved by avoiding building new infrastructures. Perhaps waste, being a city problem, should be treated inside of the city and not in the surroundings of the city, making people understand and look ar the pollution that they are making and what is necessary to solve it.

Minimizing environmental impacts from SWM is a work in progress, not existing a unique solution for it. Life cycle assessment, waste hierarchy principle, and circular economy are just some of the ways to manage waste and assess their impacts on the environment that could help on such task. Using waste as secondary resources, namely, in the building sector, can be a way to use local products and reduce environmental impacts from construction and demolition waste, at least regarding transportation of such waste to faraway areas to be recycled or disposed.

19.1.9 SDG 12: Responsible Consumption and Production

The goal most related to SWM is SG12. This goal intends to promote sustainable consumption and production by "doing more and better with less," increasing net welfare gains from economic activities by not using so many resources, and by avoiding or minimizing degradation and pollution during the lifecycle of goods produced while increasing quality of life (UN 2016):

- Implement the 10-year framework of programs on sustainable consumption and production, all countries taking action, with developed countries taking the lead, taking into account the development and capabilities of developing countries.
- By 2030, achieve the sustainable management and efficient use of natural resources.
- By 2030, halve per capita global food waste at the retail and consumer levels, and reduce food losses along production and supply chains, including postharvest losses.
- By 2020, achieve the environmentally sound management of chemicals and all wastes throughout their life cycle, by agreed international frameworks, and significantly reduce their release to air, water, and soil to minimize their adverse impacts on human health and the environment.

- By 2030, substantially reduce waste generation through prevention, reduction, recycling, and reuse.
- Encourage companies, especially large and transnational companies, to adopt sustainable practices and to integrate sustainability information into their reporting cycle.
- Promote public procurement practices that are sustainable, by national policies and priorities.
- By 2030, ensure that people everywhere have the relevant information and awareness for sustainable development and lifestyles in harmony with nature.
- Support developing countries to strengthen their scientific and technological capacity to move toward more sustainable patterns of consumption and production.
- Develop and implement tools to monitor sustainable development impacts for sustainable tourism that creates jobs and promotes local culture and products.

A system approach is needed, as well as cooperation between all stakeholders, from the producer, to the consumer and to waste managers and recyclers. The system approach thinking applied to SWM already provided by Chang and Pires (2015), together with the waste hierarchy principle, and now with the circular economy vision can contribute to reaching this SDG. The reduction of waste generated through waste prevention, preparation for reuse, and recycling, in the light of the circular economy, has the potential to promote sustainable use of resources.

19.1.10 SDG 13: Climate Action

This SDG intends to promote measures to control climate change and its impacts. People are identifying impacts of climate change, from changing weather patterns, rising sea level, and more extreme weather events, affecting national economies and lives, costing people, communities, and countries today and tomorrow (UN 2016). SWM can contribute to reducing the emissions of greenhouse gases (GHG), by implementing practices that could prevent their release. Also, the diversion of organic matter from landfills, promoted in Europe by Landfill Directive, has allowed reducing the release of methane (one GHG) in European landfills, being now new goals to manage this particular waste fraction related to GHG and climate change. Waste collection, in another hand, is consuming petroleum-derived fuel in the majority, existing few projects which uses renewable fuels. The SDG related to waste is (UN 2016):

- Integrate climate change measures into national policies, strategies, and planning.

Waste management planning, strategies, and policies need to include in their guidelines mitigation and adaptation measures for climate change. Like simplified in Fig. 19.1, GHG emissions occur during all waste life cycle, from transportation to disposal in a landfill. But also reduction of GHG emissions occurs due to processes that generate biowaste-based energy, reducing the release of GHG emissions. Recycling itself prevents the release of GHG by avoiding the extraction and consumption of raw materials. Integration of SWM measures in plans and implementing those measures in the field are necessary to reach SDG 13 goal.

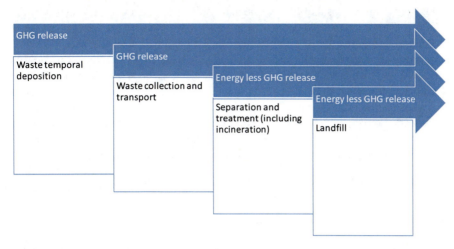

Fig. 19.1 Simplified schematic of waste management system and GHG emissions (applicable to urban waste management)

19.1.11 SDG 14: Life Below Water

This SDG intends to conserve and promote sustainable use of the oceans, seas, and marine resources. The sub-goals for this particular SDG are (UN 2016):

- By 2025, prevent and significantly reduce marine pollution of all kinds, in particular from land-based activities, including marine debris and nutrient pollution.
- Minimize and address the impacts of ocean acidification, including through enhanced scientific cooperation at all levels.

In the first sub-goal, the role of SWM to reduce marine pollution is relevant. Marine debris is waste created by humans that have been discharged into coastal or marine environments, resulting from activities on land or sea (UNEP and NOAA 2011). The majority of marine debris are made of plastics, macroplastics, and microplastics (plastic particles <5 mm in diameter including nanoparticles), where smaller macroplastics (<2.5 cm) has origin from lost bottle caps or plastic fragment (GESAMP 2016; UNEP 2016). Common macroplastics (until 1 m) originating from rivers or maritime sources such as plastic bags, food and other packaging, fishing floats, buoys, balloons, and macroplastics larger than 1 m from fishing activities or catastrophic events such as fishing nets and traps, rope, boat hulls, and plastic films from agriculture are also marine debris (GESAMP 2016; UNEP 2016). Waste management activities should be conducted to avoid littering and consequent contamination of coastal areas and rivers with this type of waste. For example, campaigns to collect bottle caps, implementation of waste management practices at the fishing sector, and extended producer responsibility principle implemented to plastic products to induce ecodesign of products to avoid marine litter are some of the possible measures to help fight marine litter. Also, due to this environmental problem, the waste collection has to be made to avoid waste loss during collection (due to the poor coverage of dumpsters and dump trucks (Sheavly and Register

2007) and avoid disposal of light waste in landfills to not get out of the landfill by the wind. The control of sources of marine litter is needed to avoid inadvertent release of waste and consequently, of marine litter (Sheavly and Register 2007).

For the second sub-goal, the one devoted to solving ocean acidification, the role of SWM to solve ocean acidification is equivalent to the role of climate change. Ocean acidification is due to the absorption of carbon dioxide emissions from human activities by the ocean, equivalent to about 30% of the total emissions for the last 200 years from nonrenewable fuel combustion, cement production, and land-use change (Sabine et al. 2004). The measures and actions which can be done to minimize climate change can also be used to minimize ocean acidification.

19.1.12   SDG 15: Life on Land

This goal intends to reduce deforestation and combat desertification caused by human activities and climate change. In specific, the sub-goal related to SWM is (UN 2016):

- By 2020, ensure the conservation, restoration, and sustainable use of terrestrial and inland freshwater ecosystems and their services, in particular, forests, wetlands, mountains, and drylands, in line with obligations under international agreements.

When SWM is managed with the purpose to avoid or minimize landfilling and when it is made to utilize existing infrastructure and not building dedicated waste treatment and recycling technologies, the land is used more efficiently. Industrial symbiosis and sharing of infrastructures based on the type of waste processed and not waste origin (what matters is the composition of the waste to be processed together) can reduce the multiplicity of infrastructures, getting scale and reducing the land needed to build them.

19.1.13   SDG 17: Partnerships

The Agenda 2030 intends to promote partnerships that could (UN 2016):

mobilize, redirect and unlock the transformative power of trillions of dollars of private resources to deliver on sustainable development objectives.

For developing countries, the long-term investments are needed to promote sustainable energy, infrastructures, and transport, as well as information and communication technologies (UN 2016), not forgetting waste management infrastructures. The sub-goal related to SWM is:

- Promote the development, transfer, dissemination, and diffusion of environmentally sound technologies to developing countries on favorable terms, including on concessional and preferential terms, as mutually agreed.

To reach this sub-goal, national entities in developing countries must implement waste management technologies. The choice of the technologies needs to respect several aspects, such as financial planning and management, strategies to ensure operational efficiency, public involvement in waste issues, to know the waste to be processed, to avoid the transfer of technology from industrialized countries to developing countries without the appropriate considerations. One example of a technology of difficult transfer from industrialized to developing countries is incineration or waste-to-energy facilities. Those infrastructures are expensive to build and to operate, require a reduce amount of employees, and do not promote other green jobs around recyclables. More interesting infrastructures are the ones which potentiate recycling (mechanical, biological), allowing to create more green jobs and promoting a circular economy in developing countries.

19.2 Final Remarks

SWM can help to achieve 13 out of 17 UN SDG, but the relationship between SWM and SDG is reciprocal: SWM can help to achieve, but to reach the goals, the SWM performance has to change. Improving the application of waste hierarchy will help to improve the living conditions and better public health of more than 2–3 billion people who currently lack services, prevent plastics entering the oceans, contribute to climate change mitigation and adaptation, and help to restore terrestrial ecosystems (Rodić and Wilson 2017). Authorities and private companies would create decent jobs, and workers lives and of their relatives will improve, increasing their economic sustainability. For developing countries, measures related to waste collection services extended to serve all citizens, closing dumpsites and ending dumpsites burning, and converting dumpsites into sanitary landfills will make considerable improvements. However, in developed countries, measures to ensure that waste jobs are decent and fair conditions are ensured to workers are determined to make waste jobs cleaner and recognized by society. Developed and developing countries' governments could apply several instruments to ensure the role of SWM in reaching SDG goals. Instruments such as direct regulation, economic instruments, voluntary agreements, and information instruments are helpful to put in practice SWM measures in light of SDG.

References

Bonatti V (2018) Taking out the garbage: Migrant women's unseen environmental work. Eur J Women's Stud 25:41–55

Chang NB, Pires A (2015) Sustainable solid waste management: a systems engineering approach. IEEE book series on systems science and engineering. Wiley/IEEE, New York

Cierjacks A, Behr F, Kowarik I (2012) Operational performance indicators for litter management at festivals in semi-natural landscapes. Ecol Indic 13:328–337

Dermatas D (2017) Waste management and research and the sustainable development goals: focus on soil and groundwater pollution. Waste Manag Res 35:453–455

European Commission (2014a) Communication from the Commission to the European Parliament, The Council, The European Economic and Social Committee and the Committee of the regions towards a circular economy: a zero waste programme for Europe COM/2014/0398 final. European Commission. Available via http://eur-lex.europa.eu/legal-content/GA/TXT/?uri=CELEX%3A52014DC0398. Accessed 15 Jan 2018

European Commission (2014b) Commission Staff working document impact assessment - Accompanying the document Proposal for a Directive of the European parliament and of the Council amending Directives 2008/98/EC on waste, 94/62/EC on packaging and packaging waste, 1999/31/EC on the landfill of waste, 2000/53/EC on end-of-life vehicles, 2006/66/EC on batteries and accumulators and waste batteries and accumulators, and 2012/19/EU on waste electrical and electronic equipment. European Commission. Available via http://register.consilium.europa.eu/doc/srv?l=EN&f=ST%2011598%202014%20ADD%203. Accessed 15 Jan 2018

European Commission (2001) Employment effects of waste management policies. European Commission, Brussels

GESAMP (2016) Sources, fate and effects of microplastics in the marine environment: part two of a global assessment. In: Kershaw PJ, Rochman CM (eds.) IMO/FAO/UNESCO-IOC/UNIDO/WMO/IAEA/UN/UNEP/UNDP joint group of experts on the scientific aspects of marine environmental protection. Rep Stud GESAMP 93, p 220

Gregson N, Crang M, Botticello J, Calestani M, Krzywoszynska A (2016) Doing the 'dirty work' of the green economy: resource recovery and migrant labour in the EU. Eur Urban Reg Stud 23:541–555

Martinho G, Gomes A, Ramos M, Santos P, Gonçalves G, Fonseca M, Pires A (2018) Solid waste prevention and management at green festivals: a case study of the Andanças Festival, Portugal. Waste Manag 71:10–18

Rodić L, Wilson DC (2017) Resolving governance issues to achieve priority: sustainable development goals related to solid waste management in developing countries. Sustain 9:404

Sabine CL, Feely RA, Gruber N, Key RM, Lee K, Bullister JL, Wanninkhof R, Wong CS, Wallace DWR, Tilbrook B, Millero FJ, Peng TH, Kozyr A, Ono T, Rios AF (2004) The oceanic sink for anthropogenic CO_2. Science 305:367–371

Sheavly SB, Register KM (2007) Marine debris & plastics: environmental concerns, sources, impacts and solutions. J Polym Environ 15:301–305

United Nations (UN) (2016) Sustainable development goals – 17 goals to transform our world. United Nations. Available via http://www.un.org/sustainabledevelopment/sustainable-development-goals/. Accessed 20 Feb 2018

United Nations Environment Programme (UNEP) (2016) Marine plastic debris and microplastics — Global lessons and research to inspire action and guide policy change. United Nations Environment Programme, Nairobi

United Nations Environment Programme (UNEP), National Oceanic and Atmospheric Administration (NOAA) (2011) Honolulu Strategy. National Oceanic and Atmospheric Administration. Available via https://marinedebris.noaa.gov/sites/default/files/publications-files/Honolulu_Strategy.pdf Accessed 20 Feb 2018

Index

A

Accurate cost data, 177
Adaptive management strategies, 132
Advanced forecast models, 146–147
Aesthetic values, 338
Altruistic behavior, 77
Ampliroll and Multilift, 40
Analytic hierarchy process (AHP), 243–244
Artificial intelligent approaches, 146
Artificial neural networks (ANNs), 146
Augmented ε-constraint method, 269, 283, 284
Autoregressive integrated moving average
(ARIMA) models, 144, 160

B

Basel Convention, 61
Batteries and accumulators, 65–66
Behavior studies, 198
Benchmarking analysis, 96
Benchmarking study, 169
Biodegradable municipal waste, 57
Biological treatment, 53
Bottom-up initiatives, 132

C

Capital or investment costs, 101
Carbon footprint
environmental impact systems, 192
GHG, 192, 193
GWP, 192
LCA, 193, 194
Catalan municipalities, 176
Cathode-ray tube (CRT), 50

Chinese postman problem (CPP), 208, 216,
220–222
Circular economy, 130
Closed-loop recycling, 50
Closed-loop supply chain (CLSP), 323
CO_2 emissions, 267, 269, 272
Collection duration rates, 293
Collection method component
container and vehicle, 41
semiautomated collection, 41
Company current operation mode
clients and containers, 290
collection site, 289
lunch break, 290
performance indicators, 290
vehicle route timings, 290
Connecticut Department of Environmental
Protection, 114
ε-constraint method, 229
Consumer price index (CPI), 103
Container classification diagram, 33
Container component, 33–37
Containers' capacity ranges, 37
Container-vehicle system, 32
Contextual factors, 75, 198
Convenience factors, 76
Coupling systems, 34
Cradle to Cradle: Remaking the Way We
Make Things, 49
Crane-based lifting devices, 39
Crane-compatible containers, 35
Crane coupling, 34
Crowd logistics
advantages, 329
definition, 328

© Springer International Publishing AG, part of Springer Nature 2019
A. Pires et al., *Sustainable Solid Waste Collection and Management*,
https://doi.org/10.1007/978-3-319-93200-2

362 Index

Crowd logistics (*cont.*)
  postal operator, 328
  stakeholders, 328
Cumulative energy demand (CED), 316
Current network, 232, 291–293
Current network expansion, 298
Current routes characteristics, 291

D
Danish Environmental Protection Agency, 104
DecideIT software, 250
Decision Lab 2000, 251
Decision support systems (DSS), 250
Dedicated collection system study, 132
Deposit-refund systems, 111
Depreciation, 104
Descriptive indicators
  collection system, 167
  waste collection, 168
Direct costs, 101
Directive 2000/53/EC, 66
Directive 2008/987EC, 54
Disposal techniques, 55
Down-cycling, 50

E
Eco-city projects, 127
Eco-design, 16
Ecological footprint (EF), 255
Economic indicators
  mechanisms and technologies, 176
  optimization models, 177
  solid waste collection systems, 177, 178
  types, 176
  waste collection system, 176
Economic instruments, 132
ELECTRE III–IV, 250
Elimination and choice expressing the reality
    (ELECTRE), 242, 243, 245
End-of-life recycling approach, 187
End-of-life vehicles (ELV), 66
End-of-waste criteria, 52
Energy recovery, 309
Environment Protection Act (EPASA), 13
Environmental and social impact, 207
Environmental behaviors, 77
Environmental context
  circular cities, 133
  circular economy, 130
  climate change, 129
  eco-city projects, 127

  economic growth and waste generation, 124
  European Commission, 130
  GHG, 129, 131
  global markets, 125
  globalization, 124, 125
  MSW, 127
  OECD Environmental Strategy, 124
  sustainability and circular economy, 130
  WEEE, 125
  zero-carbon goal, 128
  tools, 123
Environmental indicators
  LCA, 178
EU Waste Framework Directive, 309
European Union legislation, 47, 313
European Waste Framework Directive, 14
European Waste Framework Directive
    (2008/98/CE), 110
Eurostat, 55
e-waste management, 342
Exponential smoothing model, 159
Extended producer responsibility (EPR), 64

F
Facility location problem, 223
Forecast accuracy, 144–145
  MAPE, 145
  squared errors, 145
Forecast data, 141
Forecasting methods, 158–161
Forecasting waste generation, 142
Freight on Transit (FOT)
  experiences, 330
Fuel consumption, 102, 107
Full truckload (FTL), 275
Fully automated collection methods, 42

G
Geographic information systems (GIS), 97
  capacity constraint, 155
  database, 152
  demand constraints, 155
  distance and time matrices, 155
  economic perspective, 153
  environmental perspective, 154
  European project, 157
  features, 152
  forbidden turns, 155
  human resources, 156
  labor constraints, 155
  MSW, 153

Index   363

optimization process, 156
political constraints, 155
route definition, 153
social perspective, 154
stopping point, 155
vehicles, 155
Geometric mean absolute error (GMAE), 145, 161
Geometric mean relative absolute error (GMRAE), 162
Global warming potential (GWP), 192
Goal and scope definition, 184–188
Gradient boosting regression model, 147
Greenhouse gases (GHG), 129
Gross domestic product (GDP), 6, 124
Gross national income (GNI), 124

H

High-density polyethylene (HDPE), 35
HIPRE 3C, 250
Hiring fleet cost, 299
Holt's model, 159–160
Holt-Winters method, 160
Hydraulic ejection panel, 40

I

Incineration, 46, 54
Industrial symbiosis, 308
Information and communication technology (ICT), 95, 326
Initial costs, 102
Integrated solid waste management (ISWM) system, 309
Integrated waste collection (IWC), 309
   biological treatments, 313
   bio-waste, 314
   CED, 316
   Dual-stream collection, 312
   heavy metal content, 315
   landfill, 318
   MBT, 314
   optimization algorithms, 311
   organic matter, 313
   PAYT, 311
   public awareness, 312
   RDF, 316
   reuse, 311
   sort waste, 312
   sustainability, 318–319
   transportation and sorting, 313
   waste prevention, 310

WEEE, 312
WtE plants, 318
Integrated waste management system (IWMS), 95
Inter-American Development Bank, 13
International legislation
   Basel Convention, 61
   European Union waste policies, 63
Interpretation, 191
Interval-valued fuzzy (IVF) method, 256
Investment costs (CAPEX)
   depreciation, 104
   fractional residual value, 106
   literature references, 105
   MSW, 103
   municipalities and equipment suppliers, 104
   vehicle, 105
Iterative methods, 230

K

Key collection methods
   container and vehicle, 42
Kinshofer, 35
k-nearest neighbors algorithms (kNN), 146

L

Landfill tax, 111
Lexicographic method, 226–227, 267
Life cycle assessment (LCA)
   functional unit, 184
   goal and scope definition, 184–188
   inventory analysis phase, 183
   LCI, 188
   multi-functionality, 188
   multi-input process, 187
   physical/chemical classification, 187
   practitioner, 192
   reference flow, 185
   software, 192
   solid waste collection systems, 184
   waste collection, 186
   waste collection system, 185
Life cycle costing (LCC)
   cost, 195
   environmental, 193
   financial assessments, 193
   indirect costs, 194
   societal, 194
   type, 194
Lifecycle impact assessment (LCIA), 189–191
   critical steps, 189

364

Index

Lifecycle impact assessment (LCIA) (*cont.*)
  methodologies, 190
  system, 190
  waste collection system, 191
Lifecycle inventory (LCI), 188–189
Linear programming model
  constraints, 209
  disposal facilities, 209
  mathematical model, 210
  SWM, 208–210
Linear regression models, 162–163
Lisbon and Vale do Tejo (LVT), 234
Lisbon metropolitan area, 105, 106
Local governments, 108
Logistics indicators, 177
Long-term forecast, 142

M
Market-based instruments, 47
Mathematical programming models, 208
Mathematical programming problems, 215
MDVRPI extension, 282
MDVRPI relaxation, 280
MDVRP-MCO, 299
Mean absolute error (MAE), 161
Mean Absolute Percentage Error (MAPE), 162
Mean Square Error (MSE), 162
Mechanical-biological treatment (MBT),
  55, 313
Minimizing operational costs, 156–157
Mixed integer linear programming (MILP), 289
Mixed integer linear programming model
  (MILP)
  assignment problem, 212
  data, 214
  elements, 211
  LP models, 211, 212
Mobile garbage containers, 36
Monte Carlo analysis, 191
Moving-average models, 158–159
Multi-criteria analysis, 254–255
Multi-depot vehicle routing problem
  (MDVRP), 219, 264
Multidisciplinary teams, 89
Multifunctionality, 187
Multi-material system, 75
Multi-objective analysis, 234–235
Multi-objective MDPVRPI, 265, 273
Multi-objective Problem, 282
Multi-objective programming (MOP), 225
Multi-criteria decision analysis (MCDA)
  AHP, 243, 244

application, 241
comparison, 247
DecideIT software, 250
DSS, 250
ELECTRE, 245
ELECTRE III–IV, 250
environmental and social, 253
HIPRE 3C, 250
MOP techniques, 239
multi-criteria analysis, 254–255
number, 242
PROMETHEE, 245
PROMETHEE and GAIA, 251
quantitative or qualitative data, 241
SANNA 2009, 250
SMART, 243
stakeholder groups, 251
sustainable solid waste management, 240
sustainable waste management, 252
SWM, 246
techniques, 242
TOPSIS, 244
type, 242
utility score, 241
Multiple linear regression model, 163
Multiple objective programming (MOP), 239
Municipal selective waste collection schemes,
  257–258
Municipal solid waste (MSW), 25, 141, 168
  economic role, 26
  environmental role, 26
  generation, 123
  legal role, 26
  management, 27
  micro/local scale, 27
  social role, 26
  technical role, 26

N
Naïve forecast model, 158
National waste regulation
  batteries and accumulators, 65
  central/federal governments, 64
  ELV, 66
  EPR, 64
  packaging waste, 67
  PRO, 64
  waste oils, 69
  WEEE Directive, 68
Net calorific value (NCV), 54
Network expansion, 293–298
New York City Department of Sanitation, 215

Index 365

Non-linear programming, 215
Normal Period, 288
NOx emission, 254

O

Objective function (0) models, 302
One-factor-at-a-time (OFAT) methods, 191
Operating and maintenance costs (O&M)
    annual maintenance costs, 106
    collection system/circuit, 108
    "command and control" approach, 110
    container maintenance costs, 107
    economic instruments, 111
    energy consumption, 107
    fuel consumption, 107
    funding programs, 112
    human resources costs, 107
    landfill tax, 111
    maintenance costs analysis, 106
    MSW, 109, 110, 112
    municipalities and service, 106
    operating cost component, 107
    PAYT, 113
    product charges, 112
    property tax, 112
    regulatory instruments, 110
    SWM, 108
    transport system, 109
    user-charge billing, 112
    WEEE, 109
Operational indicators, 171
Optimal network, 233–234
Optimization algorithms, 176

P

Packaging waste, 64, 67
Pareto optimal solution, 225, 270
Pay-as-you-throw (PAYT)
    advantages, 113
    advantages and disadvantages, 115
    schemes, 113, 114
    system, 113
Performance indicators, 168–179
Periodic vehicle routing problem (PVRP), 264
Physical internet, 329
Plastic recycling, 352
Preference ranking organization and method
      for enrichment (PROMETHEE),
    243, 245
Producer responsibility organization (PRO), 64
Product service systems, 19, 20

Product-life extension services, 20
Products and goods reuse
    car stations, 20
    consumption, 19
    graphic representation, 20
    policy instruments, 19
Psychological factors, 198
Public and private sector financing
    European member states, 117
    European models, 117
    financing collection and transport
      systems, 116
    management model, 118
    MSW, 116
    municipalities, 117
    O&M costs, 116
    procurement process, 117
Public participation process, 197
Public-private partnership (PPP), 339

Q

Qualitative approaches, 142
Quantitative methods, 142

R

Recyclable waste collection rate, 169
Recycling
    downcycling, 50
    CRT, 50
    EoW principles, 52
    European countries, 48
    market value, 52
    materials, 81
    recovery management option, 53
    requirements, 51
    streams, 52
    upcycling, 49
    WtE, 53
Recycling behaviors
    altruistic behavior, 77, 78
    attitude-behavior relation, 80
    behavior change interventions, 83
    behavioral control, 80
    behaviors-attitudes relations, 79
    components, 82
    conceptual framework, 81
    environmental behaviors, 84
    factors, 74
    generalization, 84
    internal or external factors, 80
    meta-analysis, 74

366   Index

Recycling behaviors (*cont.*)
   normative components, 78
   organizational structure, 83
   particularism, 84
   PAYT, 85
   sociodemographic factors, 81
   sociodemographic variables, 75
   subjective norms, 79
   theories and models, 77
   useful life period, 84
Refuse collection vehicles (RCV), 230
Refuse-derived fuel (RDF), 316
Regression models, 146
   forecast variable, 146
Resource scarcity, 130
Reuse and recycling activities, 350
Reverse logistics (RL), 323
   advanced industrial products, 326
   and circular economy, 324
   and CLSC, 323
   distributor/service partner, 326
   EMF, 324–326
   front-end includes, 324
   implementation barriers, 327
   products/waste, 324
Reverse logistics system
   collection sites and depot locations, 262
   economic and environmental optimal
      plans, 267
   economic objective, 272
   economic schedule, 268
   environmental objective, 263
   environmental solution, 269
   fuel consumption, 263
   glass, 262
   historical data, 262
   lexicographic optimization, 268
   municipality perspective, 261
   plastic and metal, 266
   routes generation, 266–267
   social objective, 263
   sustainable solution, 272
Rio Declaration on Environment and
      Development, 4
Route definition, 153
Routes' characteristics, 294
Routes generation procedures, 266, 277–282
Routing problem, 264

S
SANNA 2009, 250
Satellite units, 39

Scenario comparison, 234
Schwartz model of altruistic behavior, 77–78
Seasonal exponential smoothing (SES), 148
Seasonal period, 288
Semiautomated collection methods, 42
Sensitivity analysis
   MCDA, 249
   parameter, 246
   tornado graph, 248
Separate collection, 342
Separate collection systems, 89
Separation behavior, 74
Simple addictive weighting (SAW), 243, 257
Simple linear regression model, 162–163
Sino-European Circular Economy and
      Resource Efficiency (SINCERE)
      project, 130
Social discount tax, 104
Social indicators, 179
Social life cycle assessment (S-LCA)
   definition, 196
   methodology, 195
   packaging waste collection schemes, 197
   semi-qualitative and qualitative data, 195
   UNEP/SETAC, 196
   UNEP's guideline, 196
Social psychology
   behavioral models, 86
   PAYT, 87
   recycling behaviors, 86
   separate collection system, 74
   techno-economical considerations, 74
   waste collection, 73
Socio-demographic factors, 198
Socio-demographic variables, 86, 88
Solid waste management (SWM), 349
Source separate collection, 316
Stakeholders, 251
Stochastic programming, 212–214
Subtour elimination constraints (SECs), 217
Support vector machine (SVM), 146
Sustainability assessment procedure, 7
Sustainability waste management policies, 5
Sustainable collection system, 267–272
Sustainable development
   challenges and perspectives, 9
   concept, 3–5
   conceptual framework, 7
   framework, 6–7
   role, 9
Sustainable development goals (SDG)
   affordable and clean energy, 352
   clean water and sanitation, 351–352

climate change, 356
consumption and production, 355–356
environmental impacts, 355
GHG and climate change, 356
GHG emissions, 356
good health and well-being goal, 351
human activities and climate change, 358
infrastructure, 354
no poverty goal, 350
partnerships, 358
SDG 9, 354
sustainable cities, 354–355
SWM, 356
waste sector, 350
work and economic growth, 352–353
zero hunger, 350–351
Sustainable development goals (SDGs), 3, 5
Sustainable solid waste management, 207, 240
Systems analysis, 308

T
Technical barriers and socioeconomic
challenges, 335–345
developed countries
environmental informatics, 335
ICTs, 336, 337
waste infrastructure, 337–338
developing countries
CBO/MSE, 341
dumpsite, 343, 344
e-waste, 342
financial sustainability, 339
hazardous waste, 342
individual workers, 341
informal sector, 340
municipal solid waste, 340
municipalities' staff, 345
social apathy, 344
Technical-operative indicators, 170
Technique for order preference by similarity to
ideal solution (TOPSIS), 244
Theory of planned behavior, 78–80
Theory of reasoned action, 78–80
Time series
ARIMA models, 144
components, 144
historical data, 143
MSW, 148
Time-series forecasting models, 143
Tornado graph, 248
Transboundary movements, 62
Transport simulation software, 177

Transportation problem, 222
Traveling salesman problem (TSP), 208, 216
asymmetric, 216

U
Uncertainty and sensitivity analyses, 191
United Nations Framework Convention on
Climate Change (UNFCCC), 128
Unrecoverable products
biodiesel production, 288
CO_2 emissions, 287
current operation mode, 289
fossil fuel, 288
non-specialized collection vehicles, 288
normal period, 288
recycling, 287
seasonal period, 288
vegetables oils, 288
waste cooking oil, 287, 288
Upcycling, 49
Urban population, 126
Urban symbiosis, 128
US Resource Conservation and Recovery Act
(USEPA, 2017), 13

V
Vehicle classification diagram, 38
Vehicle coupling, 34
Vehicle route timings, 290
Vehicle routing problem (VRP), 208, 216, 218
Vehicle routing problem with multiple use of
vehicles (VRPMU), 264

W
Waste collection, 9
Waste collection costs
CO_2 emissions, 98
contextual and operational/technological
variables, 96
contextual variables, 96
conventional standard approach, 98
cost values, 98
disadvantage, 97
financial variables, 102
fuel consumption, 102, 103
geographical and demographic data, 97
human resources, 102
labor, 101
MSW, 96
planning and management, 95

Waste collection costs (*cont.*)
  revenues, 98
  transport costs, 98
  transportation, 95
Waste collection routing software, 152
Waste collection system planning and selection,
    148–152
Waste collection systems (WCS), 152
  approaches, 30
  classification, 28–42
  collection method, 30
  container type, 29
  diagram, 28
  drop-off or pickup systems, 30
  local governments, 27
  pickup and drop-off arrangement, 32
  process, 27
  recovery and collection, 27
  role, 25
  service type categories and definitions, 31
  source and source separation, 30
  taxonomic classification, 32
  transport, 27
  vehicle type, 30
Waste cooking oil, 288
Waste cooking oil collection system, 289
Waste disposal destinations, 56
Waste electrical and electronic equipment
    (WEEE), 47, 68
Waste Framework Directive, 46, 47, 52, 317
Waste Framework Directive (2008/98/EC), 13
Waste generation, 168
Waste generation estimation
  legislations, 141
  long-term forecast, 142
  quantitative methods, 142

timescale, 142
time series, 143–144
Waste hierarchy, 307, 309, 319
Waste hierarchy principle (WHP)
  definitions, 16
  European, 16
  European legislation, 14
  stages, 15
  Waste Framework Directive, 15
  waste streams, 15
Waste infrastructure, 337–338
Waste management planning, 356
Waste management services, 112
Waste management system, 335, 342
Waste of electric and electronic
    equipment (WEEE), 231
Waste oils, 69
Waste prevention, 16–17
  policy instruments, 17, 18
  practices, 16
  quality and cost-effectiveness, 16
  types, 16
Waste stream, 5
Waste treatment and recycling
    technologies, 358
Waste vehicle, 151
Waste-to-energy (WtE), 53, 230
Water/electricity consumption, 113
Weighted Sum method, 227
Windy postman problem (WPP), 221
Working hours, 263, 264, 267

Z
Zero waste city, 127, 131, 133, 134
Zurich cargo tram, 330